AF766992

Stability Analysis of
Converter-Rich Power Grids

IEEE Press

445 Hoes Lane

Piscataway, NJ 08854

IEEE Press Editorial Board

Sarah Spurgeon, *Editor-in-Chief*

Moeness Amin
Jón Atli Benediktsson
Adam Drobot
James Duncan
Hugo Enrique Hernandez Figueroa
Ekram Hossain
Brian Johnson
Hai Li
James Lyke
Joydeep Mitra
Albert Wang
Desineni Subbaram Naidu
Yi Qian
Tony Quek
Behzad Razavi
Thomas Robertazzi
Patrick Chik Yue

Stability Analysis of Converter-Rich Power Grids

Jun Liang
Cardiff University, UK

Chuanyue Li
South China University of Technology, China

IEEE Press
WILEY

Copyright © 2026 by The Institute of Electrical and Electronics Engineers, Inc. All rights reserved.

IEEE Standards designations are trademarks of The Institute of Electrical and Electronics Engineers, Incorporated (www.ieee.org/).

Non-IEEE trademarks are the respective property of their owners.

Published by John Wiley & Sons, Inc., Hoboken, New Jersey.

No part of this publication may be reproduced, stored in a retrieval system, or transmitted in any form or by any means, electronic, mechanical, photocopying, recording, scanning, or otherwise, except as permitted under Section 107 or 108 of the 1976 United States Copyright Act, without either the prior written permission of the Publisher, or authorization through payment of the appropriate per-copy fee to the Copyright Clearance Center, Inc., 222 Rosewood Drive, Danvers, MA 01923, (978) 750-8400, fax (978) 750-4470, or on the web at www.copyright.com. Requests to the Publisher for permission should be addressed to the Permissions Department, John Wiley & Sons, Inc., 111 River Street, Hoboken, NJ 07030, (201) 748-6011, fax (201) 748-6008, or online at http://www.wiley.com/go/permission.

The manufacturer's authorized representative according to the EU General Product Safety Regulation is Wiley-VCH GmbH, Boschstr. 12, 69469 Weinheim, Germany, e-mail: Product_Safety@wiley.com.

Trademarks: Wiley and the Wiley logo are trademarks or registered trademarks of John Wiley & Sons, Inc. and/or its affiliates in the United States and other countries and may not be used without written permission. All other trademarks are the property of their respective owners. John Wiley & Sons, Inc. is not associated with any product or vendor mentioned in this book.

Limit of Liability/Disclaimer of Warranty: While the publisher and the authors have used their best efforts in preparing this work, including a review of the content of the work, neither the publisher nor the authors make any representations or warranties with respect to the accuracy or completeness of the contents of this work and specifically disclaim all warranties, including without limitation any implied warranties of merchantability or fitness for a particular purpose. No warranty may be created or extended by sales representatives, written sales materials or promotional statements for this work. The fact that an organization, website, or product is referred to in this work as a citation and/or potential source of further information does not mean that the publisher and authors endorse the information or services the organization, website, or product may provide or recommendations it may make. This work is sold with the understanding that the publisher is not engaged in rendering professional services. The advice and strategies contained herein may not be suitable for your situation. You should consult with a specialist where appropriate. Further, readers should be aware that websites listed in this work may have changed or disappeared between when this work was written and when it is read. Neither the publisher nor authors shall be liable for any loss of profit or any other commercial damages, including but not limited to special, incidental, consequential, or other damages.

For general information on our other products and services or for technical support, please contact our Customer Care Department within the United States at (800) 762-2974, outside the United States at (317) 572-3993 or fax (317) 572-4002.

Wiley also publishes its books in a variety of electronic formats. Some content that appears in print may not be available in electronic formats. For more information about Wiley products, visit our web site at www.wiley.com.

Library of Congress Cataloging-in-Publication Data has been applied for:

Hardback ISBN: 9781394249749
ePDF ISBN: 9781394249763
epub ISBN: 9781394249756
oBook ISBN: 9781394249770

Cover Design: Wiley
Cover Image: © Olena.07/Shutterstock

Contents

Foreword *xi*
Preface *xiii*
Acknowledgments *xvii*
Acronyms *xix*
Introduction *xxi*

1 Drives of High Penetration of Converters *1*
1.1 High-voltage Direct Current *1*
1.1.1 LCC-HVDC *1*
1.1.2 VSC-HVDC *3*
1.2 Renewable Energy *5*
1.2.1 Wind Generation *5*
1.2.2 Solar Photovoltaic *7*
1.3 Energy Storage System *9*
References *10*

2 Challenges and Future Development of Grid-connected Converters *13*
2.1 Conventional Classification of Power System Stability Based on Disturbances *13*
2.1.1 Steady-state Stability Condition (Without Considering Damping Characteristics) *14*
2.1.2 Steady-state Stability Condition (Considering Damping Characteristics) *16*
2.1.3 Subsynchronous Resonance *16*
2.1.4 Transient Stability *17*
2.2 Overview of VSC-induced Oscillation Events *18*
2.2.1 Oscillation Frequency of Different Electrical Quantities *20*

2.3 Subsynchronous Oscillations *22*
2.3.1 Subsynchronous Resonance *23*
2.3.2 Power Electronic Device Interactions *24*
2.4 Classification of Power System Stability *24*
2.4.1 Classification of Power System Stability in 1982 *24*
2.4.2 Classification of Power System Stability in 2004 *25*
2.4.3 Classification of Power System Stability in 2020 *26*
2.5 Control Interaction of CIG *27*
2.5.1 Past Experiences with Control Interactions from Power Electronic Devices *27*
2.5.2 Control Interaction of VSCs *28*
2.6 Overview of Weak-grid Caused Instabilities *31*
2.6.1 Past Experiences with Weak-grid Instabilities *32*
2.6.2 VSC: Weak-grid Instabilities *32*
References *34*

3 Fundamental Stability Criteria for Feedback Systems *37*
3.1 The Mathematical Mechanism of System Stability *37*
3.2 Stability Criterion via Pole Map *39*
3.3 Stability Criterion via Bode Plot Analysis *41*
3.4 Stability Criterion via Nyquist Plot Analysis *43*
Reference *46*

4 Modeling of Grid-connected Converters *47*
4.1 Basic Control Configurations of VSCs *47*
4.2 Linearization of VSC Control Systems *48*
4.2.1 Linearization of Frame Transformation for Single-phase Systems *50*
4.2.2 Linearization of Reference Frame Transformation for Three-phase Systems *52*
4.2.3 Summary of the Linearization of the Frame Transformation *54*
4.3 Modeling of Frame Transformation with Frequency Alignment Control *55*
4.3.1 Modeling of Phase-locked Loops *56*
4.3.2 Modeling of Power-frequency Droop Control *57*
4.4 Modeling and Stability Analysis of a VSC System *60*
4.4.1 Modeling a VSC System *60*
4.4.2 Transfer-function-based Stability Analysis Method for Converter Systems *62*

4.4.3 Case Study on Stability Analysis of the PLL *63*
4.4.4 Stability Analysis of the $P-\omega$ Droop Control *66*
Reference *67*

5 Stability Analysis Methods *69*
5.1 State-space Stability Analysis Method *69*
5.1.1 Typical State-space Stability Analysis Method *70*
5.1.2 Sensitivity Stability Analysis *70*
5.1.3 Demonstration of State-space Stability Analysis Method *71*
5.2 Impedance Stability Analysis Method *75*
5.2.1 Impedance Representation of Converters Based on Norton and Thevenin Equivalent Circuits *75*
5.2.2 Demonstration of How to Derive Converter Impedance *76*
5.2.3 Impedance Stability Analysis Method Based on Generalized Nyquist Criterion and Bode Criterion *77*
5.2.4 dq-sequence Impedance *82*
5.2.5 $\alpha\beta$ Impedance *84*
5.3 Comparison Among Stability Analysis Methods *86*
References *88*

6 Impedance Interaction Analysis for Converter-integrated Power Systems *89*
6.1 Impedance Interaction Analysis *89*
6.1.1 Negative Resistor Criterion in an Equivalent *RLC* Circuit *89*
6.1.2 Stability Impact of Negative Inductors and Capacitors in an *RLC* Circuit *91*
6.1.3 Effective Frequency Range of a Negative Resistor, Inductor, and Capacitor in an *RLC* Circuit *93*
6.2 Impedance Interaction in a High-order System *97*
6.2.1 Impedance Interaction Criterion *97*
6.2.2 Modified Impedance Interaction Criterion *98*
6.2.3 Case Study for Impedance Interaction Analysis *100*
6.2.4 Why Is the Impedance in the Form of a 2×2 Matrix? *103*
Reference *105*

7 Mirror-frequency Oscillation Due to Converters *107*
7.1 Mirror-frequency Effect of Converter Systems *107*
7.1.1 Mirror-frequency Effect in a Single-phase Control *108*
7.1.2 Mirror Frequency in a Three-phase System *109*
7.2 The Difference Between Mirror-frequency Oscillations and Positive/Negative Sequence Oscillations *115*

7.3 The Performance of Mirror-frequency Oscillations Across Different Electrical Quantities *117*
7.4 Understanding Balanced and Unbalanced Control Systems and Their Role in the Mirror-frequency Effect *119*
7.5 Why Is Converter Control an Unbalanced System? *122*

8 Impedance Measurement Techniques for Power System Stability Analysis *125*
8.1 Stability Analysis Method Based on Measurement *125*
8.1.1 Mechanism of Impedance Measurement *125*
8.2 Impedance Measurement Methods *126*
8.2.1 DC Impedance Measurement *126*
8.2.2 AC System Impedance Measurement *127*
8.3 Impact of Noise on Measurement Accuracy and Elimination Methods *130*
8.3.1 Impact of Noise on DC Impedance Measurements and Elimination Methods *131*
8.3.2 Impact of Noise on AC Impedance Measurements and Elimination Methods *133*
8.4 Impedance Measurement of dq-sequence Impedance or $\alpha\beta$ Impedance *135*

9 Stability Analysis of Grid-following Converters *139*
9.1 Instability Causes of PLL-based Control *139*
9.1.1 Stability Performance Comparison Between Current Control and PLL *139*
9.1.2 Dominant Factor of Current Control Causing PCC Voltage Fluctuation *142*
9.1.3 Instability Cause of Outer Loop *144*
9.2 A Tuning Method for PLL-based Current Control *146*
9.2.1 Solution for the Dominant Instability *146*
9.2.2 The Relationship Between Current Control and PLLs *149*
9.2.3 Tuning Strategies *153*
9.3 Overall Tuning Strategy for PLL-based *PV* Control Systems *154*
9.3.1 Outer-loop Tuning Strategy *154*
9.3.2 Coordinated Tuning of PLL and Current Control *156*
9.3.3 Overall Tuning Strategy *156*
9.4 Summary *158*
References *158*

10 Stability Analysis of Grid-forming Converters *159*
10.1 Development of Grid-forming Converters *159*
10.1.1 Connection with Synchronous Generators *159*
10.1.2 Development of GFM Control *161*
10.1.3 GFM Capabilities *162*
10.2 Definition and Variants of GFM Control *163*
10.3 Brief Comparative Analysis of Stability Characteristics: GFM Converters and SGs in Strong and Weak Grids *167*
10.3.1 Causes of GFM Converter Instability in Strong Grids *167*
10.3.2 A Brief Discussion on the Stability of SGs Under Weak Grid Conditions *169*
10.4 Comparative Analysis of PLL-based Grid-supporting Control and P–ω Droop-based GFM Controls *170*
10.4.1 Island Operation of a PLL-based Grid-supporting Converter *170*
10.4.2 Control Conflict Analysis: P–ω Droop and Inner Current Control in GFM Converters *171*
10.4.3 Comparative Analysis of Grid-supporting and GFM Control Strategies *172*
10.4.4 Challenges of Current-source-based GFM Control *173*
10.5 Comparative Analysis of Stability in GFL and GFM Converters *175*
10.5.1 Stability Analysis Under Various Grid Strength Conditions *176*
10.5.2 Stability Analysis with Various Cutoff Frequencies of the Inner Current Control Loop *177*
10.5.3 Stability Enhancement Using Virtual Impedance *178*
10.5.4 Summary *180*
References *182*

11 Transient Stability Analysis of Grid-following and Grid-forming Converters *185*
11.1 Transient Stability of Conventional Synchronous Generators Based Power Systems *185*
11.1.1 Introduction of Transient Stability *185*
11.1.2 Transient Stability Analysis Approaches *186*
11.1.3 Transient Stability Enhancement Methods *188*
11.2 Transient Stability of Grid-following Converters *189*
11.2.1 Overview of Transient Stability of Grid-following Converters *189*
11.2.2 Control Structure *191*
11.2.3 Transient Stability Analysis *192*
11.2.4 Simulation Verifications *196*

11.3 Transient Stability of Grid-forming Converters *198*
11.3.1 Overview of Transient Stability of Grid-forming Converters *198*
11.3.2 Control Structure *200*
11.3.3 Transient Stability Analysis *201*
11.3.4 Simulation Verifications *206*
11.4 Summary *211*
References *212*

Appendix A: Small-signal Model of Grid-following Converters *215*
Index *227*

Foreword

The rapid integration of renewable generation, energy storage, and modern load technologies, all interfaced through power electronic converters, has fundamentally transformed power system dynamics. Driven by the global push for decarbonization and grid modernization, these converter-based resources and transmission links introduce fast-switching control dynamics that differ markedly from traditional synchronous machines. While enabling higher efficiency and flexibility, this shift has also led to new stability challenges, such as subsynchronous oscillations, weak grid instability, and reduced system inertia. These issues, often overlooked in conventional power systems, now require advanced modeling and enhanced control strategies to ensure reliable operation in low-carbon grids.

Traditionally, power system stability research concerns focused on electromechanical transient stability and small-signal stability including subsynchronous resonance. In contrast, converter-driven systems exhibit fast control-loop interactions (μs–ms timescales), manifesting as wideband oscillations that threaten system security, which pose a particularly critical challenge.

Efforts to analyze converter-induced oscillations using traditional stability mechanisms, for both small disturbance and transient stability, are inadequate, leading to fundamental misunderstandings of grid-following and grid-forming converter behaviors, particularly under operational constraints like converter rating limits (e.g. for device current and DC voltage). A clear theoretical explanation and comprehensive reference text are necessary to deepen the understanding of these phenomena.

I have known Jun Liang for over three decades, serving as his PhD supervisor from 1995 to 1998 in China Electric Power Research Institute (CEPRI). During his doctoral research, he made outstanding achievements on power system stability, with particular excellence in conventional large-scale power systems transient stability dynamics. After completing his studies, he continued to dedicate himself to power electronics-based control and stability analysis in power systems. In recent years, he has collaborated with academia and industry in the UK and EU, including leading three EU projects under the Marie Skłodowska-Curie Programme, several major UK research council funded UK–China, UK–Australia, UK–US, and UK–Africa projects, and industrial innovation projects. His work addresses both traditional synchronous generator stability and modern VSC-induced stability issues, utilizing impedance-based analysis and exploring the physical meaning of converter control. His research is distinctly oriented toward solving real-world power system challenges through innovative solutions.

This book captures the recent achievements of Jun Liang and his team, building upon classical power system theory to address contemporary challenges in grid stability. It provides: (i) comprehension of stability characteristics between synchronous generators and voltage source converters; (ii) physical mechanisms by which converter–grid interactions induce oscillations; (iii) theoretical framework and stability analysis methods of converter-induced oscillations; and (iv) critical distinctions in stability behavior between grid-following and grid-forming converters.

By merging theoretical foundations with practical engineering insights, it provides a comprehensive framework for understanding stability in power systems with complex dynamics introduced by power electronic converters.

Tailored for the next generation of power systems professionals, the book balances academic depth with practical relevance. Researchers will appreciate its analytical rigor, while industry professionals will find value in its implementation-oriented guidance. Students, too, will benefit from its logical organization and accessible presentation of key principles. Ultimately, this book bridges the gap between theory and practice, equipping readers with the knowledge to analyze, assess, and mitigate stability challenges in modern power systems.

Prof. Xiaoxin Zhou
Life Fellow IEEE Honorary President,
China Electric Power Research Institute Academician,
Chinese Academy of Sciences
Beijing, China

Preface

Sustainable development requires a gradual transition from fossil fuel energy to renewable energy. This energy transition drives the transformation of electricity generation and expands the methods of electricity production. Conventional electricity generation employs synchronous generators (SGs) to convert mechanical energy, typically supplied by a prime mover, into electrical power. The prime mover is typically driven by fossil fuels or hydro energy. This energy conversion process, facilitated by the SG, is fundamental to power system operation and ultimately determines the system's key characteristics.

With the expansion of power systems, more requirements arise, necessitating power electronic devices to facilitate asynchronous integration and reactive power compensation applications. Renewables like wind and solar must operate asynchronously to maximize energy conversion, confining them to grid-following (GFL) mode. Operating at full capacity in this mode leaves no margin for grid-forming (GFM) capabilities. Back-to-back converters play an essential role in power interaction and operational isolation between two asynchronous power systems. Long-distance transmission requires direct current (DC) technology via converters for bulk power transfer. Static synchronous compensator (STATCOM) also relies on converters to provide voltage support and reactive power compensation.

The growing reliance on voltage source converter (VSC) as one of power electronic converters for all the above applications, particularly to accommodate increasing renewable energy integration, is changing system dynamics. Although most grid-connected converters are designed to follow the behavior of SGs, system characteristics are shifting from traditional electromechanical dynamics to electromechanical–power electronic hybrid dynamics.

This shift raises concerns about declining system inertia and voltage strength due to the characteristics of GFL VSCs. In recent years, VSCs have been further assigned a new role: sharing GFM capabilities with SGs. This enables nonsynchronous resources to take on responsibilities similar to SGs for counteracting the declined system characteristics. Consequently, additional requirements for voltage and frequency support in VSC control have been introduced, further complicating the control loops. Regarding power electronic switch current protection, power demands, and grid services, VSCs operate in multiple control loops and modes, increasing their potential interaction with power systems.

Power systems initially experienced control interaction-induced oscillations from line-commutated converters (LCCs) in high-voltage direct current (HVDC) applications, which interacted with SGs to cause torsional subsynchronous oscillations (SSOs). As VSCs have been increasingly adopted for renewables, energy storage, and HVDC, hundreds of VSC-related oscillation events have been reported across wide frequency ranges. Such oscillations may destabilize systems by triggering protective actions, potentially causing generation curtailment or complete tripping.

Although GFM VSCs are primarily designed to enhance system stability, for example, by providing inertia, the control interactions between power electronic devices (mainly VSCs) and the power system introduce various unspecified oscillations. Consequently, traditional stability issues linked to SGs, such as rotor angle, voltage, and frequency stability, now also include wideband oscillations caused by control interactions. These oscillations have become more pronounced due to the increasing integration of power electronic devices in recent decades.

Two additional power system stability categories have been proposed to better classify stability phenomena in modern grids, particularly those involving VSCs. These include: (i) electrical and torsional resonance phenomena—while traditionally associated with SGs, many recent cases are VSC-induced; and (ii) purely power electronics (PE)-driven oscillations. The complex interaction between modern power systems and VSC multi-loop control architectures leads to diverse oscillation mechanisms, presenting significant challenges in comprehensively understanding and clearly explaining the underlying dynamics of VSC-induced oscillations.

To facilitate comprehension and analysis of VSC-induced oscillations, this book is organized as follows:

Chapters 1 and 2 provide an overview of the power system implications of VSC-based resources and the different types of VSC-induced system oscillations. Various renewable energy sources, HVDC systems, and energy storage systems that rely on VSCs are discussed. The development of converter-induced oscillations is presented, covering subsynchronous torsional oscillations, electrical oscillations, and wideband oscillations, while introducing newly proposed stability categories.

Chapters 3 and 4 establish the mathematical foundation for the stability analysis of VSC-induced oscillations, including basic stability criteria, mathematical linearization, and modeling of VSC control systems.

Chapter 5 facilitates understanding and extends the application of stability analysis for grid-connected VSCs. Further typical stability analysis methods are introduced and compared to provide a clearer understanding of VSC-induced oscillations.

Chapter 6 examines the widely discussed impedance stability analysis of VSC-induced oscillations, illustrating how impedance interactions facilitate the study of control interactions between VSCs and power systems.

Chapter 7 presents the characteristics of VSC-induced oscillations. The converter-induced oscillation due to control interaction consists of two frequencies, which may include either subsynchronous and supersynchronous oscillations or supersynchronous and negative-sequence oscillations, performing in a mirror-frequency manner.

Chapter 8 presents the measurement-based stability analysis method, focusing on the impedance measurement method, which enables black-box system stability analysis. This approach is essential for complex and large power systems that are difficult to model in detail.

Chapters 9 and 10 conduct a comparative stability analysis of GFL and GFM converters, enabling a comprehensive understanding of their different stability characteristics. These chapters also discuss stability risks and oscillation mitigation methods, with a focus on recently proposed GFM converters compared to traditional GFL converters.

This book is intended for graduate students, academics, manufacturers, and system operators, providing a systematic overview of VSC-induced oscillations. It covers the understanding of VSC-induced oscillations, the mechanism of mirror-frequency behavior, stability analysis methods, and the stability characteristics of GFL and GFM converters. Mitigation measures are also discussed. The book not only serves as a foundation but also outlines the capabilities and limitations of typical methods, helping readers identify the most suitable approaches for industrial applications. It clarifies the limitations of each stability analysis method when applied in practice. Additionally, it presents recent developments in GFM converters, which remain susceptible to control interactions and may still induce oscillations. Recognizing the limitations and drawbacks of GFM inverters is crucial for making informed decisions.

Prof. Jun Liang, Fellow IEEE

Cardiff University

31 July 2025

Acknowledgments

I would like to express my deepest gratitude to all those who made this work possible: to my family, to my extraordinary research team and collaborators, to our industrial partners, and to the academic community.

Acronyms

AC	Alternating Current
BP	Band Pass
CLCP	Closed-loop Characteristic Equation
OLCP	Open-loop Characteristic Equation
CIG	Converter-interfaced Generation
DC	Direct Current
DDSSO	Device-dependent Subsynchronous Oscillation
DFIG	Doubly Fed Induction Generator
FFT	Fast Fourier Transform
GFM	Grid Forming
GFL	Grid Following
HVDC	High-voltage Direct Current
HP	High Pass
IG	Induction Generator
IGBT	Insulated Gate Bipolar Transistor
IGE	Induction Generator Effect
IP	Integral–Proportional
LCC	Line-commutated Converter
LHP	Left Half-plane
LPF	Low-pass Filter
MFE	Mirror-frequency Effect
MMC	Modular Multilevel Converter
MPPT	Maximum Power Point Tracking
NPC	Neutral-point-clamped Converter
PCC	Point of Common Coupling

PI	Proportional–Integral
PR	Proportional–Resonant
PWM	Pulse Width Modulation
PE	Power Electronics
PEDI	Power Electronic Device Interactions
PLL	Phase-locked Loop
PNF	Positive- and Negative-frequency
PQ	Power and Reactive Power
PV	Power and Voltage
RHP	Right Half-plane
RMS	Root Mean Square
ROCOF	Rate of Change of Frequency
SCR	Short-circuit Ratio
SG	Synchronous Generator
SM	Submodule
SSCI	Subsynchronous Control Interaction
SSTI	Subsynchronous Tortional Interaction
SSO	Subsynchronous Oscillation
SSR	Subsynchronous Resonance
STATCOM	Static Synchronous Compensator
SVC	Static Var Compensator
TES	Thermal Energy Storage
VSC	Voltage Source Converter
VSG	Virtual Synchronous Generator
VSM	Virtual Synchronous Machine

Introduction

The increasing penetration of power electronic converters in power systems is driven by electricity decarbonization, the growing demand for renewable energy, and advancements in high-voltage direct current (HVDC) technology. Power electronic converters, which facilitate nonsynchronous connections, flexible voltage support, and precise power control, will play an important role in modern power systems. However, these converters have introduced new wide-frequency oscillations, distributed in a mirror-frequency manner, into power systems due to control interactions, which threaten power system stability and have raised widespread concern.

This book aims to systematically present the fundamental stability criteria, the mechanisms of mirror-frequency oscillations, stability analysis methods, and stability measures for both grid-following (GFL) and grid-forming (GFM) converters. A detailed discussion on impedance-based stability analysis and criteria is provided, including comprehensive forms of impedance modeling, along with their merits and relevance for analyzing mirror-frequency oscillations. Additionally, impedance measurement methods for black-box systems to conduct stability analysis are introduced.

This book provides a solid foundation in stability criteria and analysis methods for converter-driven oscillations, offering valuable insights to researchers, engineers, and system operators for understanding and addressing these oscillations in modern power systems.

1

Drives of High Penetration of Converters

Jun Liang[1,*] *and Chuanyue Li*[2]

[1] *School of Engineering, Cardiff University, Cardiff, UK*
[2] *School of Electric Power Engineering, South China University of Technology, Guangzhou, China*

*Corresponding author: LiangJ1@cardiff.ac.uk

1.1 High-voltage Direct Current

The remote location of fossil fuel generation plants, renewable energies like hydro and offshore wind, has driven the demand of long-distance power transfer technology since last century. High-voltage direct current (HVDC) technology is widely adopted because it offers better efficiency for long-distance power transmission. After decades of development, HVDC technology has been developed with two main categories: line-commutated converters (LCCs) and voltage source converters (VSCs). HVDC also enables the interconnection among remote power grids and benefits the optimized power exchange crossing a wide-range grid. In recent years, VSCs have been suggested for their potential to offer grid-forming capabilities. Given the increasing adoption of VSC-HVDC, these capabilities can further enhance the grid's stability by supporting both voltage strength and frequency inertia. A detailed description of both LCC-HVDC and VSC-HVDC for the grid application is provided below.

1.1.1 LCC-HVDC

LCC-HVDC, the first generation of HVDC technology, operates by converting AC voltage to DC voltage. This process generates the output DC voltage by retaining either the positive or negative half of a three-phase AC voltage. In other words,

Stability Analysis of Converter-Rich Power Grids, First Edition. Jun Liang and Chuanyue Li.
© 2026 The Institute of Electrical and Electronics Engineers, Inc. Published 2026 by
John Wiley & Sons, Inc. All rights reserved.

LCC controls the DC voltage by adjusting the proportion of the positive or negative half of the AC voltage through the firing angle of thyristor-based converters, as illustrated in Figure 1.1. Typically, DC-side voltage and power control are used, and it has no capability to control the AC voltage. Additionally, a stable AC voltage source is crucial for its valves commutation. This is why LCC is also known as a current source converter.

The development of LCC-HVDC technology is summarized in Figure 1.2. In 1954, the first LCC-HVDC project, named the Gotland HVDC link, was commissioned with a capacity of 20 MW and a voltage rating of 100 kV, covering a transmission distance of 96 km via a submarine cable [1]. At that time, mercury-arc valves were used. However, with advancements in power electronics, thyristors became more reliable and required less maintenance. In 1972, thyristors were first used in the Eel River HVDC project, an LCC-HVDC system with ratings of ±80 kV and 320 MW in a back-to-back configuration for asynchronous interaction [2]. Since then, thyristors have completely replaced mercury-arc valves in LCC-HVDC applications. Up to now, LCC-HVDC technology, as applied in the Changji-Guquan project, can operate at up to ±1100 kV with a capacity of 12 GW, spanning distances of up to 3,305 km [3]. LCC-HVDC technology is the feasible option for ultra HVDC applications exceeding ±800 kV and has been widely applied globally for long-distance power transmission.

Figure 1.1 Line-commutated converter.

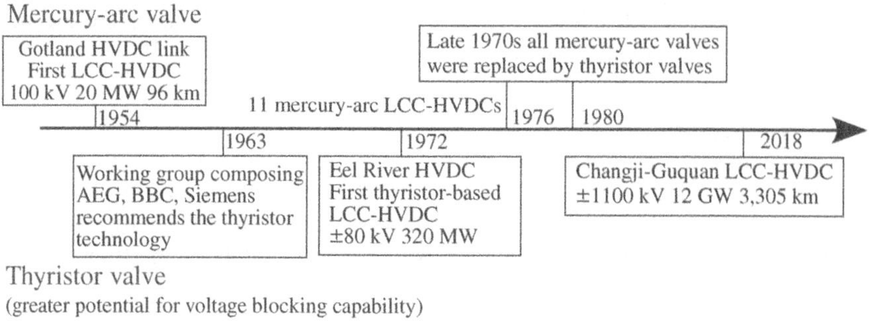

Figure 1.2 Development of LCC-HVDC.

1.1.2 VSC-HVDC

Originally, VSC technology was applied to motor drive systems. To avoid the commutation failures and the need for an AC voltage source inherent in LCC systems, VSC technology was attempted for HVDC applications. With advancements in the feasibility of series-connected Insulated Gate Bipolar Transistors (IGBTs) for high voltage withstand capability, the first VSC-HVDC system was commissioned in 1997, named Hellsjön project with ratings of ±10 kV and 3 MW.

Voltage Source Converters (VSCs) can be powered by a DC source to generate AC voltage, mimicking an AC voltage source, or vice versa. They offer several advantages over LCC, including (i) voltage and frequency support, (ii) four-quadrant power control, and (iii) black start capability. In contrast, the main advantage of LCC-HVDC is its ability to operate at higher voltage levels, making it suitable for longer distance power transfer with lower capital costs. Both HVDC technologies have their applications.

The evaluation of VSC-HVDC technology spans various topologies, from two-level converters to multilevel converters, as shown in Figure 1.3, driven by the motivations to reduce switching frequency, scale up DC voltage simply, and minimize harmonics.

Two-level converter: it represents the first generation of VSC technology applied to HVDC systems. In this topology, the DC/AC conversion operates by switching between two voltage levels at $+V_{dc}$ and $-V_{dc}$ using high-frequency pulse width modulation (PWM) to generate each instantaneous point of the AC voltage. Therefore, the switching frequency is significantly higher than the AC voltage frequency, typically ranging between 1 and 20 kHz. Since the AC voltage is created using square waves of $+V_{dc}$ and $-V_{dc}$ with specific duty cycles, the harmonics are numerous, necessitating an effective low-pass filter on the AC converter side.

Neutral point clamped (NPC) converter: it can be a diode-clamped or capacitor-clamped multilevel converter. By using additional diodes or capacitors, the two-level voltage can be clamped to achieve various predetermined voltage levels. With an organized sequence of switching, a multilevel AC voltage is created. Compared to two-level converters, in high voltage applications, diode-clamped converters avoid the difficulties of handling a large number of series IGBTs and reduce harmonics and switching frequency. However, a large number of voltage levels significantly increase the required number of high-voltage-rated diodes or capacitors, which reduces cost-effectiveness.

Modular multilevel converter (MMC): the concept of series sub-modules (SMs) eliminates the limitations of the neutral point clamped design in VSCs. It allows the converter to increase voltage levels simply by adding more SMs, benefiting from all the advantages of multilevel voltage and enhancing redundancy with additional SMs. The modular multilevel converter (MMC) has been widely adopted in VSC-HVDC application.

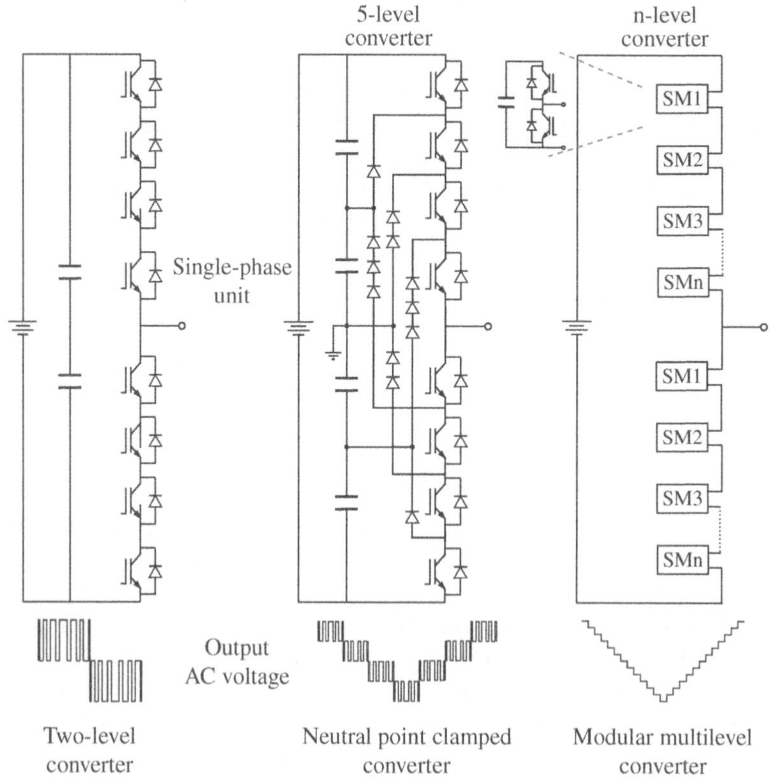

Figure 1.3 Different types of VSC technologies.

Since 1997, two-level VSC has been commercially used as an HVDC option for applications requiring AC voltage source behavior as shown in Figure 1.4. After 2000, NPC-based VSCs began to be implemented, such as the Murraylink HVDC Light Interconnection, which adopted a three-level VSC with ratings of ±150 kV and 220 MW. Since 2010, MMC technology has been applied to HVDC systems, significantly increasing the voltage and power levels of VSC-HVDC, exemplified by the Trans Bay Cable project with ratings of ±200 kV and 400 MW. The commissioned MMC station can reach a voltage level of ±525 kV, as demonstrated in the Viking Link, which transmits electricity over a distance of 767 km at 1,400 MW. By connecting two MMCs in series at a single station, the voltage level can be increased to an ultra HVDC level of ±800 kV. This approach is applied in the Kunbei-Liuzhou-Longmen (KBL) Hybrid Multi-terminal Ultra HVDC project, which boosts the power capacity to 5,000 MW and extends the transmission distance to 905 km. According to preliminary statistics, there are more than 70 VSC-HVDC projects either in operation or in the planning stages.

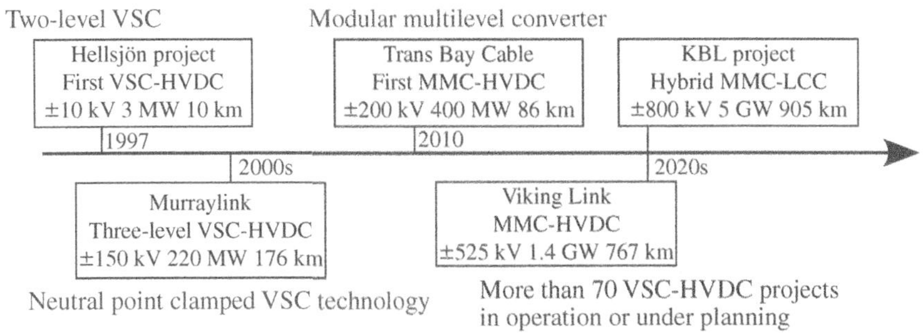

Figure 1.4 Development of VSC-HVDC.

With the advancement of MMC technology to the 800 kV voltage level, the transmission range has been extended to the ultra HVDC level. While MMCs provide stiff voltage and frequency support to the load center, the associated costs can be prohibitive when used exclusively across all HVDC stations. LCC, on the other hand, offers a more cost-effective approach for ultra long distance power transmission and boasts the capability for the longest power transmission. From an economic perspective, a hybrid configuration of LCC and MMC is widely considered at the ultra HVDC level to balance the cost impact of MMCs and benefit of the grid support capability. For instance, using LCC at the sending end and MMC at the receiving end, as implemented in the above-mentioned KBL hybrid HVDC project, ensures effective voltage and frequency support to the load center.

1.2 Renewable Energy

1.2.1 Wind Generation

When a wind turbine is connected to the grid, its operating speed is nearly fixed to the grid's frequency to deliver the necessary active power, which brings the following drawbacks:

- if wind speed and torque fluctuate, the fixed-speed drive train of a wind turbine experiences high stress, challenging its mechanical strength;
- the fixed-speed operation limits the flexibility to maximize wind energy capture;
- additionally, the induction generator used in the wind turbine may have a significant impact on the grid because it consumes reactive power for its excitation field and can draw high currents when starting directly from the grid.

Advancements in power electronics have driven the evolution of wind turbine technology, leading to improvements and solutions for the above drawbacks. To date, wind turbine technology has evolved into five types, four of which involve different combinations of power electronics, which are introduced below.

Type 1 wind turbine: this wind generator operates at a nearly fixed speed and connects directly to the grid, as shown in Figure 1.5. To limit the high inrush current resulting from connecting this wind turbine to the grid, a thyristor-based soft starter has been used since the 1980s [4]. This device mitigates the high-current effect on both the grid side and the turbine side.

Type 2 wind turbine: to increase the freedom of generator speed, a power-electronic-controlled variable resistor is used in the rotor circuit to regulate the rotor current, providing up to around 10% range of variable speed [5]. This variable-resistor-integrated induction generator is classified as the Type 2 wind turbine.

Type 3 wind turbine: in the late 1990s, back-to-back converters began to be used instead of power-electronic-controlled variable resistors, which increased the generator speed variance to about 25% [6]. This speed variance could be extended even further with a higher power-rated back-to-back converter. The converter also allows for separate active and reactive power control, providing the grid with additional services like voltage support. Currently, the Type 3 wind turbine is a popular option for wind farms, also named as double-fed induction generator.

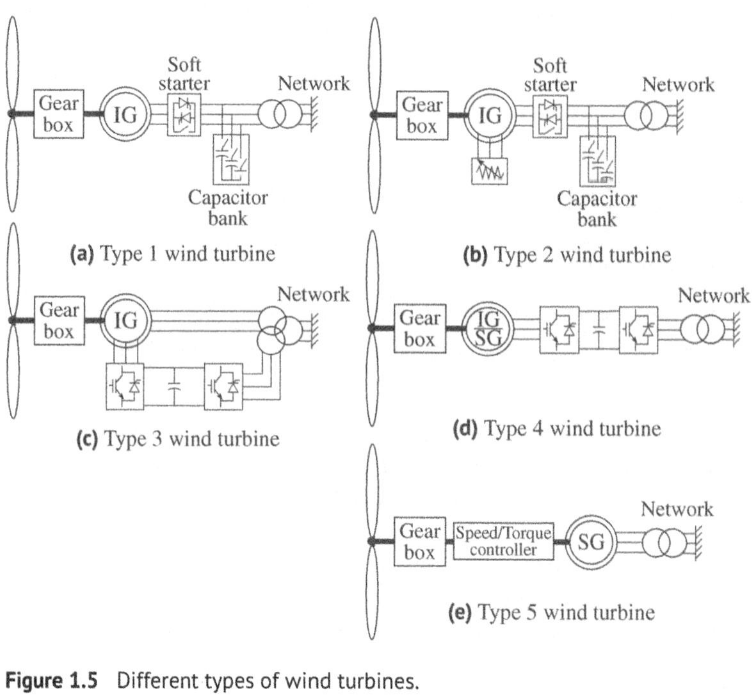

Figure 1.5 Different types of wind turbines.

Type 4 wind turbine: in the 2000s, the back-to-back converter was repositioned to the stator side to serve as the interface between the wind generator and the grid, fully decoupling the generator speed from the grid [7]. The converter's rating had to be increased to full power, but it offers a greater range of control for active and reactive power. Decoupling allows wind generators to achieve fully optimized gear safety margins and maximize wind energy capture. Additionally, both induction generators and synchronous generators can be used, offering more cost-effective options. The Type 4 wind turbine is an attractive option because a full-scale power converter decouples the wind turbine, allowing wind power to operate more like a conventional power plant.

Type 5 wind turbine: the concept of a synchronous generator is applied, featuring a typical variable-speed wind drive train as the primary driver, with the synchronous generator used for grid connection B4.62 [8]. A mechanical control box is used to change the variable speed of the wind drive train to a constant output shaft speed for the synchronous generator. Power electronic devices are barely involved in this Type 5 wind turbine.

To date, the Type 3 wind turbine has held a leading position in the market, accounting for 53.2% of it in 2022 [9]. The Type 4 wind turbine, characterized by its use of a full-scale power converter, generally entails higher capital costs. However, its capability to control the full speed of the wind generator, combined with its potential to offer additional grid services like inertia control, synchronizing power, and power system damping, is becoming more attractive and is likely to increase its market share in the future.

1.2.2 Solar Photovoltaic

Due to technological advancements in materials and manufacturing processes, photovoltaic (PV) systems become a highly affordable energy source, making them suitable for widespread deployment. Solar energy has been a critical sustainable source that supplies low-carbon energy to the grid. For grid-connected PV systems, it is required to maximize power transfer and provide auxiliary services to enhance grid stability. These two requirements commonly involve the use of DC/DC converters paired with grid-connected inverters as shown in Figure 1.6. As the power generated from a PV module primarily depends on solar irradiance and panel temperature, the DC/DC converter is responsible for maximum power point tracking (MPPT) and booting the voltage required level. The grid-connected inverter aids in supporting and stabilizing the grid. Grid-connected PV systems facilitate the overall management of PV modules and feed power to the grid through multiple entry inverters. In the solar plant, its PV modules are typically numerous and can be deployed in various configurations. These configurations are determined by factors such as power rating, voltage level, size, reliability, sustainability, and cost. The typical configurations are summarized and compared as below:

Central configuration: a large number of PV modules are connected in series to form a string. All strings are then connected in parallel to inverters. This configuration offers the lowest cost and easiest maintenance throughout its operation.

String configuration: PV module strings independently feed into the grid via their respective inverters. This configuration offers the advantage of independent operation, particularly under partial shading conditions, and enhances reliability in case of a fault in any string. Accordingly, this also results in increased costs.

Extended configuration: based on the two configurations mentioned above, additional connections among PV modules or a combination of both configurations will manifest as extended configurations. These extended configurations are highly customized and will provide the necessary flexibility and reliability with an additional cost increase; two examples are presented in Figure 1.7.

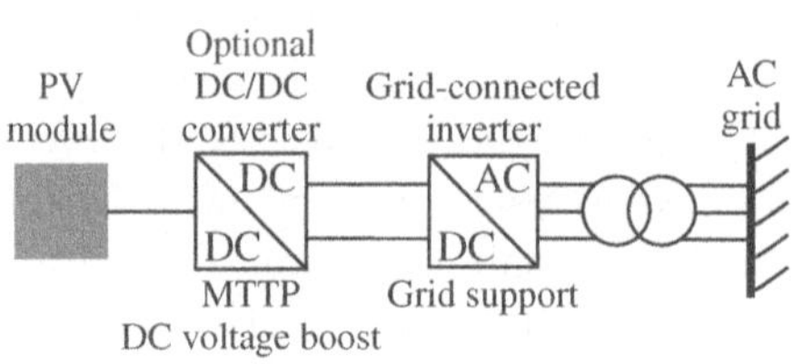

Figure 1.6 Grid-connected PV system.

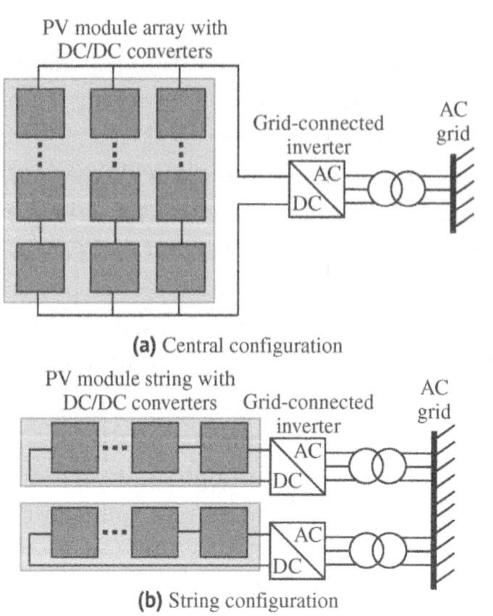

Figure 1.7 Configurations of grid-connected PV system.

Grid-connected inverters in PV systems utilize VSCs to provide grid support capabilities. The choice of VSC technology, whether it be a two-level converter, a neutral point clamped converter, or a modular multilevel converter, typically depends on the voltage level and cost considerations. As of 2023, the global solar PV capacity has reached 1,412,093 MW, while wind power capacity stands at 1,017,199 MW [10]. In 2023, solar energy accounted for nearly three-quarters of the increase in renewable energy capacity.

1.3 Energy Storage System

The growing integration of renewable energy sources puts pressure on the power system to maintain power balance and optimize power flow. The gradual decline of fossil fuel and nuclear plants due to decarbonization requirements and safety concerns further reduces the controllable power capacity of the system. Implementing energy storage systems is a viable solution to retain energy generated within the power system and return the stored energy when necessary. This allows to enhance reserve energy capacity, increase the ability to accommodate renewable energy, and enhance the power system's safety and stability margins. Additionally, energy storage systems facilitate real-time power balancing among generation, transmission, and load; help balance peak and off-peak power demands; and improve the overall efficiency of the power transfer.

Based on the types of stored energy, the energy storage systems have following categories as shown in Figure 1.8:

Mechanical energy storage system: the mechanism involves converting the excess electricity from the power system into potential energy and kinetic energy through mechanical means, including pumped hydro system, compressed air energy system, and flywheel energy system. When supplying the necessary electricity, the stored energy serves as the primary source to drive a generator. If a synchronous generator is used, a grid-connected converter may not be essential. However, if an asynchronous generator is employed to maximize energy generation, a grid-connected converter is typically required.

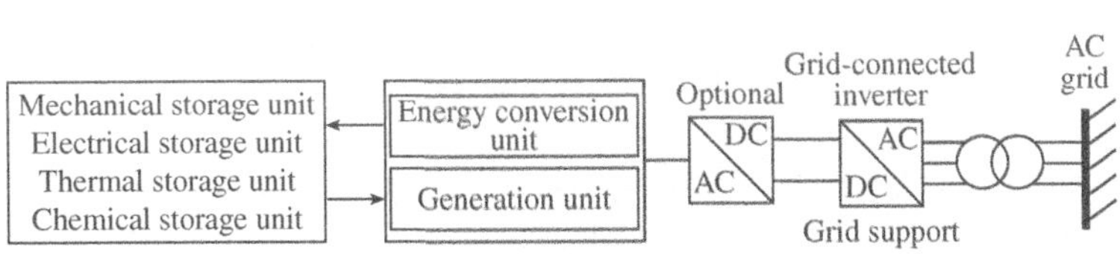

Figure 1.8 Energy storage system.

Electrochemical energy storage system: the system comprises both batteries and supercapacitors. Generally, rechargeable batteries and redox flow batteries are utilized. Rechargeable batteries include Li-ion batteries, lead-acid batteries, sodium-sulfur batteries, among others. Redox flow batteries utilize a different type of battery technology, consisting of two separate tanks that store liquid electrolytes. These electrolytes are pumped through a cell stack, where electrochemical reactions occur. Supercapacitors, also known as electrochemical double-layer capacitors, function similarly to traditional capacitors but offer a higher energy storage capacity.

Thermal energy storage: thermal energy storage (TES) systems function by accumulating energy through the heating or melting of materials, which then retain this energy in the form of either sensible or latent heat. These systems typically consist of storage media and are equipped with specialized devices engineered to facilitate both the injection and extraction of the media.

Chemical energy storage: excess electricity is utilized to drive the chemical reactions of raw materials, converting it into fuel for stored energy. When needed, this fuel is then converted back into electricity through a generator and fed back into the grid. Typical examples include hydrogen and biomass energy applications. An energy storage system commonly utilizes a VSC as the grid-connected converter. This converter serves as the interface that enables asynchronous operation of the typical energy storage or output functions. It also plays a crucial role in providing additional support services to the grid.

References

1 Olof Samuelsson. Gotland HVDC link from 1954 is awarded IEEE milestone. *IEEE Power Electronics Magazine*, pages 91–91, June 2017.

2 W.A. Patterson. The Eel River HVDC scheme — a 320 MW asynchronous interconnection between the New Brunswick Electric Power Commission and Hydro-Québec employing thyristor valves. *Canadian Electrical Engineering Journal*, 2(1): 9–16, 1977. https://doi: 10.1109/CEEJ.1977.6592831.

3 F. Yu, Y. Wu, and J. Guo. Research on the resonance characteristics of ±1100 kv UHVDC transmission system. In *The 16th IET International Conference on AC and DC Power Transmission (ACDC 2020)*, volume 2020, pages 852–857, 2020. https://doi: 10.1049/icp.2020.0183.

4 Frede Blaabjerg and Ke Ma. Future on power electronics for wind turbine systems. *IEEE Journal of Emerging and Selected Topics in Power Electronics*, 1(3): 139–152, 2013. https://doi: 10.1109/JESTPE.2013.2275978.

5 E.H. Camm, M.R. Behnke, O. Bolado, M. Bollen, M. Bradt, C. Brooks, W. Dilling, M. Edds, W.J. Hejdak, D. Houseman, S. Klein, F. Li, J. Li, P. Maibach, T. Nicolai, J. Patino, S.V. Pasupulati, N. Samaan, S. Saylors, T. Siebert, T. Smith, M. Starke,

and R. Walling. Characteristics of wind turbine generators for wind power plants. In *2009 IEEE Power & Energy Society General Meeting*, pages 1–5, 2009. https://doi: 10.1109/PES.2009.5275330.

6 A. Pena, J.C. Clare, and G.M. Asher. Doubly fed induction generator using back-to-back PWM converters and its application to variable speed wind-energy generation. *IEE Proceedings Electric Power Applications*, 143(3): 231–241, 1996.

7 F. Blaabjerg, Zhe Chen, and S.B. Kjaer. Power electronics as efficient interface in dispersed power generation systems. *IEEE Transactions on Power Electronics*, 19(5): 1184–1194, 2004. https://doi: 10.1109/TPEL.2004.833453.

8 Working Group B4.62. Connection of wind farms to weak AC networks. Technical report, CIGRE, 2016.

9 Zhao Feng, Lee Joyce, and Lathigara Anajali. Global wind market development: supply side data 2022. Technical report, Global Wind Energy Council, 2022.

10 International Renewable Energy Agency (IRENA). Renewable capacity statistics 2024. Technical report, National Grid ESO, 2024.

2

Challenges and Future Development of Grid-connected Converters

Jun Liang[1,] and Chuanyue Li[2]*

[1] *School of Engineering, Cardiff University, Cardiff, UK*
[2] *School of Electric Power Engineering, South China University of Technology, Guangzhou, China*

*Corresponding author: LiangJ1@cardiff.ac.uk

2.1 Conventional Classification of Power System Stability Based on Disturbances

The classification of converter-induced stability is mainly determined by how the converter interacts with other system components, as described in the following sections. However, the root causes differ depending on control strategies and power system device types, leading to excessive subcategories that may confuse the understanding of system stability caused by power electronic devices.

Conventional stability classification, such as those based on small and large disturbances, could provide a clearer perspective on power system stability by covering both synchronous generators and power electronic devices. Accordingly, this demonstrates that both synchronous generators and power electronic devices could induce small-disturbance and large-disturbance stability issues in power systems, though their underlying mechanisms and causes differ. Therefore, these traditional power system stability classifications will be explained first to facilitate the understanding of converter-induced instabilities.

Stability Analysis of Converter-Rich Power Grids, First Edition. Jun Liang and Chuanyue Li.
© 2026 The Institute of Electrical and Electronics Engineers, Inc. Published 2026 by John Wiley & Sons, Inc. All rights reserved.

2.1.1 Steady-state Stability Condition (Without Considering Damping Characteristics)

The steady-state stability of a power system requires it to return to (or remain close to) the pre-disturbance operating point following any small disturbance. If damping characteristics are not considered, the steady-state stability condition applies only within the system's steady-state operating limits. If the system is already at its steady-state operating limit, any small disturbance will cause instability.

In a power system, the power characteristic can be represented by:

$$P^* = \frac{V_1^* V_2^*}{X^*} \sin \delta \tag{2.1}$$

where P is the transmission power between two nodes, V_1 and V_2 are the voltages of two nodes, X represents the transmission reactance between two nodes, and δ represents the phase angle difference between the two voltages; * for P, V_1, V_2, and X denote the values in per unit.

In normal operation, where V_1^* and V_2^* are both approximately 1 p.u., the active power can be presented as:

$$P^* = \frac{1}{X^*} \sin \delta \tag{2.2}$$

The steady-state stability condition requires $\frac{dP}{d\delta} > 0$. For a given grid impedance X^*, the maximum power that can be transmitted to the grid without losing steady-state stability is achieved at $\delta = 90°$, which is denoted as steady-state power limit as follows:

$$P_{\max}^* = \frac{1}{X^*} \sin 90° = \frac{1}{X^*} \tag{2.3}$$

According to the steady-state stability criterion and Equation 2.3, when the actual power remains below $P_{\max}^*$, the system is stable (as long as the system damping is positive). However, when $P \geq P_{\max}^*$, the system becomes unstable regardless of its damping characteristics; any small power increase will induce this instability. This defines the steady-state power limit, which is the steady-state stability condition without considering damping characteristics.

The recent extensive integration of power electronic converters has introduced a new stability index for power systems, which is the short-circuit ratio (SCR), defined as:

$$\mathrm{SCR} = \frac{S_{\mathrm{sc}}}{P_{\mathrm{rating}}} \tag{2.4}$$

where P_{rating} is the converter power rating and S_{sc} is the short-circuit power.

For the purpose of simplicity in analysis and without loss of generality, the converter rating P_{rating} and the rated operating voltage V_{rating} is selected as the base value, such that:

$$S_{\text{base}} = P_{\text{rating}} \qquad V_{\text{base}} = V_{\text{rating}} \tag{2.5}$$

The base impedance is then:

$$Z_{\text{base}} = V_{\text{base}}^2 / S_{\text{base}} \tag{2.6}$$

The relationship between S_{sc} and the transmission reactance (X) is therefore defined as follows:

$$S_{\text{sc}} = V_{\text{rating}} \frac{V_{\text{rating}}}{X} \tag{2.7}$$

Converting this equation to per-unit values yields:

$$S_{\text{sc}}^* = S_{\text{sc}} / S_{\text{base}} = V_{\text{rating}} \frac{V_{\text{rating}}}{X} / S_{\text{base}} = \frac{1}{X^*} \tag{2.8}$$

Thus, the SCR can be equivalently expressed as:

$$\text{SCR} = \frac{S_{\text{sc}} / S_{\text{base}}}{P_{\text{rating}} / P_{\text{base}}} = \frac{S_{\text{sc}}^*}{P_{\text{rating}}^*} = \frac{1}{X^*} \tag{2.9}$$

where P_{rating}^* represents converter power rating in per unit (equal to 1 p.u.).

A comparison of Equations 2.3 and 2.9 yields:

$$\text{SCR} = P_{\max}^* = \frac{1}{X^*} \tag{2.10}$$

For a converter required to operate at its rated power, SCR or X^* represents the stability boundary for maintaining steady-state stability while outputting rated power. Consequently, to maintain safe operating margins below the steady-state power limit, the rated operating power of a converter must satisfy:

$$P_{\text{rating}}^* < P_{\max}^* - \Delta_{\text{margin}} \tag{2.11}$$

where Δ_{margin} represents the necessary stability margin.

Substituting $P_{\text{rating}}^* = 1$ p.u. into the equation above yields:

$$P_{\max}^* > 1 + \Delta_{\text{margin}} \tag{2.12}$$

Substituting SCR for $P_{\max}^*$ shows that, to maintain safe operating margins below the steady-state power limit, the SCR must satisfy:

$$\text{SCR} > 1 + \Delta_{\text{margin}} \tag{2.13}$$

where Δ_{margin} represents the necessary stability margin.

It should be noted that, from a mathematical point of view, if the power rating of converters could be selected higher than P_{max}, or if the transmission reactance X increases, P_{max} is reduced; then $P_{rating} > P_{max}$ is possible, leading to SCR < 1. However, in this case, if the converter actual output power is smaller than P_{max}, then the system can still maintain the steady-state stability.

It should be noted that, for renewable power generation systems and VSC-HVDC applications, converters must be ready for full rated power transfer at any time to accommodate available power variations and operational requirements. So $P_{rating} < P_{max}$. However, from a mathematical point of view, if the power rating of converter P_{rating} could be selected higher than P_{max}, or if the transmission reactance X increases, P_{max} is reduced; then $P_{rating} > P_{max}$ is possible, leading to SCR < 1. Under this case, the system can still maintain the steady-state stability, as long as the converter actual output power is smaller than P_{max}.

This steady-state stability analysis gives a general indication, which is only based on steady states, assumes 1 p.u. operation voltage, uses a simplified system structure, neglects line resistance and capacitance, and does not consider control system dynamics with damping characteristics.

2.1.2 Steady-state Stability Condition (Considering Damping Characteristics)

Steady-state stability requires a power system possess sufficient damping to suppress small disturbance-induced oscillations and can return to the pre-disturbance operating point. This is typically analyzed using small-signal modeling and stability criteria based on Eigen values, pole locations, Nyquist plot, Bode plot, damping factor, etc. While this analysis can reveal non-oscillatory instability (divergence), this mode is very rare in practice; it is usually identified and corrected during the design stage and is often caused by control design errors such as unintended positive feedback.

The various control strategies and loops of a converter, such as synchronization, phase-locked loops, power control, AC and DC voltage control, and current control, exhibit distinct oscillatory characteristics when interacting with the power system. The underlying causes of these oscillations normally vary and will be presented with examples in Section 2.5.1.

2.1.3 Subsynchronous Resonance

Subsynchronous resonance (SSR), which represents a steady-state stability phenomenon, typically occurs due to interactions between series capacitors and both the electrical and mechanical systems of synchronous generators (SGs). Line LC resonance can also interact with the rotor shafts of SGs, forming an electromechanical resonance as shown in Figure 2.1. Its oscillation frequency is normally

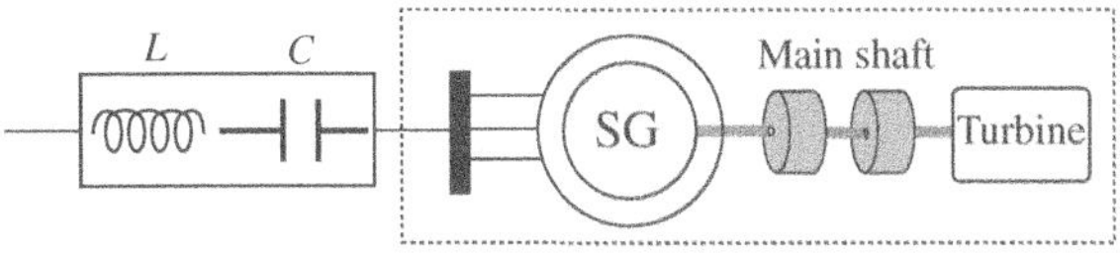

Figure 2.1 Resonance interaction between SGs and series capacitors.

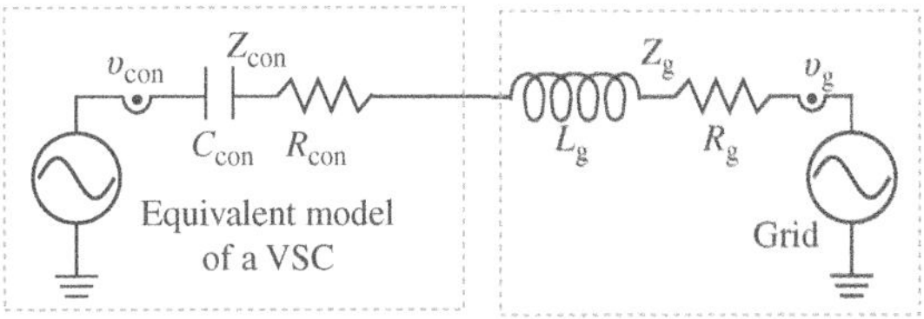

Figure 2.2 Analysis of control interaction via the impedance criterion.

below the rated system frequency. A detailed classification of SSR is presented in Section 2.3.1.

With the extensive recent integration of power electronic converters, their control systems can also induce electrical subsynchronous oscillations, which can be characterized through impedance-based electrical resonance theory. When modeling converter control as an impedance, its interaction with the system typically manifests at resonant oscillation frequencies. As illustrated in Figure 2.2, the converter control typically consists of proportional gain and integrator components, which can be represented as a resistor in series with a capacitor:

$$Z_{con} + Z_g = -R_{con} + 1/(C_{con}\omega_1 j) + R_g + L_g\omega_1 j \tag{2.14}$$

The *LC* resonance will be amplified if the overall resistance is negative or close to zero. Under these conditions, even a small disturbance can trigger resonant oscillations. In summary, the converter system exhibits small-disturbance instability if the following conditions occur:

$$\frac{1}{C_{con}\omega_1 j} + L_g\omega_1 j = 0 \quad \text{and} \quad -R_{con} + R_g \leq 0 \tag{2.15}$$

Whereas SGs induce electromechanical resonance through physical system dynamics, power electronic converters generate control-induced oscillations, which can be represented as impedance resonance phenomena.

2.1.4 Transient Stability

Transient stability refers to a power system's ability to maintain synchronism following large disturbances, where the post-disturbance steady-state operating condition may substantially differ from the pre-disturbance condition.

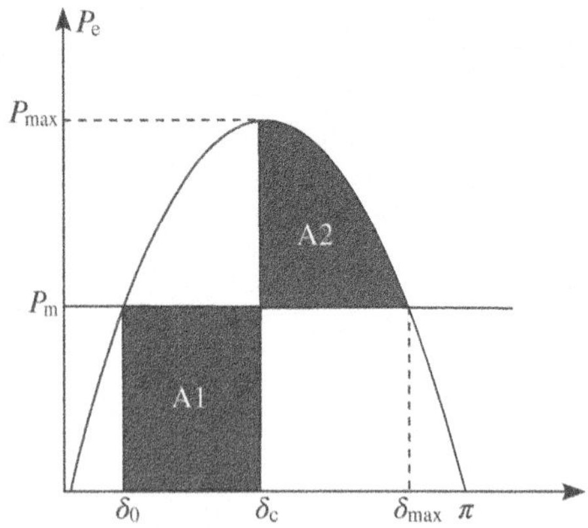

Figure 2.3 Transient stability.

Similar to the steady-state stability condition in Section 2.1.1, its stability mechanism is also governed by the power characteristics described in Equation 2.3, which is illustrated in Figure 2.3.

A classic example is an SG maintaining synchronism during severe grid faults. The Equal Area Criterion is a classical method used to assess transient stability. As shown in Figure 2.3, A1 represents the accelerating area where $P_m > P_e$ during the fault. After the fault is cleared, the decelerating area A2 forms when $P_m < P_e$. If A2 > A1, it indicates that the SG can return to synchronism, and the system is considered transiently stable. Otherwise, the SG fails to maintain synchronism, indicating transient instability. This characteristic is due to the SG inherently having mechanical inertia.

Unlike synchronous generators, converters do not rely on mechanical inertia and are not constrained by the same swing dynamics. Therefore, even if δ_c reaches to the critical power angle, the converter may still recover synchronism through fast control action, depending on the control scheme. This gives converter-based systems potential advantages in handling transient disturbances compared to conventional SGs.

2.2 Overview of VSC-induced Oscillation Events

Power system instabilities caused by power electronic devices have been known to occur since the presence of LCC-HVDC. This type of power system instability is primarily caused by the interaction between converter controls and power systems. Recently, the extensive deployment of VSCs has led to a large number of power system instability events worldwide. Moreover, the increasing use of converters is intensifying concerns about related power system instabilities.

In 2020, the "Classification of power system stability" was modified to categorize all types of instabilities caused by converters, and two new categories of "Resonance stability" and "Converter-driven stability" were proposed to include most of the VSC-induced oscillations that have occurred in recent years.

A preliminary summary of worldwide VSC-induced oscillation events is presented in Table 2.1. It has been found that in the classification of "Resonance stability," VSCs are currently causing electrical resonance due to doubly fed induction generator (DFIG) wind turbines. In the classification of "Converter-driven stability," the oscillations caused by VSCs and grid impedance (weak grid)

Table 2.1 Preliminary summary of worldwide VSC-induced oscillation events.

Year	Location	Frequency	Interacting devices	
Resonance stability				
2009	South Texas Transmission System, Texas, USA	22 Hz	DFIG wind farm Series capacitor	Electrical resonance (converter induced)
2012	Hebei wind farm, China	7 Hz	DFIG wind farm Series capacitor	
2013	Guyuan Wind farm Hebei, China	8.1 Hz	DFIG wind farm Series capacitor	
2017	South Texas Transmission System, Texas, USA	20–30 Hz	DFIG wind farm Series capacitor	
Converter-driven stability				
2011	Competitive Renewable Energy Zone Texas, USA	4 Hz	Wind converter Weak grid	Converter grid impedance
2014	First Solar's solar farm, California, USA	7 Hz	PV converter Weak grid	
2015	Hami wind farm Xinjiang, China	27–33 Hz	Wind converter Weak grid	
2016–17	Luxi HVDC, Yunan, China	1270 Hz	VSC-HVDC Weak grid	
2018 Publication year	Hydro One Networks, Canada	20, 420, 780, 2370 Hz	Converter Grid impedance	
2018	EYv back-to-back HVDC China	700, 1810 Hz	VSC-HVDC Grid impedance	
2019 Publication year	Hydro One Networks, Canada	3.5 Hz	PV converter Weak grid	
2019	Hornsea one, UK	8–9 Hz	Wind farm converter weak grid	
2021	Dominion Energy's service territory, USA	21.5–22.5 Hz	PV converter Grid impedance	
2021	SSEN-T and SPEN transmission system, Scotland, UK	8 Hz	Under investigation	
2010	Woodward Wind Oklahoma, USA	12–14 Hz	Wind farms Wind farms	Wind farm converter Other PE device (No SG)
2015 Publication year	Nanao wind farm China	20 Hz	Wind DFIG MMC	
2017	Jinxia wind farm, Xijiang, China	37.5 Hz	Wind DFIG STATCOM	

dominate the causes. Interactions with other power electronic devices, such as wind turbines, wind turbines with modular multilevel converter (MMCs), and wind turbines with static synchronous compensator (STATCOMs), have been observed. Therefore, VSC control can interact with many power system devices and circuits to induce oscillations, with oscillation frequencies ranging from several hertz to more than 1,000 Hz.

The interaction caused by the VSC control side is due to its multiple control loops and wide control bandwidth, which can interact with many of the aforementioned power system devices. These devices are only those that have been recorded; there may be additional devices causing new oscillation issues. The control interaction between VSCs and power systems involves complicated stability issues, with many causes contributing and potentially leading to severe power outages, which poses a great challenge to power system stability and places a significant hindrance on VSC-involved renewable integration.

Therefore, organizations working on power systems have been paying significant attention to these oscillations to fully understand, recognize, and resolve these issues. In the following sections, stability analysis and solutions for VSC-induced oscillations will be discussed.

2.2.1 Oscillation Frequency of Different Electrical Quantities

The oscillation frequency is a typical indicator used to classify the event as either a subsynchronous oscillation or a supersynchronous oscillation. For a given oscillation event, its frequency can be identified from various electrical quantities, such as power, root mean square (RMS) voltage, instantaneous voltage, and *dq*-frame voltage. System operators typically use power or RMS voltage to represent the oscillation frequency. However, the oscillation frequency of different electrical quantities can vary.

The oscillation frequencies of frequency-decoupled quantities, such as power, RMS voltage, and dq-frame voltage, are different from those of instantaneous values like instantaneous voltage and current. This is because these decoupled quantities are calculated based on their separation from the fundamental frequency f_0*, making them distinct from the instantaneous values. The relationship between the frequency-decoupled values (*f_1*) and the instantaneous values (*f_2*) is as follows:*

$$f_2 = f_0 \pm f_1 \tag{2.16}$$

An example with a voltage oscillation at $f_1 = 8$ Hz is presented to demonstrate the oscillation frequency characteristics of different electrical quantities. A converter system, as illustrated in Figure 2.4, is employed. The Fourier analysis results for voltage, current, and power at the measurement point are shown in Figure 2.5. Key observations include:

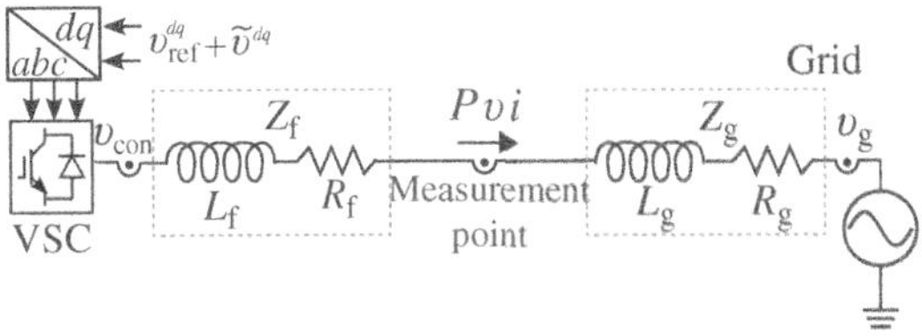

Figure 2.4 Converter system.

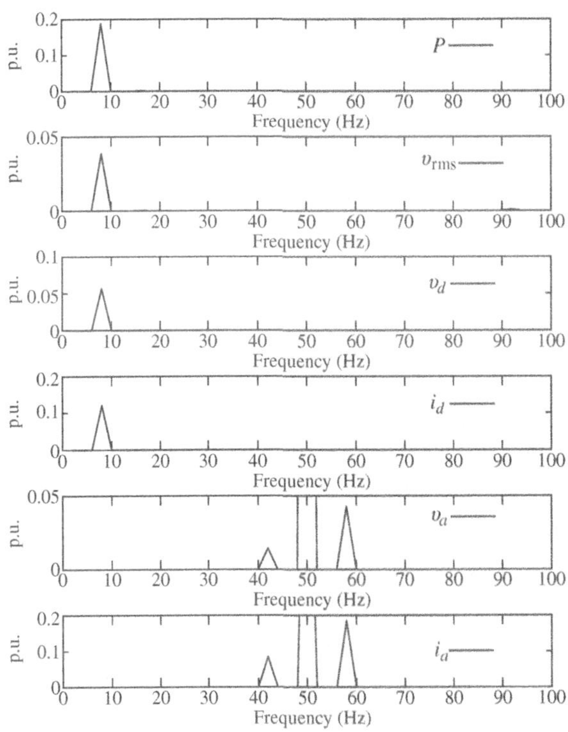

Figure 2.5 Oscillation frequency of different electrical quantities for the same oscillation event.

- 8 Hz oscillation is observed in power P, RMS value v, and dq-frame quantities such as v_d and i_d.
- 42 and 58 Hz oscillations are observed in instantaneous voltage v_a and current i_a quantities.

The mechanism behind Equation 2.16 is attributed to the mirror-frequency effect, which will be described in Chapter 7.

2.3 Subsynchronous Oscillations

Subsynchronous oscillation (SSO) is a major concern for system operators as it threatens device lifespan and causes system instability. In synchronous generator-dominated power systems, SSOs primarily arise from two sources: interactions between turbine-generators and passive system elements, such as series capacitors, or between turbine-generators and power electronic control systems, such as HVDC and static var compensators (SVCs). Accordingly, these interactions are classified into two categories: subsynchronous resonance (SSR) and device-dependent subsynchronous oscillations (DDSSO) of the System Dynamic Performance Subcommittee Power System Engineering Committee [1], of the System Dynamic Performance Subcommittee [2].

With the high penetration of VSCs in recent decades, it has been found that power electronic devices interacting with generators, or even power electronic devices alone, have induced many new subsynchronous oscillations. Consequently, a new classification of subsynchronous oscillations is updated [3]. In brief, the device-dependent subsynchronous oscillation is included in the SSR category under the subcategory of torsional resonance. A new category, referred to as power electronic device interactions (PEDIs), is proposed to encompass SSOs induced by power electronic devices without interaction with synchronous generators. A summary of the classification of SSO in 1985 and 2022 is presented in Figure 2.6, with further details provided below.

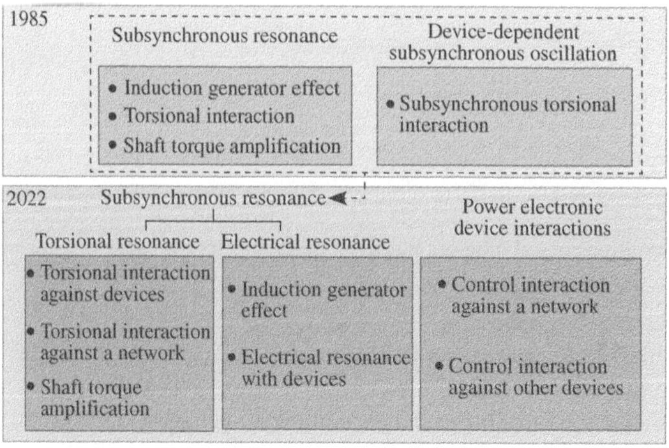

Figure 2.6 Classification of subsynchronous oscillations.

Generally, both SSR in SGs and SSO induced by VSC control interactions belong to the steady-state stability classification.

2.3.1 Subsynchronous Resonance

Subsynchronous resonance initially describes the electromechanical interaction between the generator and the series capacitor, categorized as:

1. **Induction generator effect** only involves the electrical system dynamics of the generator's self-excitation. If the equivalent resistance of the generator and networks is negative at the resonant frequency, electrical resonance occurs in the armature currents, which is either sustained or grows.
2. **Torsional interaction** is an interplay between the electrical and mechanical system dynamics of the generator. If the rotor oscillation is close to the electrical resonance frequency and the damping torque of the rotating system is insufficient at this frequency, the generator will be self-excited, producing a torque that reinforces the aforementioned rotor oscillations. This can result in sustained or growing oscillations.
3. **Shaft torque amplification** involves electrical oscillations amplifying the mechanical torsional resonance. When the electrical oscillation is close to the rotor's natural resonance frequency and the torsional damping is low, large torques may be induced in the shaft.

Besides the series capacitor, it has been observed that active devices, such as power electronic control systems, can induce the self-excitation of the generator in pure electrical dynamics or in both electrical and mechanical dynamics. These phenomena are referred to as subsynchronous control interaction (SSCI) and subsynchronous torsional interaction (SSTI), described in Sections 2.5.2.2 and 2.5.2.1, which have been categorized under a new SSO type called DDSSO. However, these DDSSOs involve the self-excitation of generators and are similar to the SSR family. Therefore, in the latest update, the classification of subsynchronous resonance has been further refined to include DDSSO, starting from electrical resonance and torsional resonance as outlined below.

1. **Electrical resonance** involves the electrical system dynamics of the generator's self-excitation at the resonant frequency, leading to the reinforcement of electrical oscillations.
   - *Electrical resonance with devices* refers to the electrical resonance induced by the active devices such as power electronic converters, for example, the SSCI phenomenon, which is described in detail in Section 2.5.2.2.
   - *Induction generator effect* refers to the aforementioned induction generator effect.

2. **Torsional resonance** involves the interplay between the electrical and mechanical system dynamics of the generator's self-excitation at the resonant frequency, ultimately leading to the amplification of torque or rotor oscillations.
   - *Torsional interaction against devices* refers to the torsional oscillation or amplification induced by active devices, such as power electronic converters. For example, the SSTI phenomenon, which is described in detail in Section 2.5.2.1.
   - *Torsional interaction against a network* refers to the aforementioned torsional interaction with passive elements of a network, such as capacitors.
   - *Shaft torque amplification* refers to the aforementioned shaft torque amplification.

2.3.2 Power Electronic Device Interactions

Apart from SSR in generators, power electronic devices have been observed to induce SSO due to control interactions with the network or other devices. The relevant oscillation events of VSCs are presented in Table 2.1. This has been proposed as a new category of SSO and is divided into two subcategories:

1. **Control interaction against a network** refers to power electronic control systems interacting with passive network elements, ultimately leading to network oscillations, such as the weak-grid oscillation events shown in Table 2.1.
2. **Control interaction against other devices** refers to power electronic control systems interacting with other network devices, ultimately leading to network oscillations, such as the oscillation events involving wind farm converters and other power electronic (PE) devices shown in Table 2.1.

2.4 Classification of Power System Stability

Classification of power system stability is critical to effectively turn the complexity into multiple representative and solvable instability cases. Relevant stability classifications for the overall power system have been widely discussed and delivered by CIGRE Committee and IEEE Committee. This section will introduce the development of the classification of power system stability and how the recently proposed stability category of power electronic devices is distributed.

2.4.1 Classification of Power System Stability in 1982

The first effort on unifying terms and definition of power system stability was made in 1982 [4], as shown in Figure 2.7. At that time, the causes of instability were

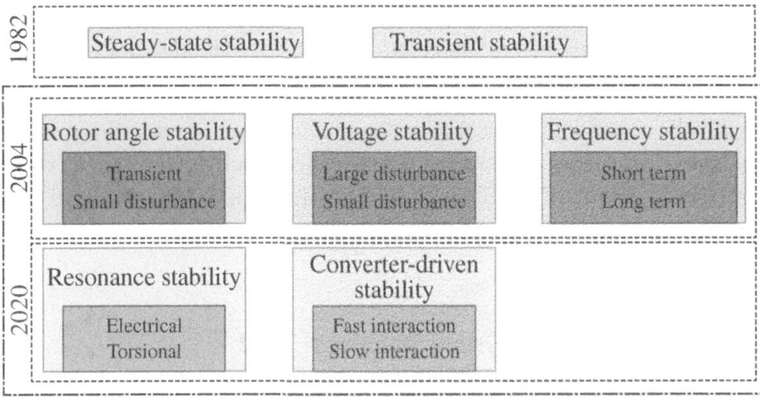

Figure 2.7 Development of classification of power system stability.

mainly from insufficient synchronizing torque and insufficient damping torque. The classification of power system stability was plain and mainly relied on the size of disturbances. Two categories were classified as steady-state stability and transient stability. The details about the classification have been explained in detail in Section 2.1.

2.4.2 Classification of Power System Stability in 2004

After continual expansion of power system and advance of technology and controls, the system instabilities have become enormously various, which led to an insufficient identification of each stability phenomenon via the previous classification.

In 2004, a new classification of power system stability was developed to encompass comprehensive instability phenomena and forms three main categories with multiple subcategories, which precisely specified each representative instability phenomenon [5], or as shown in Figure 2.7. A formal definition of power system stability is also given: *Power system stability is the ability of an electric power system, for a given initial operating condition, to regain a state of operating equilibrium after being subjected to a physical disturbance, with most system variables bounded so that practically the entire system remains intact.* A brief introduction of these three main categories are shown below:

1. **Rotor angle stability** refers to the ability of synchronous machines of an interconnected power system to remain in synchronism after being subjected to a disturbance.
2. **Voltage stability** refers to the ability of a power system to maintain steady voltages at all buses in the system after being subjected to a disturbance from a given initial operating condition.

3. **Frequency stability** refers to the ability of a power system to maintain steady frequency following a severe system upset resulting in a significant imbalance between generation and load.

This classification is more practical than the previous one for categorizing via main characters of a generator-dominant power system such as rotor angle, voltage, and frequency. Terms of steady-state stability and transient stability in the previous classification are still retained under each main category for the newly developed classification.

2.4.3 Classification of Power System Stability in 2020

With increasing penetration of converter interfaced generation (CIG) into power systems, the classification of power system stability proposed in 2004 is further extended with additional inclusion of the effect of fast-response power electronic converters in 2020 [6] and a full version [7], or as shown in Figure 2.7.

In fact, impacts of various power electronic devices on the system stability have been observed for more than 40 years. Such stability phenomenon did not raise much attention from power system due to its low adoption rate. However, in the recent decade CIG via VSC technology has been enormously penetrated to the power system for interfacing extensive sustainable energies, which leads to a great increase of related instability phenomena. Therefore, after a careful deliberation within IEEE Committee and CIGRE Committee, two more categories are defined to encompass the system stability of CIG, which are shown below:

4. **Resonance stability** refers to the ability of a power system to pose sufficient damping for system resonance especially SSR, which includes torsional resonance and electrical resonance.

   The torsional resonance manifests in the interaction between (1) series compensation and the mechanical torsional frequencies of the turbine-generator shaft and (2) power electronic devices and generator torsional system.

   The electrical resonance is due to a resonance between series compensation and the electrical characteristics of the generator. This phenomenon is mainly induced by negative damping effect from the control of connected converters.
5. **Converter-driven stability** refers to the ability of converters' control to maintain a well-damped interaction with a power system. The wide band of the control response of a CIG may lead to a power system oscillation over a wide frequency range, including subsynchronous and supersynchronous oscillations.

In this extended classification, effects of CIG on the previous three categories including rotor angle stability, voltage stability, and frequency stability are also introduced but not limited to the current extent due to the fast development of CIG.

Considerations about these effects are mainly for the trend that the form of a power system is shifting from a generator-dominant system to a generator-based system with a considerable portion of interfaced converters. The specific stability concerns of converters are the system inertia reduction, reduced fault current tolerance, and potential risk of negative control interaction.

2.5 Control Interaction of CIG

High penetration of CIG into a power system brings extensive system stability issues, which actually involve each category of the power system stability classification, as introduced previously. The main cause is due to the interaction between the control system of a converter and the power system. This phenomenon is generally termed as control interaction, which is a double-edged sword for the power system as it could worsen the stability condition or conversely help to stabilize the system oscillation. Its effectiveness highly depends on the quality of the control design. If an improper setting of the converter control is applied or the control design does not carefully consider such phenomenon, many devices in the power system would be a potential cause to induce the oscillation. To give a deep recognition and comprehensive understanding of the control interaction, experiences and recent progresses are presented.

2.5.1 Past Experiences with Control Interactions from Power Electronic Devices

Improperly designed control systems of most powerful electronic devices may interact with the power system, which leads to undesired oscillations.

Control-interaction instabilities emerged from the power system events since the 1970s in the Square Butte project of line commutated converter (LCC)-HVDCs [8]. At that time it was found that the DC control of LCC-HVDC could produce negative damping of torsional oscillations at 11.5 Hz when interacting with a turbine-generator. A term SSTI was defined to refer to this phenomenon due to control interaction.

Around 10 years later, the analyzed results at planning stage of Chester SVC application [9] showed that the voltage control of a static var compensator (SVC) could also interact with a turbine-generator and led to SSTI.

Since then a new category within Subsynchronous Resonance Group [10] is defined as DDSSO, which specially refers to the torsional interaction between turbine-generators and other power system components.

In 2021's classification of power system stability, subsynchronous resonance belongs to the category "resonance stability," and DDSSO belongs to the corresponding subcategory "torsional resonance."

2.5.2 Control Interaction of VSCs

In the recent decade, high penetration and future enormous planning of converter interfaced generation based on VSC technology have induced many instability events in the power system, and the related instability concern is persistently increasing. There are instabilities dominantly related to the control interaction of converter interfaced generation. In a power system, a wide range of devices could induce the control interaction with a VSC, which brings great difficulties to specify and classify these instabilities.

2.5.2.1 Control Interaction with Mechanical Mass System of Turbine-generators

As previously mentioned in Section 2.5.1, control interaction with mechanical mass system of turbine-generators is termed as SSTI, which was firstly found in an LCC-HVDC system. When both a VSC and a turbine-generator are integrated in the same system, as shown in Figure 2.8a, the SSTI may happen. It has been

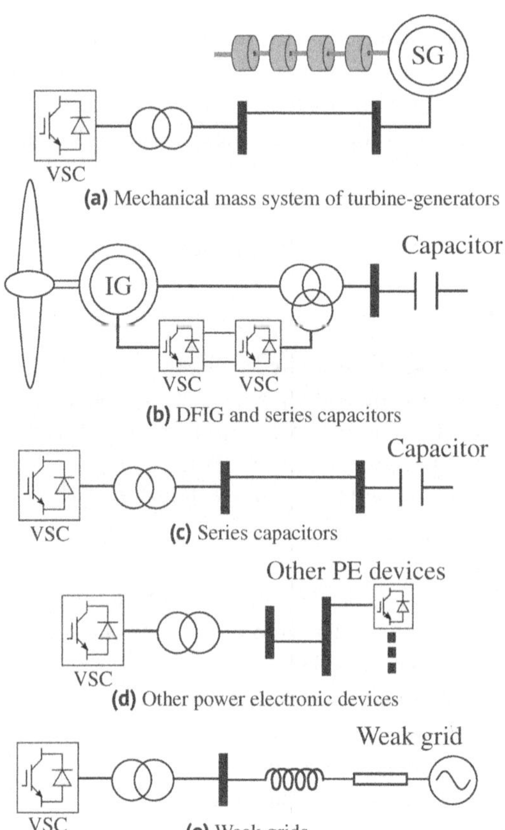

Figure 2.8 Control interaction with power system's devices.

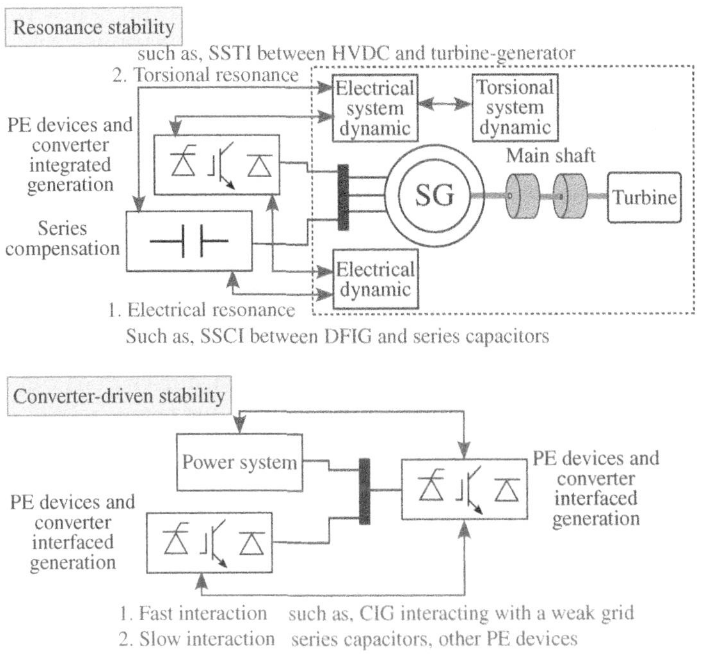

Figure 2.9 Recent Classifications of Converter-Induced Stability Issues in Power Systems

demonstrated in simulation study that this SSTI phenomenon could also happen with a VSC when connecting to a turbine-generator [11]. The current control with voltage feedforward may cause the negative-resistance effect, which could induce the torsional oscillation.

In 2014, the power system in Xinjiang China experienced an SSO event [12], which involved VSCs and torsional resonance. The investigation showed that this torsional resonance actually was induced by the similar frequency oscillation in the power system, which was primarily due to the control interaction of a VSC with a weak system. Therefore, this event is categorized into weak-grid instability rather than an SSTI event.

Although this SSTI phenomenon due to VSCs has not yet occurred in the real world, it should be well considered in the converter design and power system requirements. The SSTI phenomenon is categorized in the classification of resonance stability—torsional, as shown in Figure 2.9.

2.5.2.2 Control Interaction with Series Capacitors

The first well-known VSC instability event occurred in 2009 [13], which is due to the interaction between a doubly fed induction generator and series capacitors.

Science then, this kind of instability is termed commonly as SSCI [14] or may be known as series capacitor SSO [15]. Their definitions are presented below:

- SSCI refers to the interaction between the control system of generation resources (typically converter based) and series capacitors.
- Series capacitor SSO refers to the interaction between doubly fed induction generators (DFIGs) and series capacitors at radial connection.

Until now, the relevant SSCI events [15] are all due to the interaction between DFIGs and series capacitors. As shown in Figure 2.8b, the DFIG has both induction generators and converters. The definition of SSCI points out this interaction is due to the interaction between the control of converters and series capacitors. However, after a deep analysis of the interaction mechanism, a re-clarified conclusion [16] is given out that the dominant cause of interaction between DFIGs and series capacitors is due to induction generator effect (IGE), which is a resonance between series compensations and the electrical characteristics of generators. More specifically, the control of converters induces this IGE due to its negative-resistance effect [17].

Therefore, the current classification of power system stability puts SSCI phenomena into the category of resonance stability rather than converter-driven stability, due to IGE as the dominant cause.

Except the real-world events, in research analysis SSCI could happen purely due to the interaction between converters and series capacitors, as shown in Figure 2.8c, which should be given an appropriate attention.

The SSCI phenomenon, which happens in the interaction between DFIGs and series capacitors, is categorized into classification of resonance stability—electrical, as shown in Figure 2.9. The other SSCI phenomenon, which happens in the interaction between converters and series capacitors, is categorized into classification of converter-driven stability.

2.5.2.3 Control Interaction with Other Power Electronic Devices

The occurrence of control interaction between power electronic devices, as shown in Figure 2.8d, is also widely concerned with increasing penetration of power electronic devices. It is believed that many combinations of power electronic devices, such as VSC-STATCOM, VSC-LCC, VSC-SVC, and VSC-VSC, could suffer this control interaction.

In 2017, a 37.5 Hz oscillation occurred at Guohua wind farm in China [18], which is induced by the interaction of DFIGs and nearby STATCOM. After a detailed analysis, it was pointed out that this oscillation was caused due to the control interaction between the rotor-side converter of DFIG and the STATCOM. Therefore, this event could be considered to be a first VSC-STATCOM interaction in the real world.

HVDC links between France and the United Kingdom are the case that may risk the control interaction. Relevant transmission system operators like Réseau de Transport d'Électricité (RTE) formed the hardware-in-the-loop setup to identify the future risk of control interaction between IFA2000 and Eleclink (LCC-VSC) [19], and IFA2 and SVC Tourbe (VSC-SVC) [20].

Other analysis such as Li and Fan [21] shows that the interaction control between parallel VSCs could also happen if droop coefficients are too small.

This control interaction is categorized in the classification of converter-driven stability, as shown in Figure 2.9.

2.5.2.4 Control Interaction with Weak Power System

The commonly adopted control of VSCs is susceptible to a weak power system or named weak grid, which leads to the dominant cause of unstable control interaction of a VSC. Relying on the summarized real-world instability events due to VSC control interaction [15], it is found that almost 80% of instability events are weak-grid instability. Figure 2.8e presents the weak-grid connection, which normally refers to short-circuit ratio (SCR) < 3 or grid impedance $> 1/3$ p.u. The general explanation of this unexpected interaction is that the voltage at connection point is not stiff enough to resist the disturbance from converter control system.

Weak grid is also the dominant cause of converter-driven stability as a category of extended classification of power system stability 2021. All four subcategories, which including high-frequency oscillation, weak system stability, low-frequency oscillations, and stability issues related to power transfer limits, are highly related to this instability cause, as presented in Figure 2.9.

2.6 Overview of Weak-grid Caused Instabilities

Weak-grid connection always matters to power electronic devices across their developments. A wide range of power electronic devices such as LCC, flexible alternating current transmission systems (FACTS), and VSC suffered or are suffering the instability concern of this weak-grid connection.

Three inherent causes, which drive this unstable connection, are summarized and introduced below:

1. mismatched reactive power compensation, which leads to considerable voltage change, could cause the instability;
2. current (such as DC component, subsynchronous frequency component, fundamental component, and harmonic component) from a power electronic device, which leads to considerable voltage distortion or voltage variation

when a weak grid is connected, could destabilize control of the device or its commutation process;

3. control interaction, which the control system of a power electronic device may interact substantially with a weak grid, could lead to unexpected instability.

2.6.1 Past Experiences with Weak-grid Instabilities

During the development of LCC-HVDC, its well operation highly relies on a stiff power system. This is mainly due to its massive harmonic current injection and delicate commutation process. Therefore, weak-grid instabilities are widely concerned and the above-mentioned causes 1 and 2 are dominant factors.

Since the 1960s due attention has been paid to this concern with careful thinking and experience upgrade on filter and reactive power compensation [22]; the relevant risks of weak-grid connection have been well controlled.

At the similar time, to further enhance the stable connection to a weak grid, the inverter control of LCC-HVDCs has been carefully configured as an extinction angle control [23] to provide a safe margin for the commutation process.

With increased adoption of LCC-HVDCs, in the late 1970s, an LCC-HVDC Square Butte interacted in a negative-damping way with a turbine-generator, which caused 11.5 Hz oscillation [24]. Since then it is noticed that even at a smooth steady state, oscillations still may occur due to their control interaction with turbine-generators, which has been mentioned in Section 2.5.2.1. A similar grid configuration of CU HVDC system also raised such interaction concern. Filed tests and simulation analysis were carried out in 1979 [25]. The analysis results from both HVDC systems show that the weak grid could play an important role to substantially worsen this interaction.

In the 1980s, control-interaction oscillation due to weak-grid connection appears in the research of FACTS devices such as SVC [26]. To effectively solve this control interaction, the control systems of power electronic devices are improved based on small-signal stability analysis by parameter redesign or additional damping control, which is a cost-effective and widely accepted way.

2.6.2 VSC: Weak-grid Instabilities

Since the 1990s, the VSC is developed and warmly welcomed for grid interface due to its AC voltage control capability. In the recent decade, decarbonization of power system is propelled globally by large-scale integration of sustainable energy. The final goal of decarbonization intends to progress to the ambition of net zero. VSCs as critical interfaces of power system will widely and continually serve this ambition by integrating wind farms, PV panels, storages, and HVDCs.

Comparing to LCCs working as a current source, VSCs feature capability of voltage sources. Therefore, VSCs were believed to be more stable when connecting to a weak grid. However, control interaction is the most unpredictable cause of an unstable weak-grid connection, and it is found that VSCs may suffer this instability due to weak-grid interaction.

For such a large-scale penetration of VSCs into power system, risks of control interaction due to VSCs may be widely introduced into power system and weak-grid instability would tremendously disturb power system, which has raised a considerable concern in the world.

Over the past decade, relevant instabilities have occurred in many national power systems. As summarized in Cheng et al. [15], more than 19 events related to VSC-grid instability have occurred worldwide and around 80% events are caused by weak grids. Complexity of the VSC control system brings difficulties to explicitly understand which part of control system contributes this undesired interaction. Some of post-event analyses have pointed out which part of control system causes this weak-grid interaction, which leads to the following summary:

- voltage control: 2011 TEXAS 4 Hz oscillation [27], 2019 Hornsea one 9 Hz oscillation [28]
- phase-locked loop and current control: 2015 Xinjiang 30 and 80 Hz (dominant) oscillation [29]
- delay of the feedforward: 2017 Luxi 1270 Hz oscillation [30]

It could summarize that control interaction with a weak grid is a complex stability issue, which involves almost any part of the control system. Oscillation frequencies due to these control parts are distinctive, which is considered to be determined by their parameter design. A representative characteristic of oscillation frequency based on the above summarized cases are demonstrated in Figure 2.10, which may help to understand the dominant part of the control system when interacting with a weak grid. Oscillation frequency induced by the outer loop is the lowest due to its tuning design, which intends to response slower than the inner loop and PLL. The dynamics of delay and filter are normally much faster than the control, which causes the high-frequency oscillation.

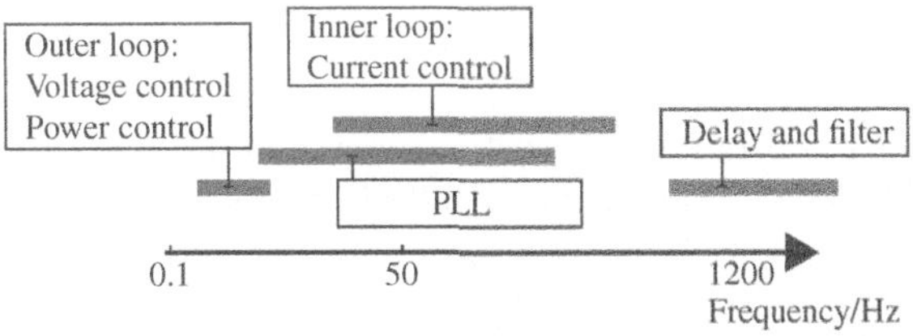

Figure 2.10 General dynamic range of different converter controls

References

1 IEEE Subsynchronous Resonance Working Group of the System Dynamic Performance Subcommittee Power System Engineering Committee. Terms, definitions and symbols for subsynchronous oscillations. *IEEE Transactions on Power Apparatus and Systems*, PAS-104(6): 1326–1334, 1985. https://doi.org/10.1109/TPAS.1985.319152.

2 IEEE Subsynchronous Resonance Working Group of the System Dynamic Performance Subcommittee. Reader's guide to subsynchronous resonance. *IEEE Transactions on Power Systems*, 7(1): 150–157, 1992. https://doi.org/10.1109/59.141698.

3 WG C4B4.52. Guidelines for subsynchronous oscillation studies in power electronics dominated power systems. Technical report, CIGRE, 2022.

4 Power System Engineering Committee. Proposed terms & definitions for power system stability. *IEEE Transactions on Power Apparatus and Systems*, PAS-101(7): 1894–1898, 1982. https://doi.org/10.1109/TPAS.1982.317476.

5 P. Kundur, J. Paserba, V. Ajjarapu, G. Andersson, A. Bose, C. Canizares, N. Hatziargyriou, D. Hill, A. Stankovic, C. Taylor, T. Van Cutsem, and V. Vittal. Definition and classification of power system stability ieee/cigre joint task force on stability terms and definitions. *IEEE Transactions on Power Systems*, 19(3): 1387–1401, 2004. https://doi.org/10.1109/TPWRS.2004.825981.

6 Nikos Hatziargyriou, Jovica Milanovic, Claudia Rahmann, Venkataramana Ajjarapu, Claudio Canizares, Istvan Erlich, David Hill, Ian Hiskens, Innocent Kamwa, Bikash Pal, Pouyan Pourbeik, Juan Sanchez-Gasca, Aleksandar Stankovic, Thierry Van Cutsem, Vijay Vittal, and Costas Vournas. Definition and classification of power system stability – revisited & extended. *IEEE Transactions on Power Systems*, 36(4): 3271–3281, 2021. https://doi.org/10.1109/TPWRS.2020.3041774.

7 Nikos Hatziargyriou, Jovica Milanovic, Claudia Rahmann, Venkataramana Ajjarapu, Claudio Canizares, Istvan Erlich, David Hill, Ian Hiskens, Innocent Kamwa, Bikash Pal, Pouyan Pourbeik, Juan Sanchez-Gasca, Aleksandar Stankovic, Thierry Van Cutsem, Vijay Vittal, and Costas Vournas. Stability definitions and characterization of dynamic behavior in systems with high penetration of power electronic interfaced technologies. Technical report, IEEE Power and Energy Society, 2020.

8 M. Bahrman, E.V. Larsen, R.J. Piwko, and H.S. Patel. Experience with HVDC – turbine-generator torsional interaction at square butte. *IEEE Transactions on Power Apparatus and Systems*, PAS-99(3): 966–975, 1980. https://doi.org/10.1109/TPAS.1980.319726.

9 N. Rostamkolai, R.J. Piwko, E.V. Larsen, D.A. Fisher, M.A. Mobarak, and A.E. Poitras. Subsynchronous interactions with static var compensators-concepts and

practical implications. *IEEE Transactions on Power Systems*, 5(4): 1324–1332, 1990. https://doi.org/10.1109/59.99384.

10 IEEE Committee-Subsynchronous Resonance Working Group. Reader's guide to subsynchronous resonance. *IEEE Transactions on Power Systems*, 7(1): 150–157, 1992. https://doi.org/10.1109/59.141698.

11 Lennart Harnefors. Analysis of subsynchronous torsional interaction with power electronic converters. *IEEE Transactions on Power Systems*, 22(1): 305–313, 2007. https://doi.org/10.1109/TPWRS.2006.889038.

12 Huakun Liu, Xiaorong Xie, Jingbo He, Tao Xu, Zhao Yu, Chao Wang, and Chuanyu Zhang. Subsynchronous interaction between direct-drive PMSG based wind farms and weak AC networks. *IEEE Transactions on Power Systems*, 32(6): 4708–4720, 2017. https://doi.org/10.1109/TPWRS.2017.2682197.

13 D. Kidd and P. Hassink. Transmission operator perspective of sub-synchronous interaction. In *PES T&D 2012*, pages 1–3, 2012. https://doi.org/10.1109/TDC.2012.6281627.

14 P.E. Lawrence and C. Gross, Jr. Sub-synchronous grid conditions: New event, new problem, and new solutions. In *37th Annual Western Protective Relay Conference*, 2010.

15 Yunzhi Cheng, Lingling Fan, Jonathan Rose, Shun-Hsien Huang, John Schmall, Xiaoyu Wang, Xiaorong Xie, Jan Shair, Jayanth R. Ramamurthy, Nilesh Modi, Chun Li, Chen Wang, Shahil Shah, Bikash Pal, Zhixin Miao, Andrew Isaacs, Jean Mahseredjian, and Jenny Zhou. Real-world subsynchronous oscillation events in power grids with high penetrations of inverter-based resources. *IEEE Transactions on Power Systems*, 38(1): 316–330, 2023. https://doi.org/10.1109/TPWRS.2022.3161418.

16 Lingling Fan, Rajesh Kavasseri, Zhixin Lee Miao, and Chanxia Zhu. Modeling of DFIG-based wind farms for SSR analysis. *IEEE Transactions on Power Delivery*, 25(4): 2073–2082, 2010. https://doi.org/10.1109/TPWRD.2010.2050912.

17 Einar V. Larsen. Wind generators and series-compensated AC transmission lines. In *2012 IEEE Power and Energy Society General Meeting*, pages 1–4, 2012. https://doi.org/10.1109/PESGM.2012.6344581.

18 Yanhui Xu, Shimeng Zhao, Yuping Cao, and Kai Sun. Understanding subsynchronous oscillations in DFIG-based wind farms without series compensation. *IEEE Access*, 7: 107201–107210, 2019. https://doi.org/10.1109/ACCESS.2019.2933156.

19 H. Saad, P. Rault, S. Dennetière, M. Schudel, C. Wikstrom, and K. Sharifabadi. Hil simulation to assess interaction risks of HVDC systems for upcoming grid development. In *IECON 2020 The 46th Annual Conference of the IEEE Industrial Electronics Society*, pages 5041–5048, 2020. https://doi.org/10.1109/IECON43393.2020.9254807.

20 Boris Bruned, Sébastien Dennetière, Yannick Vernay, Hani Saad, and Vinicius Oiring de Castro Cezar. Optimized real-time simulation setup for interaction study between VSC-HVDC and SVC on the French network. In *IECON 2022 – 48th Annual Conference of the IEEE Industrial Electronics Society*, pages 1–6, 2022. https://doi.org/10.1109/IECON49645.2022.9969119.

21 Yin Li and Lingling Fan. Stability analysis of two parallel converters with voltage–current droop control. *IEEE Transactions on Power Delivery*, 32(6): 2389–2397, 2017. https://doi.org/10.1109/TPWRD.2017.2656062.

22 J. Kauferle, R. Mey, and Y. Rogowsky. H.V.D.C. stations connected to weak A.C. systems. *IEEE Transactions on Power Apparatus and Systems*, PAS-89(7): 1610–1619, 1970. https://doi.org/10.1109/TPAS.1970.292808.

23 Eckhard Rumpf and S. Ranade. Comparison of suitable control systems for HVDC stations connected to weak AC systems part i: New control systems. *IEEE Transactions on Power Apparatus and Systems*, PAS-91(2): 549–555, 1972. https://doi.org/10.1109/TPAS.1972.293239.

24 M. Bahrman, E.V. Larsen, R.J. Piwko, and H.S. Patel. Experience with HVDC – turbine-generator torsional interaction at square butte. *IEEE Transactions on Power Apparatus and Systems*, PAS-99(3): 966–975, 1980. https://doi.org/10.1109/TPAS.1980.319726.

25 K. Mortensen, E.V. Larsen, and R.J. Piwko. Field tests and analysis of torsional interaction between the coal creek turbine-generators and the CU HVDC system. *IEEE Transactions on Power Apparatus and Systems*, PAS-100(1): 336–344, 1981. https://doi.org/10.1109/TPAS.1981.316861.

26 A.J.P. Ramos and H. Tyll. Dynamic performance of a radial weak power system with multiple static var compensators. *IEEE Transactions on Power Systems*, 4(4): 1316–1325, 1989. https://doi.org/10.1109/59.41681.

27 Shun-Hsien Huang, John Schmall, Jose Conto, John Adams, Yang Zhang, and Cathey Carter. Voltage control challenges on weak grids with high penetration of wind generation: Ercot experience. In *2012 IEEE Power and Energy Society General Meeting*, pages 1–7, 2012. https://doi.org/10.1109/PESGM.2012.6344713.

28 UK National Grid. Appendices to the technical report on the events of 9 August 2019. Technical report, National Grid ESO, 2019.

29 Jian Sun, Mingjie Li, Zhigang Zhang, Tao Xu, Jingbo He, Haijiao Wang, and Guanghui Li. Renewable energy transmission by HVDC across the continent: system challenges and opportunities. *CSEE Journal of Power and Energy Systems*, 3(4): 353–364, 2017. https://doi.org/10.17775/CSEEJPES.2017.01200.

30 Changyue Zou, Hong Rao, Shukai Xu, Yan Li, Weiwei Li, Jun Chen, Xiaobin Zhao, Yu Yang, and Bo Lei. Analysis of resonance between a VSC-HVDC converter and the AC grid. *IEEE Transactions on Power Electronics*, 33(12): 10157–10168, 2018. https://doi.org/10.1109/TPEL.2018.2809705.

3

Fundamental Stability Criteria for Feedback Systems

Jun Liang[1,*] *and Chuanyue Li*[2]

[1] *School of Engineering, Cardiff University, Cardiff, UK*
[2] *School of Electric Power Engineering, South China University of Technology, Guangzhou, China*

*Corresponding author: LiangJ1@cardiff.ac.uk

3.1 The Mathematical Mechanism of System Stability

This section will mathematically explain how to determine the stability characteristics of a system. A typical feedback system is depicted in Figure 3.1, and the following terminologies are defined:

- $u(s)$: the input such as control reference
- $y(s)$): the output of the system
- $G(s)$: forward path transfer function
- $H(s)$: feedback transfer function

Correspondingly, the transfer function is given below:

$$y(s) = \underbrace{\frac{G(s)}{1 + \underbrace{G(s)H(s)}_{L(s)}}}_{T(s)} u(s) \tag{3.1}$$

where $T(s)$ is the closed loop transfer function, $L(s)$ is the loop transfer function, and $1 + G(s)H(s)$ is the characteristic equation.

Stability Analysis of Converter-Rich Power Grids, First Edition. Jun Liang and Chuanyue Li.
© 2026 The Institute of Electrical and Electronics Engineers, Inc. Published 2026 by John Wiley & Sons, Inc. All rights reserved.

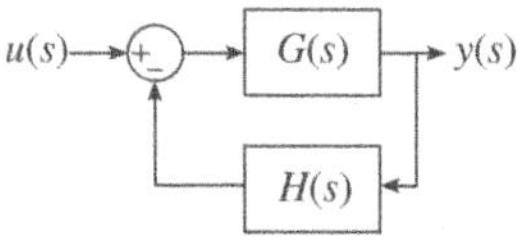

Figure 3.1 Typical feedback system.

$T(s)$ represents a linear time-invariant system that is independent of the input. Typically, it is analyzed as a continuous system in the s-domain but can be converted into the z-domain for discrete analysis.

$T(s)$ can be formulated as the form of a numerator and a denominator, and as presented below:

$$T(s) = \frac{\sum_{i=0}^{m} b_i s^i}{\sum_{i=0}^{n} a_i s^i} \tag{3.2}$$

where a_i and b_i represent the gains.

The denominator polynomial of $T(s)$ is defined as the characteristic equation:

$$C(s) = \sum_{i=0}^{n} a_i s^i = \prod_{i=1}^{n} (s - p_i) = 0 \tag{3.3}$$

Then $T(s)$ can be further expanded in partial fractions as:

$$T(s) = \frac{\prod_{i=1}^{m} (s - z_i)}{\prod_{i=1}^{n} (s - p_i)} = \sum_{i=1}^{n} \frac{R_i}{s - p_i} \tag{3.4}$$

where z_i is the zeros of $T(s)$ and p_i are the poles of $T(s)$.

In fact, the poles of $T(s)$ are the roots of $C(s)$. The time response of $T(s)$ is determined by taking the inverse Laplace transform:

$$T(t) = \sum_{i=1}^{n} R_i e^{p_i t} \tag{3.5}$$

where p_i can be represented as $\sigma_i + j\omega_{d,i}$ and σ_i is the real part of the i-th pole.

Equation 3.5 includes the term $e^{\sigma_i t}$. If $\sigma_i > 0$, $T(t)$ will not converge and the system will become unstable. This provides the mathematical basis for analyzing system stability. Consequently, system stability can be clearly visualized on the s-plane, where σ_i corresponds to the real axis and ω_i to the imaginary axis.

Key information on the stability characteristics of a system has been provided above, which is sufficient for the stability analysis of a converter system. For a rigorous mathematical proof or a more detailed derivation of system stability characteristics, please refer to Farid Golnaraghi [1].

The stability characteristics of a system are primarily determined by the positions of the poles in the transfer function's s-plane. To enable efficient stability analysis and extend its use in the frequency domain, various stability criteria based on pole maps or frequency domain have been devised. These criteria are closely aligned with the fundamental characteristic of stability, which is to form a judgment on whether there are poles close to or in the right half-plane (RHP).

3.2 Stability Criterion via Pole Map

A system is represented by a transfer function, and its stability is assessed based on the position of its poles. The criterion for determining stability include:

- *if the poles of the transfer function $T(s)$, which are the roots of its characteristic function $C(s)$ as represented by both Equations 3.4 and 3.3, are located in the RHP of the s-plane, this indicates system instability.*
- *if the poles are located on the imaginary axis of the pole maps, it indicates that the system is marginally stable.*

In general, without the aid of computer, it is difficult to reformulate $C(s)$, as represented in Equation 3.3, from $\sum_{i=0}^{n} a_i s^i$ into the form of $\prod_{i=1}^{n}(s - p_i)$. One method using the Routh-Hurwitz criterion is introduced below to analyze pole positions, which is applied to $C(s)$ as formulated in Equation 3.3. By reformulating the coefficients of its characteristic equation into a Routh array, if the first column of this array are all positive, it indicates that all the roots of $C(s)$ are in the left half-plane (LHP), confirming that the system is stable.

Based on Equation 3.3, the first two rows of the Routh array can be formed:

$$\begin{array}{llllll} \text{Row 1}: & n & a_n & a_{n-2} & a_{n-4} & \cdots \\ \text{Row 2}: & n-1 & a_{n-1} & a_{n-3} & a_{n-5} & \cdots \\ & n-2 & b1 & b2 & b3 & \cdots \\ & n-3 & c1 & c2 & c3 & \cdots \\ \cdots & \cdots & \cdots & \cdots & \cdots & \cdots \end{array} \tag{3.6}$$

The rest of the array, such as $b1$, $b2$, $c1$, $c2$, can be calculated using the relation below:

$$b1 = -\frac{\det\begin{bmatrix} a_n & a_{n-2} \\ a_{n-1} & a_{n-3} \end{bmatrix}}{a_{n-1}}$$

$$b2 = -\frac{\det\begin{bmatrix} a_n & a_{n-4} \\ a_{n-1} & a_{n-5} \end{bmatrix}}{a_{n-1}}$$

$$c1 = -\frac{\det\begin{bmatrix} a_{n-1} & a_{n-3} \\ b1 & b2 \end{bmatrix}}{b1}$$

$$c2 = -\frac{\det\begin{bmatrix} a_{n-1} & a_{n-5} \\ b1 & b3 \end{bmatrix}}{b1}$$

The Routh-Hurwitz criterion offers a technique to determine if the transfer function $T(s)$ has any poles in the RHP. However, calculating the Routh array by hand can be laborious for high-order systems.

With advancements in computer technology, the poles of $T(s)$ can be accurately and simply determined with the help of computers, such as using MATLAB, instead of the aforementioned method.

In fact, the poles can also determine a system's response in terms of damping performance and oscillation. A typical second-order system is presented below to demonstrate how poles manifest the system's stability and response:

$$T(s) = \frac{\omega_n^2}{s^2 + 2\zeta\omega_n s + \omega_n^2} \tag{3.7}$$

where ζ is the damping ratio and ω_n is the natural frequency.

The poles are calculated as the roots of the characteristic equation of $T(s)$, as shown below:

$$\text{if}\,\zeta \geq 1,\ \text{overdamped}\quad p_{1,2} = \underbrace{-\zeta\omega_n \pm \omega_n\sqrt{\zeta^2 - 1}}_{\sigma} \tag{3.8}$$

$$\text{if}\,0 < \zeta < 1,\ \text{underdamped}\quad p_{1,2} = \underbrace{-\zeta\omega_n}_{\sigma} \pm j\underbrace{\omega_n\sqrt{1 - \zeta^2}}_{\omega_d} \tag{3.9}$$

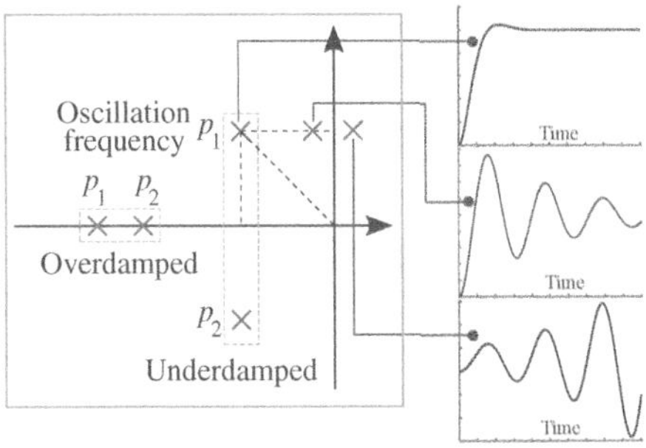

Figure 3.2 Example of the relationship between poles and output response.

The overdamped system $T(s)$ has two real poles, indicating that the system response will not exhibit any oscillation. The underdamped system $T(s)$ contains two complex poles, and the oscillation frequency is at ω_d. Pole positions of both situations are presented in Figure 3.2. The time response of the underdamped system is presented below:

$$y(t) = 1 - e^{-\zeta\omega_n t}\left(\cos(\omega_d t) + \frac{\zeta}{\sqrt{1-\zeta^2}}\sin(\omega_d t)\right) \quad \text{Step response} \tag{3.10}$$

$$y(t) = \frac{\omega_n}{\sqrt{1-\zeta^2}}e^{-\zeta\omega_n t}\sin(\omega_d t) \quad \text{Impulse response} \tag{3.11}$$

It is found that the imaginary part of the pole represents the oscillation frequency ω_d in the time domain. If the damping factor is selected properly, the step response will not manifest any oscillation. An example is presented in Figure 3.2 by moving $p_{1.1}$ to the RHP. $p_{1.1}$ and $p_{1.2}$ are located in the LHP, which lead to a stable step response. However, $p_{1.2}$ is closer to the RHP, indicating that the damping factor is smaller and the system is less stable, with oscillations manifesting at frequency ω_d. $p_{1.3}$, located in the RHP, indicates that the system is unstable at the same oscillation frequency ω_d.

3.3 Stability Criterion via Bode Plot Analysis

A transfer function can be analyzed in the frequency domain to illustrate the system's response across various frequencies. It avoids the identification of each pole of the transfer function and provides an alternative perspective of the response of the transfer function in the frequency domain. The transformation from the s-domain to the frequency domain involves substituting s with $j\omega$ in the transfer function as $T(\omega j)$. The Bode plot, which graphs the magnitude and phase of $T(\omega j)$, is one method used to analyze stability.

The same system, formulated by Equation 3.1, is used here to explain how the Bode plot illustrates system stability. Both the loop transfer function $L(s)$ and the characteristic equation $C(s)$ can be applied for the Bode plot analysis.

As presented in Equation 3.4, the characteristic equation is:

$$C(s) = 1 + G(s)H(s) = \prod_{i=1}^{n}(s - p_i) \tag{3.12}$$

After substituting s with ωj yields:

$$C(\omega j) = 1 + G(\omega j)H(\omega j) = \prod_{i=1}^{n}(\omega j - p_i) \tag{3.13}$$

where $p_i = \sigma_i + \omega_i j$.

If σ_n of p_n is 0, resulting in that:

$$\begin{gathered} C(\omega j) = \prod_{i=1}^{n-1}(\omega j - \sigma_i - \omega_i j) \times (\omega j - \omega_n j) \\ \text{when } \omega = \omega_n,\ C(\omega j) = 0 \end{gathered} \tag{3.14}$$

Stability criterion of the characteristic equation: for any roots p_i located on the imaginary axis of the pole map, where $\sigma_i = 0$, there is always a value of ω that results in $C(\omega j)$ to 0. In other words, $C(\omega j) = 0$ or close to 0 indicates that there are poles on the imaginary axis or close to the RHP, which suggests that the system is marginally stable and tends to be unstable.

If the loop transfer function $L(s)$ is analyzed for stability using a Bode plot, the situation indicating that the pole is on the imaginary axis is presented below:

$$C(\omega j) = 0 \rightarrow 1 + L(\omega j) = 0 \rightarrow |L(\omega j)| = 1 \& \angle L(\omega j) = -180^\circ \tag{3.15}$$

Stability criteria of the loop transfer function: as explained by Equation 3.15, when $|L(\omega j)| = 1 \& \angle L(\omega j) = -180^\circ$, it leads to $C(\omega j) = 0$ and indicates the system is marginally stable.

An example is given below to present if there are poles close to imaginary axis of the pole map, how Bode plot manifests the above stability criteria. Let $G(s) = [(k_i/s + k_p) \times 1/(0.1s + 1)]$ and $H(s) = 1$, $T(s)$ is yielded as:

$$T(s) = \frac{k_p s + k_i}{0.1(s^2 + \underbrace{10(1 + k_p)}_{2\zeta\omega_n}s + \underbrace{10k_i}_{\omega_n^2})} \tag{3.16}$$

Let $w_n = 100$ rad/s. When $\zeta = 0.3$, the Bode plots of both $T(s)$ and $L(s)$ are presented in Figure 3.3. If the poles are far from the imaginary axis, the above situation

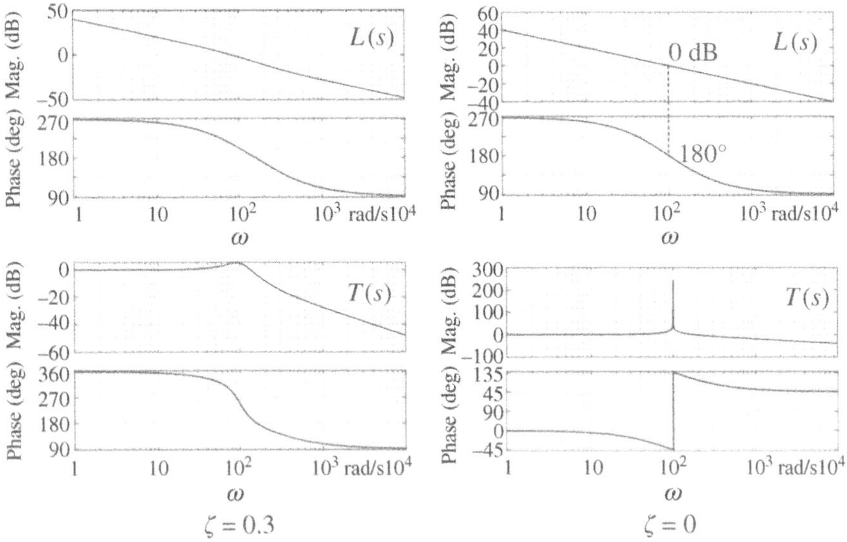

Figure 3.3 Stability analysis via Bode plots for $\zeta = 0.3$ and $\zeta = 0$.

described by the stability criteria does not occur. While changing ζ from 0.3 to 0, when $\zeta = 0$, the poles are on the imaginary axis, and the Bode plot clearly demonstrates the stability criteria. For example, the gain of $T(s)$ is extremely large at $\omega = 100$ rad/s because $C(s) = 0$, and correspondingly, $|L(\omega j)| = 1 \& \angle L(\omega j) = 180°$ at 100 rad/s.

A Bode plot serves as a practical tool for visualizing stability analysis, clearly displaying gains across different frequencies. The stability of a system is determined by the characteristics of poles along the imaginary axis in the pole map, offering straightforward identification methods. For instance, an infinite gain at a specific frequency indicates that the system is marginally stable and likely to oscillate at that frequency. Conversely, if the poles are in the RHP, the gain will not be infinite and does not carry significant implications.

3.4 Stability Criterion via Nyquist Plot Analysis

The Nyquist stability criterion is another method used in the frequency domain for the stability analysis of linear time-invariant systems. The Bode plot stability criterion relies on the characteristic of the poles on the imaginary axis. Meanwhile, the Nyquist stability criterion is to identify if there are any RHP poles. The Nyquist plot represents a transfer function, denoted as $T(\omega j)$, in terms of its real

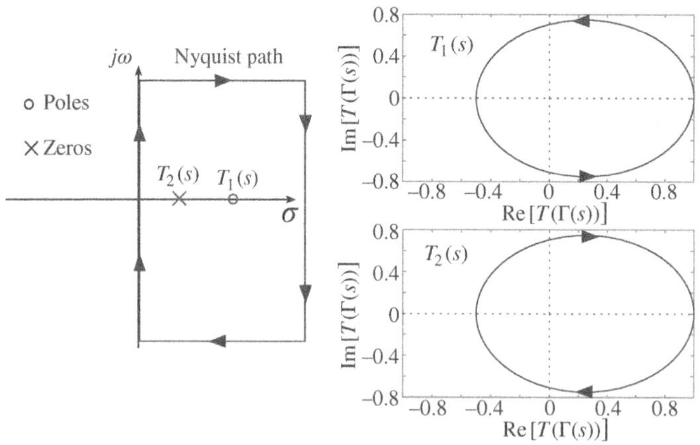

Figure 3.4 Nyquist plot based on the argument principle.

part, $\mathrm{Re}(T(\omega j))$, and imaginary part, $\mathrm{Im}(T(\omega j))$. Its mathematical mechanism for the RHP pole identification is based on the argument principle.

Argument principle: assume an arbitrary closed path $\Gamma(s)$ is selected in the s-plane, ensuring that it does not intersect any poles or zeros of $T(s)$. The corresponding path $T(\Gamma(s))$ will encircle the origin $Z - P$ times, where Z and P are the number of zeros and poles encircled by the path $T(\Gamma(s))$, respectively. The number of corresponding $T(\Gamma(s))$ path encircling the origin can be defined as:

$$N = Z - P \tag{3.17}$$

Based on the argument principle, a Nyquist path is therefore defined to cover the full RHP to identify if there are any poles and zeros inside. The Nyquist path is normally defined to encircle the RHP from $-\omega j$ to ωj. Two simple transfer functions $T_1(s)$ and $T_2(s)$ each containing one RHP pole and one RHP zero, respectively, are given below to demonstrate how the argument principle works in the Nyquist plot:

$$T_1(s) = \frac{s+1}{s-2} \tag{3.18}$$

$$T_2(s) = \frac{s-1}{s+2} \tag{3.19}$$

If a Nyquist path is applied to s of $T_1(s)$ and $T_2(s)$, the corresponding Nyquist plots are presented in Figure 3.4:

- $T_1(s)$ contains one RHP pole, indicating that $P = -1$ at $s = 2$, $Z = 0$, and $N = -1$; therefore, its Nyquist plot encircles the origin once. The presence of the pole inverts the Nyquist path, leading to an inverse direction.

- $T_2(s)$ contains one RHP zero $Z = 1$ at $s = 1$, $P = 0$, and $N = 1$; therefore, its Nyquist plot encircles the origin once, following the same direction as the Nyquist path.

A minimum-phase transfer function is defined as having no RHP poles or zeros and $N = 0$. As mentioned in Section 3.1, the characteristic equation $C(s)$ of a system in Figure 3.1 indicates that the system is unstable if there are any RHP roots/zeros. *If $C(s)$ is of a minimum-phase type, the Nyquist stability criterion can be defined for the system stability analysis as follows:*

- *if the Nyquist path $\Gamma(s)$ is applied to the characteristic equation $C(s)$, the system is stable only if the $C(\Gamma(s))$ path did not encircle the origin point.* Since $P = 0$ due to the minimum-phase transfer function, $N > 0$ indicates that the Nyquist plot encircles the origin and there are zeros in the RHP, which indicate the system is unstable.
- alternatively, loop transfer function $L(s)$ can also be applied for Nyquist stability analysis. Since $C(s) = 1 + L(s)$, the Nyquist stability criterion requires that *the system remains stable if the Nyquist plot according to $L(s)$ does not encircle the point $(-1, j0)$.*

An example, which adopts the same system described by Equation 3.16 for Bode plot stability analysis, is provided to demonstrate how Nyquist criterion is applied for stability analysis, as shown in Figure 3.5. When $\zeta = 0.3$, the system $T(s)$ exhibits no RHP poles, as evidenced by the Nyquist plots of both $C(s)$ and $L(s)$. These plots

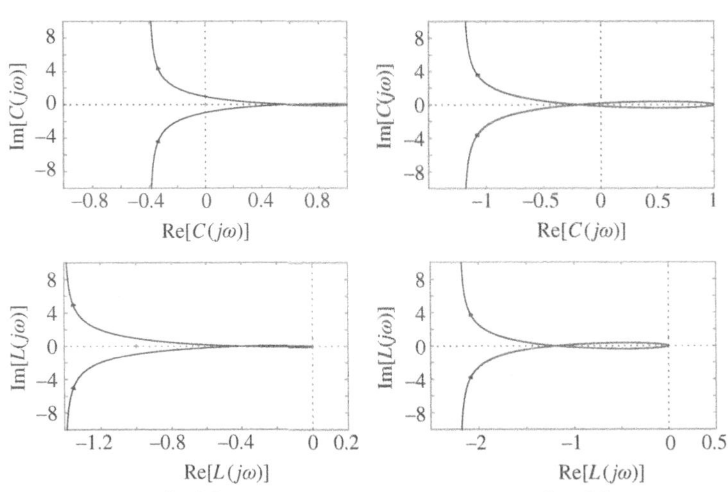

Figure 3.5 Stability analysis via Nyquist plots for $\zeta = 0.3$ and $\zeta = -0.1$ of Equation 3.16.

do not encircle either the origin or the point $(-1, 0j)$, confirming the absence of RHP poles. Conversely, when $\zeta = -0.1$, the system $T(s)$ exhibits RHP poles. This is demonstrated by the Nyquist plots of $C(s)$ and $L(s)$, which encircle both the origin and the point $(-1, 0j)$, indicating the presence of RHP poles in the system.

Reference

1 Benjamin C. Kuo and Farid Golnaraghi. *Automation Control Systems*. Wiley, Hoboken, NJ, 2009.

4

Modeling of Grid-connected Converters

Jun Liang[1,] and Chuanyue Li[2]*

[1] *School of Engineering, Cardiff University, Cardiff, UK*
[2] *School of Electric Power Engineering, South China University of Technology, Guangzhou, China*

*Corresponding author: LiangJ1@cardiff.ac.uk

4.1 Basic Control Configurations of VSCs

Voltage source converter (VSC) acts as a controllable voltage source; therefore, the VSC control primarily is to control the voltage magnitude and phase of $v_{\text{con}} \cos\theta$ to achieve the required control objectives. The modulation technology, such as the pulse-width modulation, will convert the voltage signal to the real output voltage of VSC.

The primary function of VSC control is to regulate the voltage magnitude and phase of $v_{\text{con}}^{\text{mag}} \cos\theta$ in order to meet specified control objectives. The control system is divided into two fundamental components: frequency alignment and VSC control.

Generally, the way of frequency alignment of VSCs forms their control configurations. Currently, as two well-known frequency alignment methods: phase-locked loop (PLL) and power-frequency (P–f) droop, these two VSC control configurations are widely adopted, as shown in Figure 4.1.

The P–f droop was initially adopted in microgrids for autonomous load balance, while in recent years, the P–f droop control, which enables grid-connected VSCs to work like synchronous generators, is recommended and introduced to power systems for VSCs providing grid-forming capability. This control configuration is named as grid-forming control and raises a wide interest. Currently, grid-forming converters are being tested and demonstrated by industries and system operators.

Stability Analysis of Converter-Rich Power Grids, First Edition. Jun Liang and Chuanyue Li.
© 2026 The Institute of Electrical and Electronics Engineers, Inc. Published 2026 by John Wiley & Sons, Inc. All rights reserved.

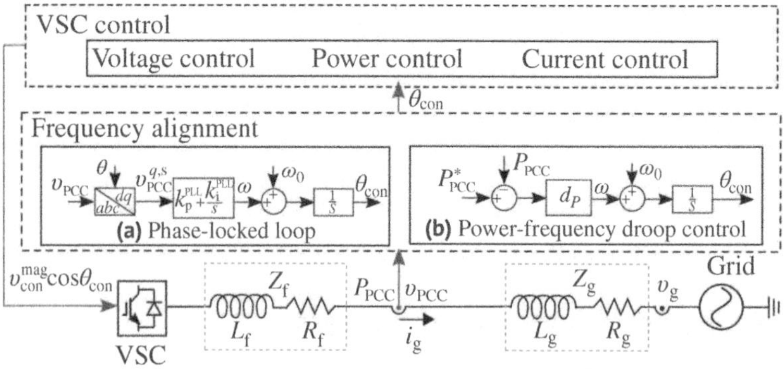

Figure 4.1 Two VSC control configurations.

Since the application of VSCs connecting to power systems, the PLL is a widely adopted way for frequency alignment, which has been sustained for many decades. PLL is a voltage-based synchronization way. It enables VSCs to align their frequency in an open-line situation, which leads to a smooth connection to the grid.

VSC control aims to manage specific objectives, such as point of common coupling (PCC) voltage, power at the PCC point, and output current. These specific objectives are controlled by adjusting the corresponding voltage reference $v_{con}^{mag}\cos\theta$ of the VSC. Since VSCs are based on power electronic switches, their current withstanding capability is low, typically around 2 p.u. for current ratings. Current control is crucial to prevent overcurrent by saturating the current control references as the inner loop. The outer loop control modes include voltage and power control. If only current control is required by a VSC, outer loop control will not be applied. However, if control modes require power or voltage control, an outer loop will be implemented, as illustrated in Figure 4.1. It is important to note that additional functions can be included in the outer loop, depending on the requirements of the VSC. Details of VSC control will be introduced in the following chapter.

4.2 Linearization of VSC Control Systems

Modeling a VSC with transfer functions is foundational for stability analysis, as described in Section 3.1. The main modeling work is on how to represent the control system as transfer functions. Other elements, such as pulse width modulation (PWM) dead time and digital signal processing steps, are modeled as delay units. Because these delays typically span very small intervals, their effects are considered negligible in this section.

Generally, the control of a VSC adopts Proportional-Integral (PI) controllers to achieve an infinite DC value gain, effectively eliminating control error. Therefore, AC signal to DC signal transformation is essential for PI-based control system. The transformation requires the frequency information via frequency alignment control, as described in Figure 4.1. Alternatively, a Proportional-Resonant (PR) controller is used for controlling AC values, but PR controllers highly depend on the specific resonant frequency. If the resonant frequency is mismatched, control errors will present. The PR controller still needs frequency alignment control for optimal results. In fact, it has been proven that the PI controller is equivalent to the PR controller [1]. Due to the effectiveness and simplicity of a PI controller, it has been widely adopted by VSCs.

For a VSC that adopts PI controllers, the control system operates according to the following sequence: "AC-DC transformation," "PI-based control system," "DC-AC transformation." Then, the output AC voltage signal is sent to the modulation stage to be converted into the VSC output voltage. The overall process is depicted in Figure 4.2. The following parameters are defined:

- v_{con} represents the converter voltage signal for VSC modulation, where the subscript $_{con}$ denotes the converter. Generally, v_{con} can also represent the output voltage of the VSC, as the modulation process transfers the $_{con}$ signal to the output voltage with only a small time delay.

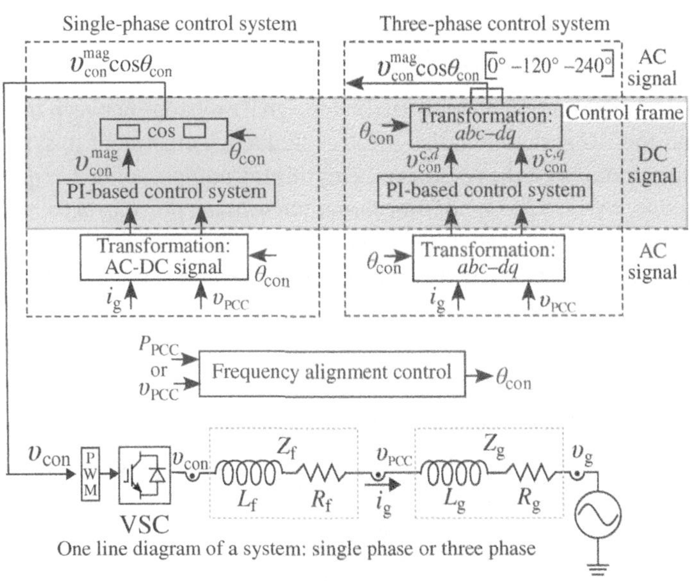

Figure 4.2 Single-phase system and three-phase system.

- $v_{\text{con}}^{\text{mag}}, v_{\text{con}}^{\text{c,mag}}$ are the magnitude of v_{con}, where the superscript ^{c} denotes the control frame, and ^{mag} indicates the magnitude. $v_{\text{con}}^{\text{mag}}$ represents the magnitude of v_{con} in the AC system; $v_{\text{con}}^{\text{c,mag}}$ represents the magnitude in the control frame.
- $v_{\text{con}}^{\text{c},d}, v_{\text{con}}^{\text{c},q}$ are the control frame voltages, where the superscript ^{c} denotes the control frame, ^{d} represents the d-axis of the control frame, and ^{q} represents the q-axis of the control frame.

A VSC can be operated using either individual single-phase control or comprehensive three-phase control, each employing a different "DC-AC transformation" method. In single-phase control, a sinusoidal function is used to convert the phase θ_{con} and magnitude $v_{\text{con}}^{\text{mag}}$ into a sinusoidal voltage signal $v_{\text{con}}^{\text{mag}} \cos\theta_{\text{con}}$, while the three-phase control utilizes an abc–dq transformation to convert $v_{\text{con}}^{\text{c},d}$ and $v_{\text{con}}^{\text{c},q}$ into sinusoidal voltage signals, as shown in Figure 4.2.

PI-based control using DC signals is already a linear system. The main challenge lies in linearizing the AC signal, including the components marked with *, as indicated below:

$$v_{\text{mag}} \underbrace{\cos(\overbrace{\omega_0 t}^{*} + \theta)}_{*} \tag{4.1}$$

where v_{mag} is the magnitude of the signal.

The linearization is to:

- decouple the AC signal from the frequency $\omega_0 t$
- linearize the frequency-decoupled signal $m\cos\theta$

The AC-DC signal transformation not only facilitates signal exchange between the control frame and the AC system but also involves frequency alignment to synchronize the control frame with the AC system. *This transformation, defined as the frame transformation, refers to the signal transformation between the control frame and the AC system.* The frame transformation is highly nonlinear and forms the basis for modeling VSC systems. The linearization of this frame transformation for both single-phase and three-phase control applications will be detailed in the subsequent sections.

4.2.1 Linearization of Frame Transformation for Single-phase Systems

As shown in Figure 4.2, a single-phase VSC system uses a sinusoidal function to convert the DC signals $v_{\text{con}}^{\text{c,mag}}$ and θ_{con} from the control frame into converter output voltage v_{con}, as formulated below:

$$v_{\text{con}}^{\text{c,mag}} \cos(\theta_{\text{con}}) = \underbrace{v_{\text{con}}^{\text{mag}} \cos(\omega_0 t + \theta_0)}_{v_{\text{con}}} \tag{4.2}$$

where $\theta_{con} = \omega_0 t + \theta_0$, the subscript ^{c} denotes the control frame.

The small-signal modeling is adopted for the linearization of $v_{con}^{mag} \cos\theta_{con}$, which is presented below:

$$(v_{con}^{c,mag} + \widetilde{v}_{con}^{c,mag}) \cos(\omega_0 t + \theta_0 + \widetilde{\theta}_{con}) = \underbrace{v_{con}^{mag} \cos(\omega_0 t + \theta_0)}_{v_{con}} + \underbrace{\widetilde{v}_{con}^{mag} \cos(\omega_0 t + \theta_0)}_{\widetilde{v}_{con}} \tag{4.3}$$

4.2.1.1 Frequency Decoupling

To transfer Equation 4.3 into the vector form, the following equation is developed:

$$\mathrm{j}(v_{con}^{c,mag} + \widetilde{v}_{con}^{c,mag}) \sin(\omega_0 t + \theta_0 + \widetilde{\theta}_{con}) = \mathrm{j}(\underbrace{v_{con}^{mag} \sin(\omega_0 t + \theta_0)}_{v_{con}} + \underbrace{\widetilde{v}_{con}^{mag} \sin(\omega_0 t + \theta_0)}_{\widetilde{v}_{con}}) \tag{4.4}$$

By combining both Equations 4.3 and 4.4, its vector form based on Euler's theorem can be formulated as:

$$(v_{con}^{c,mag} + \widetilde{v}_{con}^{c,mag}) e^{\mathrm{j}(\theta_0 + \widetilde{\theta}_{con})} e^{\mathrm{j}\omega_0 t} = v_{con}^{mag} e^{\mathrm{j}\theta_0} e^{\mathrm{j}\omega_0 t} + \widetilde{v}_{con}^{mag} e^{\mathrm{j}\theta_0} e^{\mathrm{j}\omega_0 t} \tag{4.5}$$

The frequency component $e^{\mathrm{j}\omega_0 t}$ can be eliminated on both sides, which is equivalent to the process of *abc–dq* transformation. Equation 4.5 can be rewritten as:

$$(v_{con}^{c,mag} + \widetilde{v}_{con}^{c,mag}) e^{\mathrm{j}(\theta_0 + \widetilde{\theta}_{con})} = v_{con}^{mag} e^{\mathrm{j}\theta_0} + \widetilde{v}_{con}^{mag} e^{\mathrm{j}\theta_0 t} \tag{4.6}$$

4.2.1.2 Linearization of Sinusoidal Functions

Next step is to linearize $(v_{con}^{c,mag} + \widetilde{v}_{con}^{c,mag}) e^{\mathrm{j}(\theta_0 + \widetilde{\theta}_{con})}$ of Equation 4.6 and take real part of Equation 4.6 as an example below:

$$\begin{aligned}(v_{con}^{c,mag} + \widetilde{v}_{con}^{c,mag}) \cos(\theta_0 + \widetilde{\theta}_{con}) = {} & v_{con}^{c,mag} [\cos(\theta_0) \cos\widetilde{\theta}_{con} - \sin(\theta_0) \sin\widetilde{\theta}_{con}] \\ & + \widetilde{v}_{con}^{c,mag} [\cos(\theta_0) \cos\widetilde{\theta}_{con} - \sin(\theta_0) \sin\widetilde{\theta}_{con}]\end{aligned} \tag{4.7}$$

The small-signal modeling gives following relations:

$$\sin\widetilde{\theta}_{con} = \widetilde{\theta}_{con} \qquad \widetilde{v}_{con}^{c,mag} \widetilde{\theta}_{con} = 0 \qquad \cos\widetilde{\theta}_{con} = 1 \tag{4.8}$$

Accordingly, Equation 4.7 can be rearranged as:

$$\begin{gathered}(v_{con}^{c,mag} + \widetilde{v}_{con}^{c,mag}) \cos(\theta_0 + \widetilde{\theta}_{con}) = \\ v_{con}^{c,mag} \cos\theta_0 - \widetilde{\theta}_{con} v_{con}^{c,mag} \sin\theta_0 + \widetilde{v}_{con}^{c,mag} \cos\theta_0\end{gathered} \tag{4.9}$$

The linearization of the imaginary part of $(v_{con}^{c,mag} + \widetilde{v}_{con}^{c,mag}) e^{\mathrm{j}(\theta_0 + \widetilde{\theta}_{con})}$ can be performed using the same process as detailed above.

Therefore, Equation 4.6 is linearized based on the above process and rewritten in the following matrix form:

$$\begin{bmatrix} v_{\text{con}}^{\text{mag}} \cos\theta_0 \\ v_{\text{con}}^{\text{mag}} \sin\theta_0 \end{bmatrix} + \begin{bmatrix} \widetilde{v}_{\text{con}}^{\text{mag}} \cos\theta_0 \\ \widetilde{v}_{\text{con}}^{\text{mag}} \sin\theta_0 \end{bmatrix} = \begin{bmatrix} v_{\text{con}}^{\text{c,mag}} \cos\theta_0 \\ v_{\text{con}}^{\text{c,mag}} \sin\theta_0 \end{bmatrix} + \begin{bmatrix} -\widetilde{\theta}_{\text{con}} v_{\text{con}}^{\text{c,mag}} \sin\theta_0 \\ \widetilde{\theta}_{\text{con}} v_{\text{con}}^{\text{c,mag}} \cos\theta_0 \end{bmatrix} + \begin{bmatrix} \widetilde{v}_{\text{con}}^{\text{c,mag}} \cos\theta_0 \\ \widetilde{v}_{\text{con}}^{\text{c,mag}} \sin\theta_0 \end{bmatrix} \tag{4.10}$$

The values $v_{\text{con}}^{\text{mag}}$ and $v_{\text{con}}^{\text{c,mag}}$ represent the steady-state values of converter voltage magnitude in the AC system and control frame, respectively. The control frame is well aligned with the AC system at steady-state conditions, leading to $v_{\text{con}}^{\text{mag}} = v_{\text{con}}^{\text{c,mag}}$. Therefore, Equation 4.10 can be simplified as:

$$\begin{bmatrix} \widetilde{v}_{\text{con}}^{\text{mag}} \cos\theta_0 \\ \widetilde{v}_{\text{con}}^{\text{mag}} \sin\theta_0 \end{bmatrix} = \begin{bmatrix} -\widetilde{\theta}_{\text{con}} v_{\text{con}}^{\text{c,mag}} \sin\theta_0 \\ \widetilde{\theta}_{\text{con}} v_{\text{con}}^{\text{c,mag}} \cos\theta_0 \end{bmatrix} + \begin{bmatrix} \widetilde{v}_{\text{con}}^{\text{c,mag}} \cos\theta_0 \\ \widetilde{v}_{\text{con}}^{\text{c,mag}} \sin\theta_0 \end{bmatrix} \tag{4.11}$$

4.2.2 Linearization of Reference Frame Transformation for Three-phase Systems

As shown in Figure 4.2, a three-phase VSC system employs a *dq–abc* transformation to convert the DC signals $v_{\text{con}}^{\text{c},d}$ and $v_{\text{con}}^{\text{c},q}$ from the control frame into the AC system v_{con}, along with θ_{con}. To align with the form of the *dq* frame, the three-phase voltage is transferred to its $\alpha\beta$ frame. The *dq–abc* transformation can be formulated as follows:

$$\begin{bmatrix} v_{\text{con}}^{\alpha} \\ v_{\text{con}}^{\beta} \end{bmatrix} = \begin{bmatrix} \cos\theta_{\text{con}} & -\sin\theta_{\text{con}} \\ \sin\theta_{\text{con}} & \cos\theta_{\text{con}} \end{bmatrix} \begin{bmatrix} v_{\text{con}}^{\text{c},d} \\ v_{\text{con}}^{\text{c},q} \end{bmatrix} \tag{4.12}$$

where v_{con}^{α} and v_{con}^{β} represent the values of the three-phase v_{con} in the $\alpha-\beta$ frame; $\theta_{\text{con}} = \omega_0 t$.

The small-signal equation of Equation 4.12 is as below:

$$\begin{bmatrix} v_{\text{con}}^{\alpha} + \widetilde{v}_{\text{con}}^{\alpha} \\ v_{\text{con}}^{\beta} + \widetilde{v}_{\text{con}}^{\beta} \end{bmatrix} = \begin{bmatrix} \cos(\omega_0 t + \widetilde{\theta}_{\text{con}}) & -\sin(\omega_0 t + \widetilde{\theta}_{\text{con}}) \\ \sin(\omega_0 t + \widetilde{\theta}_{\text{con}}) & \cos(\omega_0 t + \widetilde{\theta}_{\text{con}}) \end{bmatrix} \begin{bmatrix} v_{\text{con}}^{\text{c},d} + \widetilde{v}_{\text{con}}^{\text{c},d} \\ v_{\text{con}}^{\text{c},q} + \widetilde{v}_{\text{con}}^{\text{c},q} \end{bmatrix} \tag{4.13}$$

4.2.2.1 Frequency Decoupling

To extract the frequency component from v_{con}^{α}, the mechanism of *abc–dq* transformation is applied, and v_{con}^{α} can be represented as below:

$$\begin{bmatrix} v_{\text{con}}^{\alpha} + \widetilde{v}_{\text{con}}^{\alpha} \\ v_{\text{con}}^{\beta} + \widetilde{v}_{\text{con}}^{\beta} \end{bmatrix} = \begin{bmatrix} \cos(\omega_0 t) & -\sin(\omega_0 t) \\ \sin(\omega_0 t) & \cos(\omega_0 t) \end{bmatrix} \begin{bmatrix} v_{\text{con}}^{d} + \widetilde{v}_{\text{con}}^{d} \\ v_{\text{con}}^{q} + \widetilde{v}_{\text{con}}^{q} \end{bmatrix} \tag{4.14}$$

Substituting v_{con}^{α} and v_{con}^{β} in Equation 4.13 with Equation 4.14 yields:

$$\underbrace{\begin{bmatrix} \cos(\omega_0 t) & -\sin(\omega_0 t) \\ \sin(\omega_0 t) & \cos(\omega_0 t) \end{bmatrix}}_{e^{j\omega_0 t}} \begin{bmatrix} v_{\text{con}}^{d} + \widetilde{v}_{\text{con}}^{d} \\ v_{\text{con}}^{q} + \widetilde{v}_{\text{con}}^{q} \end{bmatrix} = \underbrace{\begin{bmatrix} \cos(\omega_0 t + \widetilde{\theta}_{\text{con}}) & -\sin(\omega_0 t + \widetilde{\theta}_{\text{con}}) \\ \sin(\omega_0 t + \widetilde{\theta}_{\text{con}}) & \cos(\omega_0 t + \widetilde{\theta}_{\text{con}}) \end{bmatrix}}_{e^{j(\omega_0 t + \widetilde{\theta}_{\text{con}})}} \begin{bmatrix} v_{\text{con}}^{c,d} + \widetilde{v}_{\text{con}}^{c,d} \\ v_{\text{con}}^{c,q} + \widetilde{v}_{\text{con}}^{c,q} \end{bmatrix} \tag{4.15}$$

In the vector form of Equation 4.15, it is clear that the frequency component of $e^{j\omega_0 t}$ can be eliminated on both sides, as shown below:

$$\begin{bmatrix} v_{\text{con}}^{d} + \widetilde{v}_{\text{con}}^{d} \\ v_{\text{con}}^{q} + \widetilde{v}_{\text{con}}^{q} \end{bmatrix} = \begin{bmatrix} \cos\widetilde{\theta}_{\text{con}} & -\sin\widetilde{\theta}_{\text{con}} \\ \sin\widetilde{\theta}_{\text{con}} & \cos\widetilde{\theta}_{\text{con}} \end{bmatrix} \begin{bmatrix} v_{\text{con}}^{c,d} + \widetilde{v}_{\text{con}}^{c,d} \\ v_{\text{con}}^{c,q} + \widetilde{v}_{\text{con}}^{c,q} \end{bmatrix} \tag{4.16}$$

4.2.2.2 Linearize Sinusoidal Functions

A small signal of $\widetilde{\theta}$ simplifies Equation 4.8, leading to the following simplification of Equation 4.16:

$$\begin{bmatrix} v_{\text{con}}^{d} + \widetilde{v}_{\text{con}}^{d} \\ v_{\text{con}}^{q} + \widetilde{v}_{\text{con}}^{q} \end{bmatrix} = \begin{bmatrix} 1 & -\widetilde{\theta}_{\text{con}} \\ \widetilde{\theta}_{\text{con}} & 1 \end{bmatrix} \begin{bmatrix} v_{\text{con}}^{c,d} + \widetilde{v}_{\text{con}}^{c,d} \\ v_{\text{con}}^{c,q} + \widetilde{v}_{\text{con}}^{c,q} \end{bmatrix} \tag{4.17}$$

Expanding Equation 4.17 and assuming $\widetilde{\theta}\widetilde{v}_{\text{con}}^{c,d} = 0$, $\widetilde{\theta}\widetilde{v}_{\text{con}}^{c,q} = 0$ yields:

$$\begin{bmatrix} v_{\text{con}}^{d} \\ v_{\text{con}}^{q} \end{bmatrix} + \begin{bmatrix} \widetilde{v}_{\text{con}}^{d} \\ \widetilde{v}_{\text{con}}^{q} \end{bmatrix} = \begin{bmatrix} v_{\text{con}}^{c,d} \\ v_{\text{con}}^{c,q} \end{bmatrix} + \begin{bmatrix} -\widetilde{\theta}_{\text{con}} v_{\text{con}}^{c,q} \\ \widetilde{\theta}_{\text{con}} v_{\text{con}}^{c,d} \end{bmatrix} + \begin{bmatrix} \widetilde{v}_{\text{con}}^{c,d} \\ \widetilde{v}_{\text{con}}^{c,q} \end{bmatrix} \tag{4.18}$$

$[v_{\text{con}}^{d}; v_{\text{con}}^{q}]$ and $[v_{\text{con}}^{c,d}; v_{\text{con}}^{c,q}]$ represent the steady-state values of converter voltage in the AC system and control frame, respectively. Under steady-state conditions, the control frame and the *AC* system are well aligned, leading to $[v_{\text{con}}^{d}; v_{\text{con}}^{q}] = [v_{\text{con}}^{c,d}; v_{\text{con}}^{c,q}]$. Therefore, Equation 4.18 can be simplified as:

$$\begin{bmatrix} \widetilde{v}_{\text{con}}^{d} \\ \widetilde{v}_{\text{con}}^{q} \end{bmatrix} = \begin{bmatrix} -\widetilde{\theta}_{\text{con}} v_{\text{con}}^{c,q} \\ \widetilde{\theta}_{\text{con}} v_{\text{con}}^{c,d} \end{bmatrix} + \begin{bmatrix} \widetilde{v}_{\text{con}}^{c,d} \\ \widetilde{v}_{\text{con}}^{c,q} \end{bmatrix} \tag{4.19}$$

The above equation is the final linearized equation of the *abc–dq* transformation applied in the VSC control.

4.2.2.3 Comparison of DC to AC Signal Transformation in Single-phase and Three-phase Systems

The linearization process has been presented above, and the small-signal equations for DC to AC signal transformation in single-phase and three-phase VSC systems have been developed, as shown in Equations 4.10 and 4.17. In fact, both equations are equivalent, as demonstrated by the following relations:

$$\begin{bmatrix} \widetilde{v}_{\text{con}}^{\text{mag}} \cos\theta_0 \\ \widetilde{v}_{\text{con}}^{\text{mag}} \sin\theta_0 \end{bmatrix} = \begin{bmatrix} \widetilde{v}_{\text{con}}^{d} \\ \widetilde{v}_{\text{con}}^{q} \end{bmatrix} \quad \begin{bmatrix} \widetilde{v}_{\text{con}}^{\text{c,mag}} \cos\theta_0 \\ \widetilde{v}_{\text{con}}^{\text{c,mag}} \sin\theta_0 \end{bmatrix} = \begin{bmatrix} \widetilde{v}_{\text{con}}^{\text{c},d} \\ \widetilde{v}_{\text{con}}^{\text{c},q} \end{bmatrix} \tag{4.20}$$

Therefore, it can be concluded that the frame transformation, whether via a sinusoidal function or *dq–abc* transformation as shown in Figure 4.2, is the same in the small-signal equation. If both the PI-based control systems and the AC-DC signal transformations are the same in both single-phase and three-phase systems, they share the same small-signal equations and exhibit the same stability dynamics.

4.2.3 Summary of the Linearization of the Frame Transformation

As most VSCs adopt a PI controller for the control system, which is inherently linear, the frame transformation between the control frame and the AC system, which requires converting AC signals to DC signals, is a crucial part of linearization.

As demonstrated above, the frame transformation that adopts a sinusoidal function and the dq–abc transformation in both single-phase and three-phase systems are equivalent in their linearized equations. The transformation matrix as shown in Equation 4.20 in the three-phase system is used to represent the frame transformation, and vice versa as shown in Equation 4.21, applicable to both single-phase and three-phase systems.

Based on Equation 4.20, the frame transformation from the AC system to the control frame is as follows:

$$\begin{bmatrix} \widetilde{v}_{\text{con}}^{\text{c},d} \\ \widetilde{v}_{\text{con}}^{\text{c},q} \end{bmatrix} = \underbrace{\begin{bmatrix} \widetilde{\theta}_{\text{con}} v_{\text{con}}^{\text{c},q} \\ -\widetilde{\theta}_{\text{con}} v_{\text{con}}^{\text{c},d} \end{bmatrix}}_{\boldsymbol{T}_{\text{con}}^{v}} + \begin{bmatrix} \widetilde{v}_{\text{con}}^{d} \\ \widetilde{v}_{\text{con}}^{q} \end{bmatrix} \tag{4.21}$$

where $\boldsymbol{T}_{\text{con}}^{v}$ is defined as the transformation matrix between $\widetilde{v}_{\text{con}}$ and $\widetilde{v}_{\text{con}}^{\text{c}}$.

Correspondingly, the frame transformation from the AC system to the control frame of i_{g} is presented below:

$$\begin{bmatrix} \widetilde{i}_{\text{g}}^{\text{c},d} \\ \widetilde{i}_{\text{g}}^{\text{c},q} \end{bmatrix} = \underbrace{\begin{bmatrix} \widetilde{\theta}_{\text{con}} i_{\text{g}}^{\text{c},q} \\ -\widetilde{\theta}_{\text{con}} i_{\text{g}}^{\text{c},d} \end{bmatrix}}_{\boldsymbol{T}_{\text{g}}^{i}} + \begin{bmatrix} \widetilde{i}_{\text{g}}^{d} \\ \widetilde{i}_{\text{g}}^{q} \end{bmatrix} \tag{4.22}$$

where $\boldsymbol{T}_{\text{g}}^{i}$ is defined as the transformation matrix between $\widetilde{i}_{\text{g}}$ and $\widetilde{i}_{\text{g}}^{\text{c}}$.

The transformation of v_{PCC} from the AC system to the control frame is presented below:

$$\begin{bmatrix} \tilde{v}^{\text{c},d}_{\text{PCC}} \\ \tilde{v}^{\text{c},q}_{\text{PCC}} \end{bmatrix} = \underbrace{\begin{bmatrix} \tilde{\theta}_{\text{con}} v^{\text{c},q}_{\text{PCC}} \\ -\tilde{\theta}_{\text{con}} v^{\text{c},d}_{\text{PCC}} \end{bmatrix}}_{T^{v}_{\text{PCC}}} + \begin{bmatrix} \tilde{v}^{d}_{\text{PCC}} \\ \tilde{v}^{q}_{\text{PCC}} \end{bmatrix} \tag{4.23}$$

where T^{v}_{PCC} is defined as the transformation matrix between $\tilde{v}_{\text{PCC}}$ and $\tilde{v}^{\text{c}}_{\text{PCC}}$.

Using v_{con} as an example, the inversed frame transformation from the control frame to the AC system can be presented as $-T^{v}_{\text{con}}$, or as follows:

$$\begin{bmatrix} \tilde{v}^{d}_{\text{con}} \\ \tilde{v}^{q}_{\text{con}} \end{bmatrix} = \underbrace{\begin{bmatrix} -\tilde{\theta}_{\text{con}} v^{\text{c},q}_{\text{con}} \\ \tilde{\theta}_{\text{con}} v^{\text{c},d}_{\text{con}} \end{bmatrix}}_{-T^{v}_{\text{con}}} + \begin{bmatrix} \tilde{v}^{\text{c},d}_{\text{con}} \\ \tilde{v}^{\text{c},q}_{\text{con}} \end{bmatrix} \tag{4.24}$$

4.3 Modeling of Frame Transformation with Frequency Alignment Control

The control frame and AC system are well aligned via the frequency alignment control under steady state, leading to equal steady-state values on both sides. For example, $[v^{d}_{\text{PCC}}, v^{q}_{\text{PCC}}] = [v^{\text{c},d}_{\text{PCC}}, v^{\text{c},q}_{\text{PCC}}]$. However, the small disturbance $\tilde{\theta}_{\text{con}}$ causes differences, which are presented as linearized frame transformation between the AC system and the control frame, as shown in Figure 4.3. v_{PCC} and i_{g} are transferred to the control frame as the control input via the frame transformations T^{v}_{PCC} and T^{i}_{g}. v_{con}, as the control output, is transferred back to the AC system for the VSC modulation via the inversed frame transformation $-T^{v}_{\text{con}}$. These frame transformations are shown in Equations 4.22, 4.23, and 4.24.

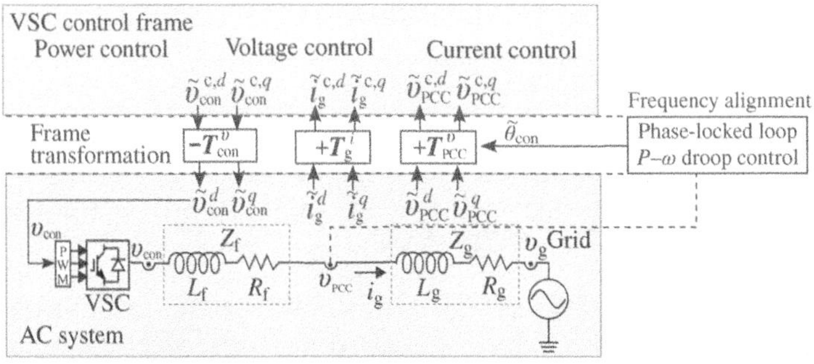

Figure 4.3 Modeling of the frame transformation with frequency alignment.

In this section, the dynamics of the frequency alignment control will be modeled into the frame transformation. The frequency alignment, including PLL and P–ω droop control, is the foundation of a VSC control system, determining its characteristics. The PLL requires the VSC to work as a current source, while P–ω droop control enables the VSC to work as a voltage source. Modeling frequency alignment control is the foundation and first step toward developing the overall VSC system model, which is typically challenging to understand and develop. The modeling of both PLL and P–ω droop control will be introduced in detail.

4.3.1 Modeling of Phase-locked Loops

The control structure of a PLL is presented in Figure 4.4. The mechanism of PLL operates by utilizing the PCC voltage based on the relationship $\underbrace{v_{\text{con}}^{\text{mag}} \sin(\theta_{\text{PCC}} - \theta_{\text{con}})}_{v_{\text{PCC}}^{\text{c},q}} = 0$. This is achieved by applying a PI controller to seamlessly track θ_{PCC} through adjustments to θ_{con}. The small-signal relation of the PLL can be represented as below:

$$\widetilde{v}_{\text{PCC}}^{\text{c},q} \times \left(k_{\text{p}}^{\text{PLL}} + \frac{k_{\text{i}}^{\text{PLL}}}{s} \right) \times \frac{1}{s} = \widetilde{\theta}_{\text{con}} \tag{4.25}$$

$\widetilde{v}_{\text{PCC}}^{\text{c},q}$ is the intermediate parameter, and the primary input is $\widetilde{v}_{\text{PCC}}$. Their relationship can be represented as shown below by referring to the linearized Equation 4.21 of v_{con}'s abc–dq transformation of:

$$\begin{bmatrix} \widetilde{v}_{\text{PCC}}^{\text{c},d} \\ \widetilde{v}_{\text{PCC}}^{\text{c},q} \end{bmatrix} = \begin{bmatrix} \widetilde{\theta}_{\text{con}} v_{\text{PCC}}^{\text{c},q} \\ -\widetilde{\theta}_{\text{con}} v_{\text{PCC}}^{\text{c},d} \end{bmatrix} + \begin{bmatrix} \widetilde{v}_{\text{PCC}}^{d} \\ \widetilde{v}_{\text{PCC}}^{q} \end{bmatrix} \tag{4.26}$$

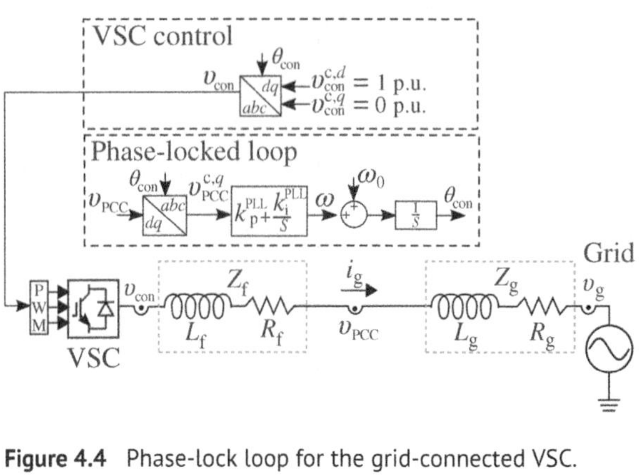

Figure 4.4 Phase-lock loop for the grid-connected VSC.

Substituting $\widetilde{v}_{\mathrm{PCC}}^{\mathrm{c},q}$ in Equation 4.25 with Equation 4.26 yields:

$$(\widetilde{v}_{\mathrm{PCC}}^{q} - \widetilde{\theta}_{\mathrm{con}} v_{\mathrm{PCC}}^{d}) \times \left(k_{\mathrm{p}}^{\mathrm{PLL}} + \frac{k_{\mathrm{i}}^{\mathrm{PLL}}}{s} \right) \times \frac{1}{s} = \widetilde{\theta}_{\mathrm{con}} \tag{4.27}$$

Rearranging Equation 4.27 yields:

$$\widetilde{\theta}_{\mathrm{con}} = \underbrace{\frac{k_{\mathrm{p}}^{\mathrm{PLL}} s + k_{\mathrm{i}}^{\mathrm{PLL}}}{s^2 + v_{\mathrm{PCC}}^{d} k_{\mathrm{p}}^{\mathrm{PLL}} s + v_{\mathrm{PCC}}^{d} k_{\mathrm{i}}^{\mathrm{PLL}}}}_{G_{\mathrm{PLL}}} \widetilde{v}_{\mathrm{PCC}}^{q} \tag{4.28}$$

The above Equation 4.27 represents the small-signal dynamics of the PLL. Substituting $\widetilde{\theta}_{\mathrm{con}}$ with Equation 4.28, the frame transformations via Equations 4.22, 4.23, and 4.24 can be further represented as below by integrating the PLL's dynamics:

$$\begin{bmatrix} \widetilde{v}_{\mathrm{PCC}}^{\mathrm{c},d} \\ \widetilde{v}_{\mathrm{PCC}}^{\mathrm{c},q} \end{bmatrix} = \underbrace{\underbrace{\begin{bmatrix} 0 & G_{\mathrm{PLL}} v_{\mathrm{PCC}}^{\mathrm{c},q} \\ 0 & -G_{\mathrm{PLL}} v_{\mathrm{PCC}}^{\mathrm{c},d} \end{bmatrix}}_{G_{\mathrm{PCC}}^{v}} \begin{bmatrix} \widetilde{v}_{\mathrm{PCC}}^{d} \\ \widetilde{v}_{\mathrm{PCC}}^{q} \end{bmatrix} + \begin{bmatrix} \widetilde{v}_{\mathrm{PCC}}^{d} \\ \widetilde{v}_{\mathrm{PCC}}^{q} \end{bmatrix}}_{\widetilde{T}_{\mathrm{PCC}}^{v}} \tag{4.29}$$

$$\begin{bmatrix} \widetilde{i}_{\mathrm{g}}^{\mathrm{c},d} \\ \widetilde{i}_{\mathrm{g}}^{\mathrm{c},q} \end{bmatrix} = \underbrace{\underbrace{\begin{bmatrix} 0 & G_{\mathrm{PLL}} i_{\mathrm{g}}^{\mathrm{c},q} \\ 0 & -G_{\mathrm{PLL}} i_{\mathrm{g}}^{\mathrm{c},d} \end{bmatrix}}_{G_{\mathrm{g}}^{i}} \begin{bmatrix} \widetilde{v}_{\mathrm{PCC}}^{d} \\ \widetilde{v}_{\mathrm{PCC}}^{q} \end{bmatrix} + \begin{bmatrix} \widetilde{i}_{\mathrm{g}}^{d} \\ \widetilde{i}_{\mathrm{g}}^{q} \end{bmatrix}}_{\widetilde{T}_{\mathrm{g}}^{i}} \tag{4.30}$$

$$\begin{bmatrix} \widetilde{v}_{\mathrm{con}}^{d} \\ \widetilde{v}_{\mathrm{con}}^{q} \end{bmatrix} = \underbrace{\underbrace{\begin{bmatrix} 0 & -G_{\mathrm{PLL}} v_{\mathrm{con}}^{\mathrm{c},q} \\ 0 & G_{\mathrm{PLL}} v_{\mathrm{con}}^{\mathrm{c},d} \end{bmatrix}}_{-G_{\mathrm{con}}^{v}} \begin{bmatrix} \widetilde{v}_{\mathrm{PCC}}^{d} \\ \widetilde{v}_{\mathrm{PCC}}^{q} \end{bmatrix} + \begin{bmatrix} \widetilde{v}_{\mathrm{con}}^{\mathrm{c},d} \\ \widetilde{v}_{\mathrm{con}}^{\mathrm{c},q} \end{bmatrix}}_{-\widetilde{T}_{\mathrm{con}}^{v}} \tag{4.31}$$

With the frequency alignment using the PLL, the frame transformation in Figure 4.3 of $\widetilde{T}_{\mathrm{PCC}}^{v}$, $\widetilde{T}_{\mathrm{g}}^{i}$, $-\widetilde{T}_{\mathrm{con}}^{v}$ can be represented as above Equations 4.29, 4.30, and 4.31.

4.3.2 Modeling of Power-frequency Droop Control

The P–ω droop control utilizes the relationship between active power and phase difference in an inductive system. Furthermore, employing frequency instead of

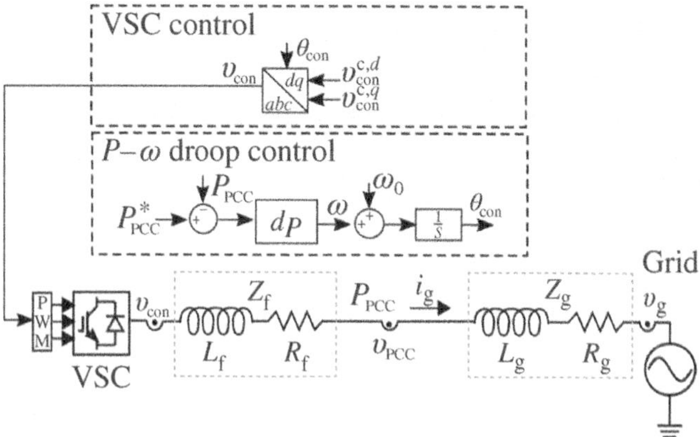

Figure 4.5 P–ω droop control for the grid-connected VSC.

phase can additionally enable autonomous power balance in a grid. This frequency alignment is achieved via active power, enabling the VSC to function like voltage sources. A P–ω droop control is presented in Figure 4.5, which has the following relation:

$$\theta_{\text{con}} = \int [\omega_0 + d_P(P^* - P)]dt \tag{4.32}$$

where the superscript * represents the reference of P and P^* is the active power reference.

To linearize the above Equation 4.32, P is represented in the form of voltage and current via small-signal modeling, as shown below:

$$\begin{aligned} P_{\text{PCC}} + \widetilde{P}_{\text{PCC}} &= 1.5[(v^d_{\text{PCC}} + \widetilde{v}^d_{\text{PCC}})(i^d_g + \widetilde{i}^d_g) + (v^q_{\text{PCC}} + \widetilde{v}^q_{\text{PCC}})(i^q_g + \widetilde{i}^q_g)] \\ &= 1.5(v^d_{\text{PCC}} i^d_g + v^q_{\text{PCC}} i^q_g) + 1.5\widetilde{v}^d_{\text{PCC}}\widetilde{i}^d_g + 1.5\widetilde{v}^q_{\text{PCC}}\widetilde{i}^q_g \\ &+1.5(i^d_g \widetilde{v}^d_{\text{PCC}} + v^d_{\text{PCC}}\widetilde{i}^d_g + i^q_g \widetilde{v}^q_{\text{PCC}} + v^q_{\text{PCC}}\widetilde{i}^q_g) \end{aligned} \tag{4.33}$$

where $P_{\text{PCC}} = 1.5(v^d_{\text{PCC}} i^d_g + v^q_{\text{PCC}} i^q_g)$ under stead state; $v^d_{\text{PCC}} i^d_g \approx 0$, $v^q_{\text{PCC}} i^q_g \approx 0$.

Rearranging Equation 4.33 yields:

$$\widetilde{P}_{\text{PCC}} = 1.5(i^d_g \widetilde{v}^d_{\text{PCC}} + v^d_{\text{PCC}}\widetilde{i}^d_g + i^q_g \widetilde{v}^q_{\text{PCC}} + v^q_{\text{PCC}}\widetilde{i}^q_g) \tag{4.34}$$

The small-signal model of Equation 4.32 in the s-domain is as follows:

$$\widetilde{\theta}_{\text{con}} = \frac{d_P(\widetilde{P}^* - \widetilde{P})}{s} \tag{4.35}$$

Substituting $\widetilde{P}$ in Equation 4.35 with Equation 4.34 yields:

$$\widetilde{\theta}_{\mathrm{con}} = \frac{d_P\widetilde{P}^*}{s} - \frac{1.5d_P}{s}(i_{\mathrm{g}}^d\widetilde{v}_{\mathrm{PCC}}^d + v_{\mathrm{PCC}}^d\widetilde{i}_{\mathrm{g}}^d + i_{\mathrm{g}}^q\widetilde{v}_{\mathrm{PCC}}^q + v_{\mathrm{PCC}}^q\widetilde{i}_{\mathrm{g}}^q) \quad (4.36)$$

The above Equation 4.36 represents the small-signal dynamics of the P–ω droop control. Substituting $\widetilde{\theta}_{\mathrm{con}}$ with Equation 4.36, the frame transformations via Equations 4.22, 4.23, and 4.24 can be further represented as below by integrating the P–ω droop control's dynamics:

$$\begin{bmatrix}\widetilde{v}_{\mathrm{PCC}}^{\mathrm{c},d}\\ \widetilde{v}_{\mathrm{PCC}}^{\mathrm{c},q}\end{bmatrix} = \begin{bmatrix}\widetilde{v}_{\mathrm{PCC}}^{d}\\ \widetilde{v}_{\mathrm{PCC}}^{q}\end{bmatrix} + \begin{bmatrix}\frac{d_P v_{\mathrm{PCC}}^{\mathrm{c},q}}{s} & 0\\ 0 & -\frac{d_P v_{\mathrm{PCC}}^{\mathrm{c},d}}{s}\end{bmatrix}\begin{bmatrix}\widetilde{P}^* & 0\\ 0 & \widetilde{P}^*\end{bmatrix} + \underbrace{1.5d_P\left(\underbrace{\begin{bmatrix}-\frac{v_{\mathrm{PCC}}^{d}v_{\mathrm{PCC}}^{\mathrm{c},q}}{s} & -\frac{v_{\mathrm{PCC}}^{q}v_{\mathrm{PCC}}^{\mathrm{c},q}}{s}\\ \frac{v_{\mathrm{PCC}}^{d}v_{\mathrm{PCC}}^{\mathrm{c},d}}{s} & \frac{v_{\mathrm{PCC}}^{q}v_{\mathrm{PCC}}^{\mathrm{c},d}}{s}\end{bmatrix}}_{\boldsymbol{H}_{\mathrm{PCC}}^{i}}\begin{bmatrix}\widetilde{i}_{\mathrm{g}}^{d}\\ \widetilde{i}_{\mathrm{g}}^{q}\end{bmatrix} + \underbrace{\begin{bmatrix}-\frac{i_{\mathrm{g}}^{d}v_{\mathrm{PCC}}^{\mathrm{c},q}}{s} & -\frac{i_{\mathrm{g}}^{q}v_{\mathrm{PCC}}^{\mathrm{c},q}}{s}\\ \frac{i_{\mathrm{g}}^{d}v_{\mathrm{PCC}}^{\mathrm{c},d}}{s} & \frac{i_{\mathrm{g}}^{q}v_{\mathrm{PCC}}^{\mathrm{c},d}}{s}\end{bmatrix}}_{\boldsymbol{H}_{\mathrm{PCC}}^{v}}\begin{bmatrix}\widetilde{v}_{\mathrm{PCC}}^{d}\\ \widetilde{v}_{\mathrm{PCC}}^{q}\end{bmatrix}\right)}_{\widetilde{\boldsymbol{T}}_{\mathrm{PCC}}^{v}} \quad (4.37)$$

$$\begin{bmatrix}\widetilde{i}_{\mathrm{g}}^{\mathrm{c},d}\\ \widetilde{i}_{\mathrm{g}}^{\mathrm{c},q}\end{bmatrix} = \begin{bmatrix}\widetilde{i}_{\mathrm{g}}^{d}\\ \widetilde{i}_{\mathrm{g}}^{q}\end{bmatrix} + \begin{bmatrix}\frac{d_P i_{\mathrm{g}}^{\mathrm{c},q}}{s} & 0\\ 0 & -\frac{d_P i_{\mathrm{g}}^{\mathrm{c},d}}{s}\end{bmatrix}\begin{bmatrix}\widetilde{P}^* & 0\\ 0 & \widetilde{P}^*\end{bmatrix} + \underbrace{1.5d_P\left(\underbrace{\begin{bmatrix}-\frac{v_{\mathrm{PCC}}^{d}i_{\mathrm{g}}^{\mathrm{c},q}}{s} & -\frac{v_{\mathrm{PCC}}^{q}i_{\mathrm{g}}^{\mathrm{c},q}}{s}\\ \frac{v_{\mathrm{PCC}}^{d}i_{\mathrm{g}}^{\mathrm{c},d}}{s} & \frac{v_{\mathrm{PCC}}^{q}i_{\mathrm{g}}^{\mathrm{c},d}}{s}\end{bmatrix}}_{\boldsymbol{H}_{\mathrm{g}}^{i}}\begin{bmatrix}\widetilde{i}_{\mathrm{g}}^{d}\\ \widetilde{i}_{\mathrm{g}}^{q}\end{bmatrix} + \underbrace{\begin{bmatrix}-\frac{i_{\mathrm{g}}^{d}i_{\mathrm{g}}^{\mathrm{c},q}}{s} & -\frac{i_{\mathrm{g}}^{q}i_{\mathrm{g}}^{\mathrm{c},q}}{s}\\ \frac{i_{\mathrm{g}}^{d}i_{\mathrm{g}}^{\mathrm{c},d}}{s} & \frac{i_{\mathrm{g}}^{q}i_{\mathrm{g}}^{\mathrm{c},d}}{s}\end{bmatrix}}_{\boldsymbol{H}_{\mathrm{g}}^{v}}\begin{bmatrix}\widetilde{v}_{\mathrm{PCC}}^{d}\\ \widetilde{v}_{\mathrm{PCC}}^{q}\end{bmatrix}\right)}_{\widetilde{\boldsymbol{T}}_{\mathrm{g}}^{i}} \quad (4.38)$$

$$\begin{bmatrix} \tilde{v}_{\text{con}}^{d} \\ \tilde{v}_{\text{con}}^{q} \end{bmatrix} = \begin{bmatrix} \tilde{v}_{\text{con}}^{\text{c},d} \\ \tilde{v}_{\text{con}}^{\text{c},q} \end{bmatrix} + \underbrace{\begin{bmatrix} -\frac{d_P v_{\text{con}}^{\text{c},q}}{s} & 0 \\ 0 & \frac{d_P v_{\text{con}}^{\text{c},d}}{s} \end{bmatrix}}_{H_P} \begin{bmatrix} \tilde{P}^* & 0 \\ 0 & \tilde{P}^* \end{bmatrix} +$$

$$\underbrace{1.5d_P \left(\underbrace{\begin{bmatrix} \frac{v_{\text{PCC}}^{d} v_{\text{con}}^{\text{c},q}}{s} & \frac{v_{\text{PCC}}^{q} v_{\text{con}}^{\text{c},q}}{s} \\ -\frac{v_{\text{PCC}}^{d} v_{\text{con}}^{\text{c},d}}{s} & -\frac{v_{\text{PCC}}^{q} v_{\text{con}}^{\text{c},d}}{s} \end{bmatrix}}_{-H_{\text{con}}^{i}} \begin{bmatrix} \tilde{i}_{\text{g}}^{d} \\ \tilde{i}_{\text{g}}^{q} \end{bmatrix} + \underbrace{\begin{bmatrix} \frac{i_{\text{g}}^{d} v_{\text{con}}^{\text{c},q}}{s} & \frac{i_{\text{g}}^{q} v_{\text{con}}^{\text{c},q}}{s} \\ -\frac{i_{\text{g}}^{d} v_{\text{con}}^{\text{c},d}}{s} & -\frac{i_{\text{g}}^{q} v_{\text{con}}^{\text{c},d}}{s} \end{bmatrix}}_{-H_{\text{con}}^{v}} \begin{bmatrix} \tilde{v}_{\text{PCC}}^{d} \\ \tilde{v}_{\text{PCC}}^{q} \end{bmatrix} \right)}_{-\tilde{T}_{\text{con}}^{v}} \tag{4.39}$$

With the frequency alignment using the P–ω droop control, the frame transformation in Figure 4.3 of $\tilde{T}_{\text{PCC}}^{v}$, $\tilde{T}_{\text{g}}^{i}$, $-\tilde{T}_{\text{con}}^{v}$ can be represented as above Equations 4.37, 4.38, and 4.39.

4.4 Modeling and Stability Analysis of a VSC System

The modeling of a VSC system encompasses three parts: VSC control, the AC system, and the frame transformation between them. The challenging part of frame transformation modeling is detailed in Section 4.3. The VSC control utilizes dq-frame control, and the modeling of the AC system is straightforward, as described below. Additionally, examples of stability analysis for frequency alignment, which includes both PLL and P–ω droop control, are provided, respectively.

4.4.1 Modeling a VSC System

As explained in Section 4.2, the PI-based VSC control system is a linear system in which $\boldsymbol{v}_{\text{PCC}}^{\text{c}}$ and $\boldsymbol{i}_{\text{g}}^{\text{c}}$ are the inputs, and the output is $\boldsymbol{v}_{\text{con}}^{\text{c}}$. The AC system is linearized in the dq frame to decouple the frequency ω_0 from the AC voltage and AC. Consequently, the three-phase values are rewritten in the dq form, for example, $\boldsymbol{v} = [v^d; v^q]$.

The detailed control system of a VSC will be explained in the following section. In this section, the control system is generalized and modeled as:

$$\tilde{\boldsymbol{v}}_{\text{con}}^{\text{c}} = \boldsymbol{G}_{\text{cur}}^{\text{c}} \tilde{\boldsymbol{i}}_{\text{g}}^{\text{c}} + \boldsymbol{G}_{\text{vol}}^{\text{c}} \tilde{\boldsymbol{v}}_{\text{PCC}}^{\text{c}} + \boldsymbol{G}_{\text{ref}} \tilde{\boldsymbol{u}} \tag{4.40}$$

where $\boldsymbol{G}^{\text{c}}_{\text{cur}}$ is the transfer function of the control system relevant to $\tilde{\boldsymbol{i}}^{\text{c}}_{\text{g}}$; $\boldsymbol{G}^{\text{c}}_{\text{vol}}$ is the transfer function of the control system relevant to $\tilde{\boldsymbol{v}}^{\text{c}}_{\text{PCC}}$; $\tilde{\boldsymbol{u}}$ represents the references of the control, which could be voltage, current, or power references; $\boldsymbol{G}_{\text{ref}}$ is the equivalent gain from the control system.

Transferring Equation 4.40 of the control system to the AC system via the frame transformation based on Equations 4.22, 4.23, and 4.24 yields:

$$\tilde{\boldsymbol{v}}_{\text{con}} = -\tilde{\boldsymbol{T}}^{v}_{\text{con}} + \underbrace{[\boldsymbol{G}^{\text{c}}_{\text{cur}}(\tilde{\boldsymbol{i}}_{\text{g}} + \tilde{\boldsymbol{T}}^{i}_{\text{g}}) + \boldsymbol{G}^{\text{c}}_{\text{vol}}(\tilde{\boldsymbol{v}}_{\text{PCC}} + \tilde{\boldsymbol{T}}^{v}_{\text{PCC}}) + \boldsymbol{G}_{\text{ref}}\tilde{\boldsymbol{u}}]}_{\tilde{\boldsymbol{v}}^{\text{c}}_{\text{con}}} \tag{4.41}$$

Rearranging yields:

$$\tilde{\boldsymbol{v}}_{\text{con}} = \underbrace{\boldsymbol{G}^{\text{c}}_{\text{cur}}\tilde{\boldsymbol{i}}_{\text{g}} + \boldsymbol{G}^{\text{c}}_{\text{vol}}\tilde{\boldsymbol{v}}_{\text{PCC}} + \underbrace{\boldsymbol{G}^{\text{c}}_{\text{cur}}\tilde{\boldsymbol{T}}^{i}_{\text{g}} + \boldsymbol{G}^{\text{c}}_{\text{vol}}\tilde{\boldsymbol{T}}^{v}_{\text{PCC}} - \tilde{\boldsymbol{T}}^{v}_{\text{con}}}_{\tilde{\boldsymbol{T}}(\tilde{\boldsymbol{i}}_{\text{g}},\tilde{\boldsymbol{v}}_{\text{PCC}})}}_{\boldsymbol{G}_{\text{g}}\tilde{\boldsymbol{i}}_{\text{g}}+\boldsymbol{G}_{\text{PCC}}\tilde{\boldsymbol{v}}_{\text{PCC}}} + \boldsymbol{G}_{\text{ref}}\tilde{\boldsymbol{u}} \tag{4.42}$$

where as shown in Equations 4.22, 4.23, and 4.24, $\tilde{\boldsymbol{T}}^{v}_{\text{con}}$, $\tilde{\boldsymbol{T}}^{v}_{\text{PCC}}$, and $\tilde{\boldsymbol{T}}^{i}_{\text{g}}$ are functions of $\tilde{\boldsymbol{i}}_{\text{g}}$ and $\tilde{\boldsymbol{v}}_{\text{PCC}}$ (denoted as $\tilde{\boldsymbol{T}}(\tilde{\boldsymbol{i}}_{\text{g}}, \tilde{\boldsymbol{v}}_{\text{PCC}})$).

Using $\boldsymbol{G}^{i}_{\text{col}}$ and $\boldsymbol{G}^{v}_{\text{col}}$ to represent the overall control gains of $\tilde{\boldsymbol{i}}_{\text{g}}$ and $\tilde{\boldsymbol{v}}_{\text{PCC}}$ of Equation 4.42 yields:

$$\tilde{\boldsymbol{v}}_{\text{con}} = \boldsymbol{G}^{i}_{\text{col}}\tilde{\boldsymbol{i}}_{\text{g}} + \boldsymbol{G}^{v}_{\text{col}}\tilde{\boldsymbol{v}}_{\text{PCC}} + \boldsymbol{G}_{\text{ref}}\tilde{\boldsymbol{u}} \tag{4.43}$$

The modeling of converter's filter impedance $\boldsymbol{Z}_{\text{f}}$ requires decoupling the system frequency, as described in Section 4.2.2.1, which leads to the following functions:

$$\tilde{\boldsymbol{v}}_{\text{con}} = \boldsymbol{Z}_{\text{f}}\tilde{\boldsymbol{i}}_{\text{g}} + \tilde{\boldsymbol{v}}_{\text{PCC}} \tag{4.44}$$

where $\boldsymbol{Z}_{\text{f}} = \begin{bmatrix} L_{\text{f}}s + R_{\text{f}} & -\omega_0 L_{\text{f}} \\ \omega_0 L_{\text{f}} & L_{\text{f}}s + R_{\text{f}} \end{bmatrix}$.

Substituting $\tilde{\boldsymbol{v}}_{\text{con}}$ in Equation 4.43 with Equation 4.44 and rearranging yields:

$$(\boldsymbol{Z}_{\text{f}} - \boldsymbol{G}^{i}_{\text{col}})\tilde{\boldsymbol{i}}_{\text{g}} = (\boldsymbol{G}^{v}_{\text{col}} - \boldsymbol{I})\tilde{\boldsymbol{v}}_{\text{PCC}} + \boldsymbol{G}_{\text{ref}}\tilde{\boldsymbol{u}} \tag{4.45}$$

Equation 4.45 represents the overall model of the converter system.

A simple AC system using an AC voltage source and an impedance is applied. The modeling of the AC system requires decoupling the system frequency, as described in Section 4.2.2.1, which leads to the following functions:

$$\tilde{\boldsymbol{v}}_{\text{PCC}} = \boldsymbol{Z}_{\text{g}}\tilde{\boldsymbol{i}}_{\text{g}} + \tilde{\boldsymbol{v}}_{\text{g}} \tag{4.46}$$

where $\boldsymbol{Z}_{\text{g}} = \begin{bmatrix} L_{\text{g}}s + R_{\text{g}} & -\omega_0 L_{\text{g}} \\ \omega_0 L_{\text{g}} & L_{\text{g}}s + R_{\text{g}} \end{bmatrix}$.

Substituting $\tilde{\boldsymbol{v}}_{\mathrm{PCC}}$ in Equation 4.45 with Equation 4.46 and rearranging yields:

$$\underbrace{(\boldsymbol{Z}_{\mathrm{f}} + \boldsymbol{Z}_{\mathrm{g}} - \boldsymbol{G}^{i}_{\mathrm{col}} - \boldsymbol{G}^{v}_{\mathrm{col}}\boldsymbol{Z}_{\mathrm{g}})}_{G_{\mathrm{sys}}}\tilde{\boldsymbol{i}}_{\mathrm{g}} = -\underbrace{(\boldsymbol{I} - \boldsymbol{G}^{v}_{\mathrm{col}})}_{G_{v}}\tilde{\boldsymbol{v}}_{\mathrm{g}} + \boldsymbol{G}_{\mathrm{ref}}\tilde{\boldsymbol{u}} \tag{4.47}$$

Using $\boldsymbol{G}_{\mathrm{sys}}$ and $\boldsymbol{G}_{\mathrm{v}}$ to represent the overall s-function gains of $\tilde{\boldsymbol{i}}_{\mathrm{g}}$ and $\tilde{\boldsymbol{v}}_{\mathrm{g}}$ of Equation 4.47 yields:

$$\boldsymbol{G}_{\mathrm{sys}}\tilde{\boldsymbol{i}}_{\mathrm{g}} = \boldsymbol{G}_{\mathrm{ref}}\tilde{\boldsymbol{u}} - \boldsymbol{G}_{\mathrm{v}}\tilde{\boldsymbol{v}}_{\mathrm{g}} \tag{4.48}$$

Equation 4.48 is the system model based on the transfer functions. $\tilde{\boldsymbol{u}}$ and $\tilde{\boldsymbol{v}}_{\mathrm{g}}$ can be treated as inputs to a system model, while $\tilde{\boldsymbol{i}}_{\mathrm{g}}$ represents the outputs of state variables.

4.4.2 Transfer-function-based Stability Analysis Method for Converter Systems

The stability criteria described in Chapter 4 are applied to a single-input, single-output transfer function. However, a converter system is represented by a two-input, two-output transfer function matrix, as shown in Equation 4.48. To solve Equation 4.48 as two decoupled single-input and single-output transfer functions, rearranging Equation 4.48 as follows:

$$\tilde{\boldsymbol{i}}_{\mathrm{g}} = \boldsymbol{G}_{\mathrm{sys}}^{-1}(\boldsymbol{G}_{\mathrm{ref}}\tilde{\boldsymbol{u}} - \boldsymbol{G}_{\mathrm{v}}\tilde{\boldsymbol{v}}_{\mathrm{g}}) \tag{4.49}$$

$\tilde{\boldsymbol{u}}$ and $\tilde{\boldsymbol{v}}_{\mathrm{g}}$ are the reference values for the converter control and the grid-side AC voltage source, respectively. It is assumed that $\tilde{\boldsymbol{v}}_{\mathrm{g}} = 0$ when a deviation is applied to $\tilde{\boldsymbol{u}}$.

The factor in Equation 4.49 that represents system stability will be identified below by referring to the state-space representation of the system, as described by Equation 5.1 in Section 5.1. The corresponding relationships with the state-space representation are as follows:

- $\tilde{\boldsymbol{y}}$ represents $\tilde{\boldsymbol{i}}_{\mathrm{g}}$;
- $\tilde{\boldsymbol{x}}$ represents the state variables, including $\tilde{\boldsymbol{i}}_{\mathrm{g}}$, which leads to $\boldsymbol{D} = 0$ in Equation 5.1.

Rearranging Equation 5.1 yields the following relation:

$$\tilde{\boldsymbol{i}}_{\mathrm{g}} = \boldsymbol{C}(\boldsymbol{I}s - \boldsymbol{A})^{-1}\boldsymbol{B}\tilde{\boldsymbol{u}} \tag{4.50}$$

Since the state-space representation describes the same system, Equation 4.50 is identical to Equation 4.49, and the following relation can be obtained:

$$\boldsymbol{C}\,\underbrace{(\boldsymbol{I}s - \boldsymbol{A})^{-1}}\,\boldsymbol{B} = \boldsymbol{G}_{\mathrm{sys}}^{-1}\boldsymbol{G}_{\mathrm{ref}} \tag{4.51}$$

Further expanding yields:

$$\frac{1}{\det(\boldsymbol{I}s-\boldsymbol{A})}\boldsymbol{C}(\mathrm{adj}(\boldsymbol{I}s-\boldsymbol{A}))\boldsymbol{B}=\frac{1}{\det(\boldsymbol{G}_{\mathrm{sys}})}\mathrm{adj}(\boldsymbol{G}_{\mathrm{sys}})\boldsymbol{G}_{\mathrm{ref}} \tag{4.52}$$

where det and adj represent the determinant and adjugate matrix operations, respectively.

Equation 4.52 presents following things:

- $\det(\boldsymbol{I}s-\boldsymbol{A})$ is the characteristic equation of the system, which represents the system stability.
- $\boldsymbol{C}$, $\boldsymbol{B}$ are numerical matrices; $\mathrm{adj}(\boldsymbol{I}s-\boldsymbol{A})$ has no poles.
- $\boldsymbol{G}_{\mathrm{sys}}^{-1}\boldsymbol{G}_{\mathrm{ref}}$ contains the system's characteristic equation, which can be applied for stability analysis.

Therefore, the system stability analysis via the transfer-function-based method is as follows:

$\boldsymbol{G}_{\mathrm{sys}}^{-1}\boldsymbol{G}_{\mathrm{ref}}$ contains the system's characteristic equation as indicated in Equation 4.52. $\boldsymbol{G}_{\mathrm{ref}}$ represents the gains for the converter control references and generally does not contain any RHP poles in any element of its matrix. Therefore, the stability analysis process for a converter system is simplified to analyze $\det(\boldsymbol{G}_{\mathrm{sys}})$; $\det(\boldsymbol{G}_{\mathrm{sys}})$ serves as the system's characteristic equation to represent its stability.

4.4.3 Case Study on Stability Analysis of the PLL

Based on the above VSC system modeling and stability analysis method, an example of a PLL-based VSC system is given in this section, as shown in Figure 4.4. Only PLL is applied to a VSC with constant values of $[v_{\mathrm{con}}^{\mathrm{c},d}, v_{\mathrm{con}}^{\mathrm{c},q}]$; therefore, the modeling of VSC control based on Equation 4.41 is simplified as:

$$\tilde{\boldsymbol{v}}_{\mathrm{con}}=-\tilde{\boldsymbol{T}}_{\mathrm{con}}^{v}+\underbrace{\tilde{\boldsymbol{v}}_{\mathrm{con}}^{\mathrm{c}}}_{\tilde{\boldsymbol{u}}} \tag{4.53}$$

By applying the frame transformation with the dynamics of the PLL via Equation 4.31, the above equation can be presented as below:

$$\tilde{\boldsymbol{v}}_{\mathrm{con}}=-\boldsymbol{G}_{\mathrm{con}}^{v}\tilde{\boldsymbol{v}}_{\mathrm{PCC}}+\tilde{\boldsymbol{v}}_{\mathrm{con}}^{\mathrm{c}} \tag{4.54}$$

Substituting $\tilde{\boldsymbol{v}}_{\mathrm{con}}$ and $\tilde{\boldsymbol{v}}_{\mathrm{PCC}}$ based on the AC system model of Equations 4.44 and 4.46 yields:

$$\tilde{\boldsymbol{v}}_{\mathrm{g}}+(\boldsymbol{Z}_{\mathrm{f}}+\boldsymbol{Z}_{\mathrm{g}})\tilde{\boldsymbol{i}}_{\mathrm{g}}=-\boldsymbol{G}_{\mathrm{con}}^{v}\boldsymbol{Z}_{\mathrm{g}}\tilde{\boldsymbol{i}}_{\mathrm{g}}+\tilde{\boldsymbol{v}}_{\mathrm{con}}^{\mathrm{c}} \tag{4.55}$$

Rearranging Equation 4.53 yields:

$$\underbrace{(\boldsymbol{Z}_{\mathrm{f}}+\boldsymbol{Z}_{\mathrm{g}}+\boldsymbol{G}_{\mathrm{con}}^{v}\boldsymbol{Z}_{\mathrm{g}})}_{\boldsymbol{G}_{\mathrm{sys}}}\tilde{\boldsymbol{i}}_{\mathrm{g}}=\underbrace{\tilde{\boldsymbol{v}}_{\mathrm{con}}^{\mathrm{c}}}_{\tilde{\boldsymbol{u}}}-\underbrace{\tilde{\boldsymbol{v}}_{\mathrm{g}}}_{=0} \tag{4.56}$$

By using the stability analysis method in Section 4.4.2, det($\boldsymbol{G}_{\text{sys}}$) of Equation 4.56 can represent the stability of the PLL-based VSC system in Figure 4.4. An example is given in Figure 4.6 to demonstrate how to conduct the relevant stability analysis. The converter system uses the data of Table 4.1. SCR = 1 is used in this case to simplify the process of tuning the PLL into an unstable state. By drawing the poles of det($\boldsymbol{G}_{\text{sys}}$), it is found that when $\omega_{\text{PLL}}^{\text{n}} = 200 \times 2\pi$, the system is unstable, as shown by the poles in close proximity to x=0 on the RHP. The oscillation frequency, indicated by the poles' *y*-axis value, is 41 Hz. The simulation results are also presented in Figure 4.6. Before 1 s, the PLL is tuned with $\omega_{\text{PLL}}^{\text{n}} = 180 \times 2\pi$. After 1 s, $\omega_{\text{PLL}}^{\text{n}}$ is changed to $200 \times 2\pi$, the system starts to oscillate and becomes unstable. The unstable situation and oscillation frequency in Figure 4.6 are all matched to the pole map analysis.

4.4.3.1 Case Study on Stability Analysis of the Single-phase PLL

As explained in Section 4.2.3, both single-phase and three-phase systems can adopt the same modeling method and follow the same stability analysis method as described above. However, an additional low-pass filter is used in the single-phase

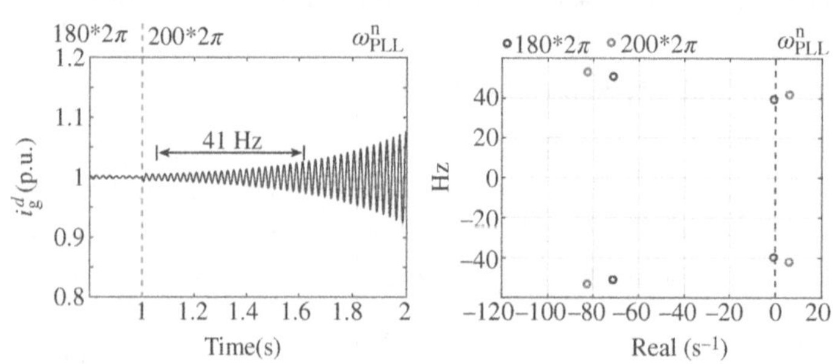

Figure 4.6 Stability analysis of the PLL-based VSC.

Table 4.1 Parameters of the converter system.

Symbol	Parameter	Value
$V_{\text{l–l}}$	L-L AC RMS voltage	320 kV
S	power rating	1000 MVA
R_{f}	filter resistor	0.05 p.u.
L_{f}	filter inductor	0.15 p.u.
SCR(L_{g} : $R_{\text{g}} = 10:1$)	short-circuit ratio	0.2
L_{g}	grid inductor	0.1818 p.u.
R_{g}	grid resistor	0.018 p.u.

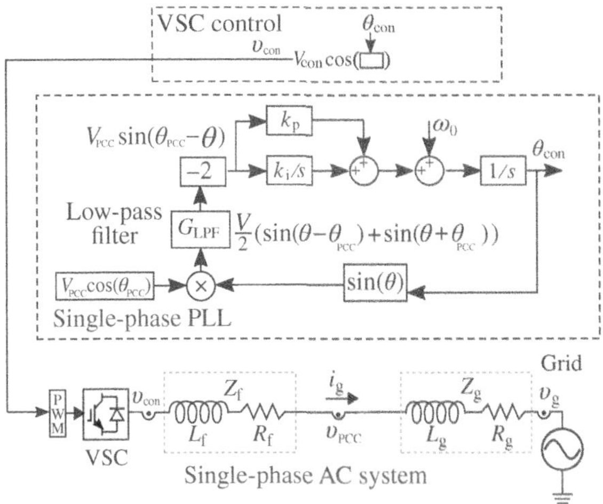

Figure 4.7 Single-phase PLL-based VSC system.

PLL to equivalently form the mathematical relationship of the *abc–dq* transformation in a three-phase PLL, as shown in both Figures 4.7 and 4.4. Therefore, the modeling of a single-phase PLL can be developed as below by referring to Equation 4.27:

$$(\tilde{v}_{\mathrm{PCC}}^{q} - \tilde{\theta}_{\mathrm{con}} v_{\mathrm{PCC}}^{d}) \times \underbrace{\frac{\omega_c}{s+\omega_c}}_{G_{\mathrm{LPF}}} \times \left(k_{\mathrm{p}}^{\mathrm{PLL}} + \frac{k_{\mathrm{i}}^{\mathrm{PLL}}}{s}\right) \times \frac{1}{s} = \tilde{\theta}_{\mathrm{con}} \tag{4.57}$$

where G_{LPF} is the transfer function of the low-pass filter in Figure 4.7.

Rearranging Equation 4.57 yields:

$$\tilde{\theta}_{\mathrm{con}} = \underbrace{\frac{G_{\mathrm{LPF}} k_{\mathrm{p}}^{\mathrm{PLL}} s + G_{\mathrm{LPF}} k_{\mathrm{i}}^{\mathrm{PLL}}}{s^2 + v_{\mathrm{PCC}}^{d} G_{\mathrm{LPF}} k_{\mathrm{p}}^{\mathrm{PLL}} s + v_{\mathrm{PCC}}^{d} G_{\mathrm{LPF}} k_{\mathrm{i}}^{\mathrm{PLL}}}}_{G_{\mathrm{PLL}}} \tilde{v}_{\mathrm{PCC}}^{q} \tag{4.58}$$

Equation 4.58 is the fundamental relation of the single-phase PLL, and it is the only difference from the three-phase PLL. The other modeling processes are exactly the same as the above-mentioned three-phase PLL, following Sections 4.3.1 and 4.4.1. The single-phase VSC in Figure 4.7 is used as an example for stability analysis, and the results are shown in Figure 4.8. Parameters adopted in the simulation and analysis are shown in Table 4.2. The poles of the developed model clearly show that the system is unstable, with a resonant frequency of 2.5 Hz.

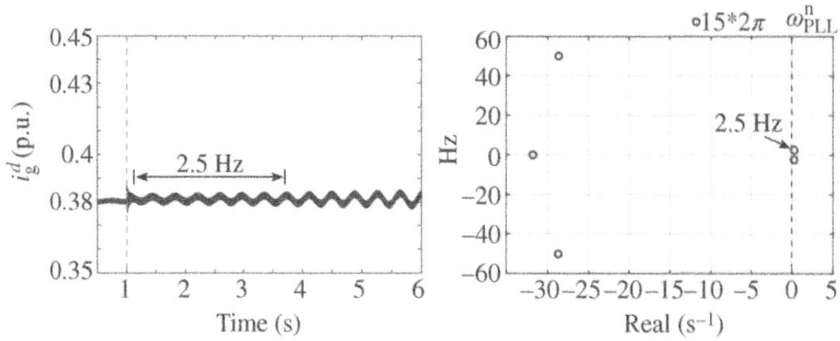

Figure 4.8 Stability analysis of the single-phase PLL-based VSC.

Table 4.2 Parameters of single-phase PLL.

Symbol	Parameter	Value
L_g	grid inductor	0.5 p.u.
R_g	grid resistor	0.005 p.u.
ω_c	cut-off frequency of LPF	$10 \times 2\pi$
ω_{PLL}^n	natural frequency of PLL	$15 \times 2\pi$

4.4.4 Stability Analysis of the *P*–*ω* Droop Control

A VSC adopting P–ω droop control has been presented in Figure 4.5. Similarly, only droop control is applied to a VSC with constant values of $[v_{con}^{c,d}, v_{con}^{c,q}]$. Therefore, the modeling of VSC control based on Equation 4.41 is simplified as:

$$\widetilde{\boldsymbol{v}}_{con} = -\widetilde{\boldsymbol{T}}_{con}^{v} + \underbrace{\widetilde{\boldsymbol{v}}_{con}^{c}}_{\widetilde{\boldsymbol{u}}} \tag{4.59}$$

$-\widetilde{\boldsymbol{T}}_{con}^{v}$ refers to the frame transformation via the P–ω droop control, which is presented in Equation 4.39. Therefore, Equation 4.59 can be represented as below:

$$\widetilde{\boldsymbol{v}}_{con} = -\boldsymbol{H}_{con}^{i}\widetilde{\boldsymbol{i}}_{g} - \boldsymbol{H}_{con}^{v}\widetilde{\boldsymbol{v}}_{PCC} + \widetilde{\boldsymbol{v}}_{con}^{c} + \boldsymbol{H}_{P}\widetilde{P}^{*} \tag{4.60}$$

Substituting $\widetilde{\boldsymbol{v}}_{con}$ and $\widetilde{\boldsymbol{v}}_{PCC}$ based on the AC system model of Equation 4.44 and 4.46, and assuming the power reference P^* will not change, yields:

$$\underbrace{(\boldsymbol{Z}_f + \boldsymbol{Z}_g + \boldsymbol{H}_{con}^{v}\boldsymbol{Z}_g + \boldsymbol{H}_{con}^{i})}_{G_{sys}}\widetilde{\boldsymbol{i}}_{g} = \underbrace{\widetilde{\boldsymbol{v}}_{con}^{c} + \boldsymbol{H}_{c}\underbrace{\widetilde{P}^{*}}_{=0}}_{\widetilde{\boldsymbol{u}}} - (\boldsymbol{H}_{con}^{v} + \boldsymbol{I})\underbrace{\widetilde{\boldsymbol{v}}_{g}}_{=0} \tag{4.61}$$

By using the stability analysis method in Section 4.4.2, det($\boldsymbol{G}_{sys}$) of Equation 4.61 can represent the stability of the P–ω droop-based VSC system in Figure 4.5. An example of stability analysis is presented based on the parameters in Table 4.3. By increasing d_P to 9.8, the system becomes unstable as poles cross into the RHP, as

Table 4.3 The Parameters of a Droop Control System.

Symbol	Parameter	Value
L_g	grid inductor	0.2 p.u.
R_g	grid resistor	0.02 p.u.
d_P	droop coefficient	9.8

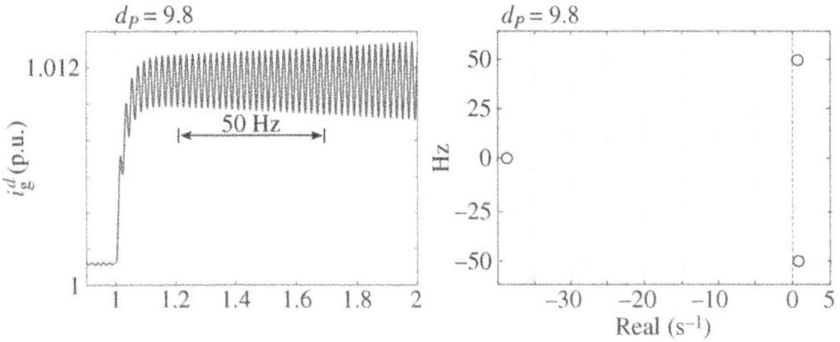

Figure 4.9 Stability analysis of the P–ω droop-based VSC system.

shown in Figure 4.9. The simulation results confirm the pole map stability analysis. When $t = 1$ s, increasing d_p to 9.8 leads to instability at an oscillation frequency of 50 Hz.

Reference

1 D.N. Zmood and D.G. Holmes. Stationary frame current regulation of PWM inverters with zero steady-state error. *IEEE Transactions on Power Electronics*, 18(3): 814–822, 2003. https://doi.org/10.1109/TPEL.2003.810852.

5

Stability Analysis Methods

Jun Liang[1,*] *and Chuanyue Li*[2]

[1] *School of Engineering, Cardiff University, Cardiff, UK*
[2] *School of Electric Power Engineering, South China University of Technology, Guangzhou, China*

*Corresponding author: LiangJ1@cardiff.ac.uk

5.1 State-space Stability Analysis Method

The state-space stability analysis method is a commonly used and widely accepted approach for analyzing linear time-invariant systems. This method decomposes the system model into a state-space form, represented by a transfer function matrix. The eigenvalues of its coefficient matrix capture all the system poles, providing a numerical matrix-based technique for identifying system poles. Unlike the process of deriving a system model into one complex transfer function to analyze all system poles, the numerical matrix method generally simplifies the calculation of these poles. However, with the aid of modern computers, this simplicity offered by state-space representation over the transfer function method is less of an advantage.

The primary benefit of the state-space stability analysis method is that it offers a deeper understanding of the system through a numerical matrix, enabling further exploration of how this matrix influences the distribution of poles.

A typical state-space stability analysis method, along with its corresponding sensitivity analysis approach, will be introduced below. Accordingly, the application of these methods to a converter system will be demonstrated.

Stability Analysis of Converter-Rich Power Grids, First Edition. Jun Liang and Chuanyue Li.
© 2026 The Institute of Electrical and Electronics Engineers, Inc. Published 2026 by
John Wiley & Sons, Inc. All rights reserved.

5.1.1 Typical State-space Stability Analysis Method

As linearizing the system has been described in Section 4.2 using small-signal modeling, the transformation of this system model into a state-space representation is presented below. A normalized state-space representation via matrices is shown as follows:

$$s\widetilde{\boldsymbol{x}} = \boldsymbol{A}\widetilde{\boldsymbol{x}} + \boldsymbol{B}\widetilde{\boldsymbol{u}} \tag{5.1a}$$

$$\widetilde{\boldsymbol{y}} = \boldsymbol{C}\widetilde{\boldsymbol{x}} + \boldsymbol{D}\widetilde{\boldsymbol{u}} \tag{5.1b}$$

A detailed description of Equation 5.1 is given below:

- $\widetilde{\ }$ represents a small deviation;
- $\widetilde{\boldsymbol{u}}$ is the input vector, including $[\widetilde{u}_1; ...; \widetilde{u}_r]$ input variables;
- $\widetilde{\boldsymbol{x}}$ is the state vector, including $[\widetilde{x}_1; ...; \widetilde{x}_n]$ state variables;
- $\widetilde{\boldsymbol{y}}$ is the output vector, including $[\widetilde{y}_1; ...; \widetilde{y}_m]$ output variables;
- $\boldsymbol{A}$ is the $n \times n$ numerical state matrix;
- $\boldsymbol{B}$ is the $n \times r$ numerical input matrix;
- $\boldsymbol{C}$ is the $m \times n$ numerical output matrix;
- $\boldsymbol{D}$ is the $m \times r$ numerical feedforward matrix.

The system stability is represented by the state matrix $\boldsymbol{A}$, whose eigenvalues $\lambda_1, ...\lambda_n$ are the system poles, which can be calculated as shown below for a nontrivial solution:

$$\det(\boldsymbol{A} - \lambda \boldsymbol{I}) = 0 \tag{5.2}$$

Equation 5.2 is the characteristic equation of the state matrix $\boldsymbol{A}$, which contains the same poles as the characteristic equation obtained from the transfer function method, as shown in Equation 3.1.

Therefore, the stability criterion in Chapter 3 can be applied, where the pole map is generally adopted. If there are RHP poles, the system will lose stability. More details about this method and the mechanism behind the criterion can be found in Prabha [1].

5.1.2 Sensitivity Stability Analysis

The numerical matrix $\boldsymbol{A}$ offers a straightforward way to explore how the elements a_{ik} of $\boldsymbol{A}$ impact the eigenvalues, a process known as sensitivity stability analysis, as defined below:

$$\frac{\widetilde{\lambda}_i}{\widetilde{a}_{jk}} \tag{5.3}$$

A sensitivity stability analysis method based on $\boldsymbol{A}$ is provided below to simplify the overall process.

For an eigenvalue λ_i, the following relation is obtained:

$$\boldsymbol{A\Phi}_i = \lambda_i \boldsymbol{\Phi}_i \tag{5.4a}$$

$$\boldsymbol{\Psi}_i^{\mathrm{T}} \boldsymbol{A} = \lambda_i \boldsymbol{\Psi}_i^{\mathrm{T}} \tag{5.4b}$$

where $\boldsymbol{\Phi}_i$ and $\boldsymbol{\Psi}_i$ are the corresponding right and left eigenvectors of λ_i; $\boldsymbol{\Psi}_i^{\mathrm{T}}$ denotes the transpose of $\boldsymbol{\Psi}_i$; $\boldsymbol{\Phi}_i = [\phi_{1i}; \phi_{ji}; \ldots; \phi_{ni}]$, $\boldsymbol{\Psi}_i^{\mathrm{T}} = [\psi_{i1}, \psi_{ik}, \ldots, \psi_{in}]$.

Applying the small deviation to Equation 5.4 yields:

$$(\boldsymbol{A} + \widetilde{\boldsymbol{A}})(\boldsymbol{\Phi}_i + \widetilde{\boldsymbol{\Phi}}_i) = (\boldsymbol{\Phi}_i + \widetilde{\boldsymbol{\Phi}}_i)(\lambda_i + \widetilde{\lambda}_i) \tag{5.5}$$

Assuming $\widetilde{\boldsymbol{A}}\widetilde{\boldsymbol{\Phi}}_i = 0$ and $\widetilde{\boldsymbol{\Phi}}_i\widetilde{\lambda}_i = 0$ and rearranging Equation 5.4 yields:

$$\boldsymbol{A}\widetilde{\boldsymbol{\Phi}}_i + \widetilde{\boldsymbol{A}}\boldsymbol{\Phi}_i = \boldsymbol{\Phi}\widetilde{\lambda}_i + \widetilde{\boldsymbol{\Phi}}_i\lambda_i \tag{5.6}$$

Multiplying $\boldsymbol{\Psi}_i^{\mathrm{T}}$ to Equation 5.6 and rearranging yields:

$$(\boldsymbol{\Psi}_i^{\mathrm{T}}\boldsymbol{A} - \lambda_i\boldsymbol{\Psi}_i^{\mathrm{T}})\widetilde{\boldsymbol{\Phi}}_i + \boldsymbol{\Psi}_i^{\mathrm{T}}\widetilde{\boldsymbol{A}}\boldsymbol{\Phi}_i = \boldsymbol{\Psi}_i^{\mathrm{T}}\boldsymbol{\Phi}\widetilde{\lambda}_i \tag{5.7}$$

Equation 5.4b indicates that $(\boldsymbol{\Psi}_i^{\mathrm{T}}\boldsymbol{A} - \lambda_i\boldsymbol{\Psi}_i^{\mathrm{T}}) = 0$, and Equation 5.7 is further simplified as:

$$\boldsymbol{\Psi}_i^{\mathrm{T}}\widetilde{\boldsymbol{A}}\boldsymbol{\Phi}_i = \boldsymbol{\Psi}_i^{\mathrm{T}}\boldsymbol{\Phi}_i\widetilde{\lambda}_i \tag{5.8}$$

Equation 5.8 provides a simple method to conduct the sensitivity analysis of Equation 5.3. When the deviation of each element $\widetilde{a}_{jk}$ is considered while the rest of the elements are 0, Equation 5.8 can be rewritten as:

$$\frac{d\lambda_i}{da_{kj}} = \frac{\psi_{ik}\phi_{ji}}{\boldsymbol{\Psi}_i^{\mathrm{T}}\boldsymbol{\Phi}_i} \tag{5.9}$$

where $\boldsymbol{\Psi}_i^{\mathrm{T}}\boldsymbol{\Phi}_i$ is a constant value.

The sensitivity analysis of a_{jk} on each eigenvalue λ_i provides deep insight into how a_{jk} impacts the system poles, which can help improve system stability.

5.1.3 Demonstration of State-space Stability Analysis Method

A PLL-based VSC system, as shown in Figure 4.4, is adopted for a state-space stability analysis. The system has been modeled in Section 4.3, which is used to demonstrate how to decompose this system model into a state-space representation. This VSC model, based on Equations 4.27, 4.44, and 4.46, is relisted below:

$$\widetilde{\theta}_{\mathrm{con}} = \frac{1}{s}\left(k_{\mathrm{p}}^{\mathrm{PLL}} + \frac{k_{\mathrm{i}}^{\mathrm{PLL}}}{s}\right)(\widetilde{v}_{\mathrm{PCC}}^{q} - v_{\mathrm{PCC}}^{d}\widetilde{\theta}_{\mathrm{con}}) \tag{5.10a}$$

$$\widetilde{v}_{\mathrm{con}}^{d} - \widetilde{v}_{\mathrm{g}}^{d} = (L_{\mathrm{f}} + L_{\mathrm{g}})s\widetilde{i}_{\mathrm{g}}^{d} - \omega_0(L_{\mathrm{f}} + L_{\mathrm{g}})\widetilde{i}_{\mathrm{g}}^{q} + (R_{\mathrm{f}} + R_{\mathrm{g}})\widetilde{i}_{\mathrm{g}}^{d} \tag{5.10b}$$

$$\widetilde{v}_{\mathrm{con}}^{q} - \widetilde{v}_{\mathrm{g}}^{q} = (L_{\mathrm{f}} + L_{\mathrm{g}})s\widetilde{i}_{\mathrm{g}}^{q} + \omega_0(L_{\mathrm{f}} + L_{\mathrm{g}})\widetilde{i}_{\mathrm{g}}^{d} + (R_{\mathrm{f}} + R_{\mathrm{g}})\widetilde{i}_{\mathrm{g}}^{q} \tag{5.10c}$$

v_g is considered the input vector on the system side and is assumed to be constant during the dynamics of the converters, which leads to $\tilde{v}_g^d = 0$ and $\tilde{v}_g^q = 0$. Rearranging Equation 5.10 into the form of a state-space representation yields:

$$s\tilde{\theta}_{con} = k_p^{PLL}(\tilde{v}_{PCC}^q - v_{PCC}^d\tilde{\theta}_{con}) + \frac{k_i^{PLL}}{s}(\tilde{v}_{PCC}^q - v_{PCC}^d\tilde{\theta}_{con}) \tag{5.11a}$$

$$s\tilde{i}_g^d = \frac{\tilde{v}_{con}^d - \tilde{v}_g^d + \omega_0(L_f + L_g)\tilde{i}_g^q - (R_f + R_g)\tilde{i}_g^d}{L_f + L_g} \tag{5.11b}$$

$$s\tilde{i}_g^q = \frac{\tilde{v}_{con}^q - \tilde{v}_g^q - \omega_0(L_f + L_g)\tilde{i}_g^d - (R_f + R_g)\tilde{i}_g^q}{L_f + L_g} \tag{5.11c}$$

To eliminate $1/s$ in Equation 5.11a, an additional state variable x_1 is proposed as the following relation:

$$s\tilde{x}_1 = \tilde{v}_{PCC}^q - v_{PCC}^d\tilde{\theta}_{con} \tag{5.12}$$

Substituting Equation 5.12 into Equation 5.11a and substituting $\tilde{v}_{con}^d$ and $\tilde{v}_{con}^q$ of Equation 5.11b and 5.11c using Equation 4.21 yields:

$$s\tilde{\theta}_{con} = k_p^{PLL}(\tilde{v}_{PCC}^q - v_{PCC}^d\tilde{\theta}_{con}) + k_i^{PLL}\tilde{x}_1 \tag{5.13a}$$

$$s\tilde{x}_1 = \tilde{v}_{PCC}^q - v_{PCC}^d\tilde{\theta}_{con} \tag{5.13b}$$

$$s\tilde{i}_g^d = -\frac{v_{con}^{c,q}}{L}\tilde{\theta}_{con} + \omega_0\tilde{i}_g^q - \frac{R}{L}\tilde{i}_g^d + \frac{1}{L}\tilde{v}_{con}^{c,d} - \frac{1}{L}\tilde{v}_g^d \tag{5.13c}$$

$$s\tilde{i}_g^q = \frac{v_{con}^{c,d}}{L}\tilde{\theta}_{con} - \omega_0\tilde{i}_g^d - \frac{R}{L}\tilde{i}_g^q + \frac{1}{L}\tilde{v}_{con}^{c,q} - \frac{1}{L}\tilde{v}_g^q \tag{5.13d}$$

where $L = L_f + L_g$ and $R = R_f + R_g$.

$\tilde{v}_{PCC}^q$ has the following relation:

$$\tilde{v}_{PCC}^q = L_g \underbrace{s\tilde{i}_g^q}_{\text{Equation 5.13d}} + \omega_0 L_g\tilde{i}_g^d + R_g\tilde{i}_g^q \tag{5.14}$$

Substituting $\tilde{v}_{PCC}^q$ in Equations 5.13a and 5.13b with Equation 5.14 and rearranging yields:

$$\begin{aligned} s\tilde{\theta}_{con} &= \left(k_p^{PLL}\left(L_g\frac{v_{con}^{c,d}}{L}\right) - k_p^{PLL}v_{PCC}^d\right)\tilde{\theta}_{con} + k_i^{PLL}\tilde{x}_1 \\ &\quad + k_p^{PLL}\left(R_g - L_g\frac{R}{L}\right)\tilde{i}_g^q + \frac{k_p^{PLL}L_g}{L}(\tilde{v}_{con}^{c,q} - \tilde{v}_g^q) \end{aligned} \tag{5.15a}$$

$$s\tilde{x}_1 = \left(L_g\frac{v_{con}^{c,d}}{L} - v_{PCC}^d\right)\tilde{\theta}_{con} + \left(R_g - L_g\frac{R}{L}\right)\tilde{i}_g^q + \frac{L_g}{L}(\tilde{v}_{con}^{c,q} - \tilde{v}_g^q) \tag{5.15b}$$

$$s\tilde{i}_g^d = -\frac{v_{con}^{c,q}}{L}\tilde{\theta}_{con} + \omega_0\tilde{i}_g^q - \frac{R}{L}\tilde{i}_g^d + \frac{1}{L}(\tilde{v}_{con}^{c,d} - \tilde{v}_g^d) \tag{5.15c}$$

$$s\tilde{i}_g^q = \frac{v_{con}^{c,d}}{L}\tilde{\theta}_{con} - \omega_0\tilde{i}_g^d - \frac{R}{L}\tilde{i}_g^q + \frac{1}{L}(\tilde{v}_{con}^{c,q} - \tilde{v}_g^q) \tag{5.15d}$$

State vector $\tilde{\boldsymbol{x}}$ is $[\tilde{\theta};\tilde{x}_1;\tilde{i}_g^d;\tilde{i}_g^q]$. Based on Equation 5.15, the state matrix $\boldsymbol{A}$ is identified as:

$$\begin{bmatrix} k_p^{PLL}\left(L_g\frac{v_{con}^{c,d}}{L}\right)-k_p^{PLL}v_{PCC}^d & k_i^{PLL} & 0 & k_p^{PLL}\left(R_g-L_g\frac{R}{L}\right) \\ L_g\frac{v_{con}^{c,d}}{L}-v_{PCC}^d & 0 & 0 & R_g-L_g\frac{R}{L} \\ -\frac{v_{con}^{c,q}}{L} & -\frac{R}{L} & 0 & \omega_0 \\ \frac{v_{con}^{c,d}}{L} & 0 & -\omega_0 & -\frac{R}{L} \end{bmatrix} \tag{5.16}$$

Both $[\tilde{v}_{con}^{c,d};\tilde{v}_{con}^{c,q}]$ and $[\tilde{v}_g^d;\tilde{v}_g^q]$ can be the input vector $\tilde{\boldsymbol{u}}$, assuming that only v_{con} or v_g is changed at one time. Based on Equation 5.15, the input matrices $\boldsymbol{B}_1$ and $\boldsymbol{B}_2$ for $[\tilde{v}_{con}^{c,d};\tilde{v}_{con}^{c,q}]$ and $[\tilde{v}_g^d;\tilde{v}_g^q]$ are presented below, respectively:

$$\boldsymbol{B}_1=\begin{bmatrix} 0 & \frac{k_p^{PLL}L_g}{L} \\ 0 & \frac{L_g}{L} \\ \frac{1}{L} & 0 \\ 0 & \frac{1}{L} \end{bmatrix};\boldsymbol{B}_2=\begin{bmatrix} 0 & -\frac{k_p^{PLL}L_g}{L} \\ 0 & -\frac{L_g}{L} \\ -\frac{1}{L} & 0 \\ 0 & -\frac{1}{L} \end{bmatrix} \tag{5.17}$$

If the output vector $\tilde{\boldsymbol{y}}$ is the grid current $\tilde{\boldsymbol{i}}_g$, since $\tilde{i}_g^d$ and $\tilde{i}_g^q$ belong to $\tilde{\boldsymbol{x}}$, the output matrix is $\boldsymbol{C} = [0\,0\,1\,0; 0\,0\,0\,1]$. Accordingly, the feedforward matrix $\boldsymbol{D}$ is 0.

The case study, which is the same as in Section 4.4.3, is presented below to demonstrate the equivalence between the transfer function method in Section 4.4 and the state-space method for pole-based stability analysis. To present the poles of the PLL-based converter system in Figure 4.4, a pole map is used to display both stable and unstable cases, as shown in Figure 5.1, where the poles match exactly for both methods.

The main advantage of the state-space method is that it allows for the numerical analysis of the system poles via the state matrix $\boldsymbol{A}$ rather than a transfer function, thereby reducing the computational burden. The sensitivity analysis of system poles can be simply applied via Equation 5.9, which helps to address instabilities. A demonstration case to resolve the unstable poles (λ_3, λ_4 of $\boldsymbol{A}$) at ω_{PLL}^n in Figure 5.1 is presented below.

The element $a_{12} = k_{PLL}^i$ of the state matrix, as shown in Equation 5.16, is related to the PI controller of the PLL and can be adjusted to shift λ_3 and λ_4 back to the left

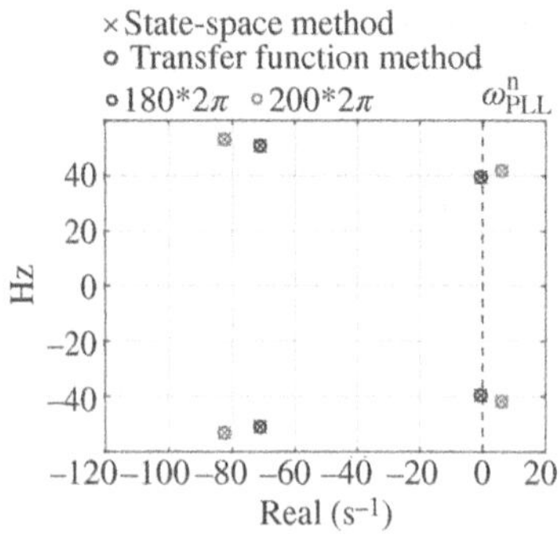

Figure 5.1 Comparison between the transfer function method and the state-space method for analyzing a PLL-based converter system.

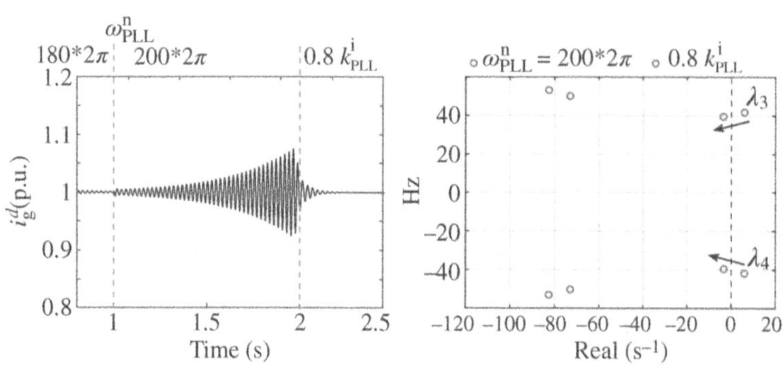

Figure 5.2 PLL stabilization based on sensitivity and stability analysis.

half-plane. According to the sensitivity analysis of a_{12} in Equation 5.9, it is found that $d\lambda_3/da_{12} = (2.1358 + 3.5411\text{j}) \times 10^{-5}$. This indicates that decreasing a_{12} can reduce the real part of λ_3, moving it further into the left half-plane of the pole map.

A validation of the sensitivity analysis of Equation 5.9 was conducted by applying $\widetilde{a}_{12} = 0.001$ in $\boldsymbol{A}$. The eigenvalue calculation of $\boldsymbol{A}$ shows that $\widetilde{\lambda}_3/\widetilde{a}_{12} = (2.1358 + 3.5411\text{j}) \times 10^{-5}$, which confirms the sensitivity analysis derived from Equation 5.9.

Based on the sensitivity analysis, a reduction of $0.8k_{\text{i}}^{\text{PLL}}$ is applied to stabilize the PLL-based converter system. The results are shown in Figure 5.2. It is found that the pole map confirms the sensitivity analysis; λ_3 and λ_4 return to the left half-plane when $k_{\text{i}}^{\text{PLL}}$ is reduced to 80%. A simulation case study is also presented, which confirms the solution. At $t = 2$ s, the system is stabilized by applying $0.8k_{\text{i}}^{\text{PLL}}$.

5.2 Impedance Stability Analysis Method

The transfer-function-based and state-space stability analysis methods discussed in Sections 4.4.2 and 5.1.1 consider the converter system as a whole model to analyze its stability. As control interaction is considered the dominant factor for instability in a converter-integrated power system, these two methods cannot reflect the interaction characteristics between the converter and the power system. In contrast, the impedance method models the converter and the power system as two separate impedances. Stability analysis is then conducted by analyzing the impedance interactions, thereby providing insight into the physical stability characteristics of control interaction.

The impedance stability analysis method is to reformulate the system into two parts, both represented as an impedance and a current/voltage source. The system's stability is determined by how these two impedances interact with each other. It is critical to identify the factors of the impedance model for stability analysis, which adopts the fundamental stability criteria introduced in Section 3.1.

Both Norton and Thevenin equivalent circuits will be applied and compared for impedance stability analysis. Then, stability analysis methods for the impedance representation will be introduced and recommended.

5.2.1 Impedance Representation of Converters Based on Norton and Thevenin Equivalent Circuits

As explained in Chapter 4, the converter system is linearized through frequency decoupling, which forms a 2×1 variable $\boldsymbol{v}$, such as $[v_d; v_q]$. Accordingly, the impedance is represented as a 2×2 impedance matrix, denoted in bold as $\boldsymbol{Z}$.

Based on the theory of Norton and Thevenin equivalent circuits, as shown in Figure 5.3, a VSC is equivalently represented as a current source, denoted as $\tilde{\boldsymbol{i}}_{\text{con,s}}$ and a parallel impedance $\boldsymbol{Z}_{\text{con}}$, or as a voltage source denoted as $\tilde{\boldsymbol{v}}_{\text{con,s}}$ and a series impedance $\boldsymbol{Z}_{\text{con}}$.

The power system is modeled as a voltage source $\tilde{\boldsymbol{v}}_{\text{g}}$ and a series impedance $\boldsymbol{Z}_{\text{g}}$.

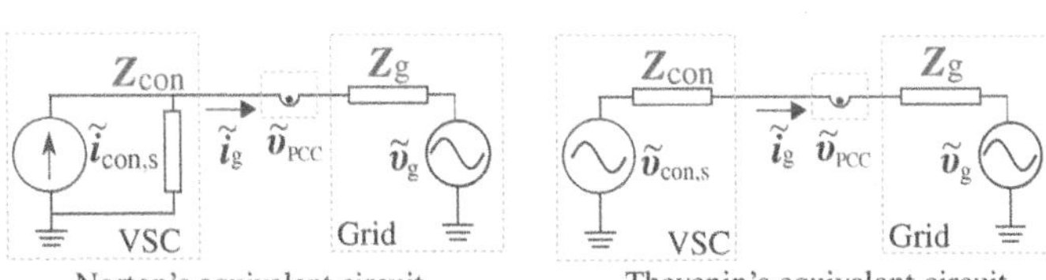

Figure 5.3 Impedance representation for a converter-integrated power system.

For grid-connected converters that are normally controlled as current sources, the Norton equivalent circuit is recommended for stability analysis. In fact, the Thevenin equivalent circuit is equally applicable. A detailed description is given below.

The Norton equivalent circuit of a converter is formulated as:

$$\tilde{\boldsymbol{v}}_{\text{PCC}} = \boldsymbol{Z}_{\text{con}}(\tilde{\boldsymbol{i}}_{\text{con,s}} - \tilde{\boldsymbol{i}}_{\text{g}}) \tag{5.18}$$

Rearranging yields:

$$\boldsymbol{Z}_{\text{con}}\tilde{\boldsymbol{i}}_{\text{con,s}} - \tilde{\boldsymbol{v}}_{\text{PCC}} = \boldsymbol{Z}_{\text{con}}\tilde{\boldsymbol{i}}_{\text{g}} \tag{5.19}$$

The Thevenin equivalent circuit of a converter is formulated as:

$$\tilde{\boldsymbol{v}}_{\text{con,s}} - \tilde{\boldsymbol{v}}_{\text{PCC}} = \boldsymbol{Z}_{\text{con}}\tilde{\boldsymbol{i}}_{\text{g}} \tag{5.20}$$

As $\tilde{\boldsymbol{i}}_{\text{con,s}}$ and $\tilde{\boldsymbol{v}}_{\text{con,s}}$ are both relevant to converter control reference, which can be represented as below based on the converter model in Equation 4.45:

$$\tilde{\boldsymbol{v}}_{\text{con,s}} = \boldsymbol{Z}_{\text{con}}\tilde{\boldsymbol{i}}_{\text{con,s}} = (\boldsymbol{I} - \boldsymbol{G}_{\text{col}}^{v})^{-1}\boldsymbol{G}_{\text{ref}}\tilde{\boldsymbol{u}} \tag{5.21}$$

If a small disturbance occurs in the converter system with a constant $\boldsymbol{u}$, where $\tilde{\boldsymbol{u}} = 0$, it leads to $\tilde{\boldsymbol{v}}_{\text{con,s}} = \boldsymbol{Z}_{\text{con}}\tilde{\boldsymbol{i}}_{\text{con,s}} = 0$. Therefore, the output impedance of the converter, $\boldsymbol{Z}_{\text{con}}$, will not be impacted regardless of whether it is based on Thevenin or Norton equivalent circuits, as shown in Equations 5.19 and 5.20.

The impedance representation of a converter can be based on both Norton and Thevenin equivalent circuits, and their output impedances are equivalent. However, for further stability analysis, it is assumed that the converter is stable under unloaded conditions, which requires:

- *for a Norton equivalent circuit,* $\boldsymbol{Z}_{\text{con}}\tilde{\boldsymbol{i}}_{\text{con,s}}$ *does not have RHP poles;*
- *for a Thevenin equivalent circuit,* $\tilde{\boldsymbol{v}}_{\text{con,s}}$ *does not have RHP poles.*

5.2.2 Demonstration of How to Derive Converter Impedance

As described in Section 4.4.1, for any converters that can be modeled as Equation 4.45, rearranging Equation 4.45 as the form of impedance representation, such as Thevenin equivalent circuit of Equation 5.20, yields:

$$\underbrace{(\boldsymbol{I} - \boldsymbol{G}_{\text{col}}^{v})^{-1}(\boldsymbol{Z}_{\text{f}} - \boldsymbol{G}_{\text{col}}^{i})}_{\boldsymbol{Z}_{\text{con}}}\tilde{\boldsymbol{i}}_{\text{g}} = -\tilde{\boldsymbol{v}}_{\text{PCC}} + \underbrace{(\boldsymbol{I} - \boldsymbol{G}_{\text{col}}^{v})^{-1}\boldsymbol{G}_{\text{ref}}\tilde{\boldsymbol{u}}}_{\tilde{\boldsymbol{v}}_{\text{con,s}}} \tag{5.22}$$

Equation 5.22 is the general form of an impedance representation of a converter.

To demonstrate how to derive the converter impedance representation, a PLL-based VSC system, as shown in Figure 4.4, is adopted. The detailed process of impedance derivation is given below. This is also a comparison between the impedance representation and transfer-function-based representation in Section 4.4.3.

The same example of a PLL-based VSC system in Section 4.4.3 is adopted, as shown in Figure 4.4. Only PLL is applied to a VSC with constant values of $[v_{\mathrm{con}}^{\mathrm{c},d}, v_{\mathrm{con}}^{\mathrm{c},q}]$, which represents $\widetilde{\boldsymbol{ref}}$. The modeling of VSC control based on Equation 4.41 is simplified as:

$$\widetilde{\boldsymbol{v}}_{\mathrm{con}} = -\widetilde{\boldsymbol{T}}_{\mathrm{con}}^{v} + \widetilde{\boldsymbol{v}}_{\mathrm{con}}^{\mathrm{c}} \tag{5.23}$$

By applying the frame transformation with the dynamics of the PLL via Equation 4.31, $\widetilde{\boldsymbol{T}}_{\mathrm{con}}^{v}$ is expanded and Equation 5.23 is represented as follows:

$$\widetilde{\boldsymbol{v}}_{\mathrm{con}} = -\boldsymbol{G}_{\mathrm{con}}^{v}\widetilde{\boldsymbol{v}}_{\mathrm{PCC}} + \widetilde{\boldsymbol{v}}_{\mathrm{con}}^{\mathrm{c}} \tag{5.24}$$

Substituting $\widetilde{\boldsymbol{v}}_{\mathrm{con}}$ and $\widetilde{\boldsymbol{v}}_{\mathrm{PCC}}$ based on the AC system model of Equation 4.44 yields:

$$\boldsymbol{Z}_{\mathrm{f}}\widetilde{\boldsymbol{i}}_{\mathrm{g}} + \widetilde{\boldsymbol{v}}_{\mathrm{PCC}} = -\boldsymbol{G}_{\mathrm{con}}^{v}\widetilde{\boldsymbol{v}}_{\mathrm{PCC}} + \widetilde{\boldsymbol{v}}_{\mathrm{con}}^{\mathrm{c}} \tag{5.25}$$

Rearranging Equation 5.25 yields:

$$\underbrace{(\boldsymbol{I} + \boldsymbol{G}_{\mathrm{con}}^{v})^{-1}\boldsymbol{Z}_{\mathrm{f}}}_{\boldsymbol{Z}_{\mathrm{con}}}\widetilde{\boldsymbol{i}}_{\mathrm{g}} = \underbrace{(\boldsymbol{I} + \boldsymbol{G}_{\mathrm{con}}^{v})^{-1}\widetilde{\boldsymbol{v}}_{\mathrm{con}}^{\mathrm{c}} - \widetilde{\boldsymbol{v}}_{\mathrm{PCC}}}_{\widetilde{\boldsymbol{u}}} \tag{5.26}$$

Therefore, for a PLL-based VSC as shown in Figure 4.4, its Thevenin equivalent impedance can be represented by Equation 5.26.

5.2.3 Impedance Stability Analysis Method Based on Generalized Nyquist Criterion and Bode Criterion

The impedance stability analysis method is to identify the factor that represents the system stability, and then apply any stability criteria described in Chapter 3. The overall system impedance representation is introduced below and explains how to identify the impedance factor.

To specify the impedance relation with the power system, the grid impedance $\widetilde{\boldsymbol{v}}_{\mathrm{PCC}}$ is formulated as:

$$\widetilde{\boldsymbol{v}}_{\mathrm{PCC}} = \boldsymbol{Z}_{\mathrm{g}}\widetilde{\boldsymbol{i}}_{\mathrm{g}} + \widetilde{\boldsymbol{v}}_{\mathrm{g}} \tag{5.27}$$

Thevenin equivalent circuit is adopted for the converter impedance modeling. Substituting $\widetilde{\boldsymbol{v}}_{\mathrm{PCC}}$ in Equation 5.20 with Equation 5.27 and rearranging yields:

$$(\boldsymbol{Z}_{\mathrm{con}} + \boldsymbol{Z}_{\mathrm{g}})\widetilde{\boldsymbol{i}}_{\mathrm{g}} = (\widetilde{\boldsymbol{v}}_{\mathrm{con,s}} - \widetilde{\boldsymbol{v}}_{\mathrm{g}}) \tag{5.28}$$

Equation 5.28 is the impedance representation of a converter-integrated power system.

5.2.3.1 Impedance Stability Criterion via Generalized Nyquist Criterion

The impedance model described by Equation 5.28 is a 2×2 system, which is a multivariable system. The Generalized Nyquist Criterion, as developed by I. Postlethwaite [2], was designed for such a multivariable system. To fulfill the form of a typical feedback system for the Generalized Nyquist Criterion, Equation 5.28 is further rearranged as follows:

$$(\boldsymbol{I} + \boldsymbol{Z}_{\text{con}}^{-1}\boldsymbol{Z}_{\text{g}})\widetilde{\boldsymbol{i}}_{\text{g}} = \underbrace{[\boldsymbol{Z}_{\text{con}}^{-1}(\widetilde{\boldsymbol{v}}_{\text{con,s}} - \widetilde{\boldsymbol{v}}_{\text{g}})]}_{\widetilde{\boldsymbol{r}}} \tag{5.29}$$

where $(\boldsymbol{Z}_{\text{g}} + \boldsymbol{I})$ is the impedance interaction matrix; $\boldsymbol{I} + \boldsymbol{Z}_{\text{con}}^{-1}\boldsymbol{Z}_{\text{g}}$ is the impedance return-ratio matrix, which is also the open-loop gain in Figure 5.4.

An assumption is made as: *$\boldsymbol{Z}_{\text{con}}^{-1}(\widetilde{\boldsymbol{v}}_{\text{con,s}} - \widetilde{\boldsymbol{v}}_{\text{g}})$ is stable and does not contain any RHP poles. It is denoted as the input vector $\widetilde{\boldsymbol{r}}$ for the feedback system in Figure 5.4.*

For such a typical feedback system as shown in Figure 5.4, the following relation is identified by I. Postlethwaite [2]:

$$\det(\boldsymbol{I} + \boldsymbol{Z}_{\text{con}}^{-1}\boldsymbol{Z}_{\text{g}}) = \frac{\text{CLCP}(s)}{\text{OLCP}(s)} \tag{5.30}$$

- CLCP(s) is closed-loop characteristic equation, which contains all the poles of the system in Figure 5.4.
- OLCP(s) is open-loop characteristic equation, which contains all the poles of open-loop gain $[\boldsymbol{I} + \boldsymbol{Z}_{\text{con}}^{-1}\boldsymbol{Z}_{\text{g}}]^{-1}$.

Therefore, the system stability can be specified based on the Generalized Nyquist Criterion for impedance interaction matrix $(\boldsymbol{I} + \boldsymbol{Z}_{\text{con}}^{-1}\boldsymbol{Z}_{\text{g}})$ as follows:

If the system modeled as Equation 5.29 is open-loop stable, where OLCP(s) *of impedance return-ratio matrix $\boldsymbol{I} + \boldsymbol{Z}_{\text{con}}^{-1}\boldsymbol{Z}_{\text{g}}$ has no RHP zeros, the system is closed-loop stable if and only if the net sum of encirclements of the critical point* $(0, 0\text{j})$ *by the set of inverse loci of the impedance interaction matrix* $\det(\boldsymbol{I} + \boldsymbol{Z}_{\text{con}}^{-1}\boldsymbol{Z}_{\text{g}})$.

A case study, which is a PLL-based converter system as described in Section 4.4.3, is adopted for impedance stability analysis based on the Generalized Nyquist Criterion to demonstrate the process of stability analysis. Two cases at different natural frequencies of the PLL are applied for a stable and an unstable scenario, respectively, as indicated in Figure 5.5a, including:

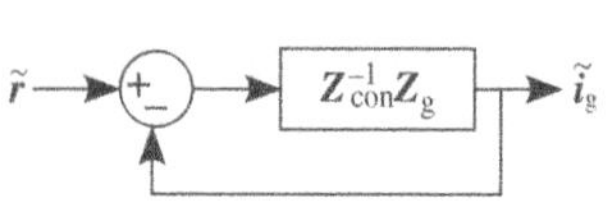

Figure 5.4 The impedance-represented feedback configuration.

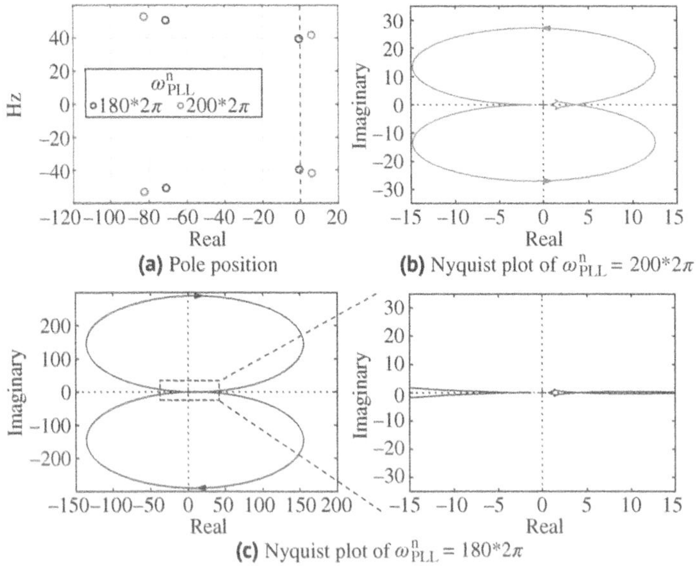

Figure 5.5 Impedance Analysis for Assessing Stability in a Converter System Using a Nyquist Plot

1. when $\omega^{n}_{PLL} = 180 \times 2\pi$, the system is stable;
2. when $\omega^{n}_{PLL} = 200 \times 2\pi$, the system is stable.

After drawing the Nyquist plot of the impedance factor $1/[\det(\boldsymbol{I} + \boldsymbol{Z}^{-1}_{con}\boldsymbol{Z}_g)]$, it accurately identifies the stability of the system, including:

1. as shown in Figure 5.5b, the Nyquist path encircles the point (0,0) counterclockwise, indicating the RHP poles, which matches the pole map in Figure 5.5a;
2. as shown in Figure 5.5c, the Nyquist path encircles the point (0,0) clockwise, indicating no RHP poles, which matches the pole map in Figure 5.5a.

Note that since $1/[\det(\boldsymbol{I} + \boldsymbol{Z}^{-1}_{con}\boldsymbol{Z}_g)]$ contains one RHP zero, the Nyquist path encircles the point (0,0) in a clockwise direction, where $Z = 1$, as shown in Figure 5.5c. Therefore, with $\omega^{n}_{PLL} = 200 \times 2\pi$, there are two RHP poles, and the Nyquist path should encircle the point (0,0) twice in a counterclockwise direction, where $P = 2$. In fact, the encirclement is $N = Z - P = -1$, which is the actual encirclement shown in Figure 5.5b.

Note that the same case has been applied in the transfer-function-based stability analysis method and the state-space stability analysis method in both Sections 4.4.2 and 5.1.3. To avoid repetition, the simulation validation will not be presented.

5.2.3.2 Impedance Stability Criterion via Bode Plot

Comparing to impedance stability analysis via $(\boldsymbol{I} + \boldsymbol{Z}_{\text{con}}^{-1}\boldsymbol{Z}_{\text{g}})$ based on the Generalized Nyquist Criterion, stability analysis via $\boldsymbol{Z}_{\text{con}} + \boldsymbol{Z}_{\text{g}}$ is an alternative method for system stability analysis, which is more consistent with the impedance form. The detailed proof of $\det(\boldsymbol{Z}_{\text{con}} + \boldsymbol{Z}_{\text{g}})$ representing system stability is given below.

Rearranging Equation 5.28 yields:

$$\tilde{\boldsymbol{i}}_{\text{g}} = (\boldsymbol{Z}_{\text{con}} + \boldsymbol{Z}_{\text{g}})^{-1}(\tilde{\boldsymbol{v}}_{\text{con,s}} - \tilde{\boldsymbol{v}}_{\text{g}}) \tag{5.31}$$

where $\boldsymbol{Z}_{\text{con}} + \boldsymbol{Z}_{\text{g}}$ is noted as the impedance interaction matrix.

The converter-integrated power system can also be modeled based on the state-space representation as described in Section 5.1.1, where $\tilde{\boldsymbol{i}}_{\text{g}}$ is the output vector $\tilde{\boldsymbol{y}}$ and is also one pair of variables of the input vector $\tilde{\boldsymbol{u}}$. $\tilde{\boldsymbol{v}}_{\text{con,s}}$ and $\tilde{\boldsymbol{v}}_{\text{g}}$ are the input vectors $\tilde{\boldsymbol{u}}_1$ and $\tilde{\boldsymbol{u}}_2$. Therefore, the relevant state-space representation is derived as:

$$s\tilde{\boldsymbol{x}} = \boldsymbol{A}\tilde{\boldsymbol{x}} + \boldsymbol{B}_1\tilde{\boldsymbol{u}} + \boldsymbol{B}_2\tilde{\boldsymbol{v}}_{\text{g}} \tag{5.32a}$$

$$\tilde{\boldsymbol{i}}_{\text{g}} = \boldsymbol{C}\tilde{\boldsymbol{x}} \tag{5.32b}$$

where $\boldsymbol{u}$ is the reference of the converter control.

It is assumed only one input of $\tilde{\boldsymbol{u}}$ and $\tilde{\boldsymbol{v}}_{\text{g}}$ is changed at a time. Therefore,

$$\tilde{\boldsymbol{i}}_{\text{g}} = \boldsymbol{C}(\boldsymbol{I}s - \boldsymbol{A})^{-1}\boldsymbol{B}_1\tilde{\boldsymbol{u}} \tag{5.33a}$$

$$\tilde{\boldsymbol{i}}_{\text{g}} = \boldsymbol{C}(\boldsymbol{I}s - \boldsymbol{A})^{-1}\boldsymbol{B}_2\tilde{\boldsymbol{v}}_{\text{g}} \tag{5.33b}$$

Taking Equation 5.33a as the example, comparing it to Equation 5.31, the following relations are found:

$$(\boldsymbol{Z}_{\text{con}} + \boldsymbol{Z}_{\text{g}})^{-1}\tilde{\boldsymbol{v}}_{\text{s,con}} = \boldsymbol{C}(\boldsymbol{I}s - \boldsymbol{A})^{-1}\boldsymbol{B}_1\tilde{\boldsymbol{u}} \tag{5.34}$$

The relation between $\tilde{\boldsymbol{v}}_{\text{s,con}}$ and $\tilde{\boldsymbol{u}}$ is presented in Equation 5.21; Equation 5.34 can be rewritten as:

$$(\boldsymbol{Z}_{\text{con}} + \boldsymbol{Z}_{\text{g}})^{-1}(\boldsymbol{G}_{\text{col}}^{v} - \boldsymbol{I})^{-1}\boldsymbol{G}_{\text{ref}} = \boldsymbol{C}(\boldsymbol{I}s - \boldsymbol{A})^{-1}\boldsymbol{B}_1 \tag{5.35}$$

$\boldsymbol{C}(\boldsymbol{I}s - \boldsymbol{A})^{-1}\boldsymbol{B}_1$ contains the characteristic equation of the system $\det(\boldsymbol{I}s - \boldsymbol{A})$ and does not contain any other poles, as shown below:

$$\boldsymbol{C}(\boldsymbol{I}s - \boldsymbol{A})^{-1}\boldsymbol{B}_1 = \frac{1}{\det(\boldsymbol{I}s - \boldsymbol{A})}\boldsymbol{C}\text{adj}(\boldsymbol{I}s - \boldsymbol{A})\boldsymbol{B}_1 \tag{5.36}$$

The following assumption is made:

- *the converter's equivalent output voltage $\tilde{\boldsymbol{v}}_{\text{s,con}}$ has no RHP poles, which equivalently requires that $(\boldsymbol{G}_{\text{col}}^{v} - \boldsymbol{I})^{-1}\boldsymbol{G}_{\text{ref}}$ has no RHP poles.*

Therefore, $(\boldsymbol{Z}_{\mathrm{con}} + \boldsymbol{Z}_{\mathrm{g}})^{-1}$ contains the system characteristic equation; furthermore, $\det(\boldsymbol{Z}_{\mathrm{con}} + \boldsymbol{Z}_{\mathrm{g}})$ contains system characteristic equation, which can represent the system stability. The stability of control interaction in the converter-integrated system can be analyzed via the impedance interaction matrix $(\boldsymbol{Z}_{\mathrm{con}} + \boldsymbol{Z}_{\mathrm{g}})$. The stability criterion via the Bode plot is therefore developed as:

If the Bode plot is applied, referring to Section 3.3, for any poles locate on the imaginary axis of the pole map, there is always a value of ω that results in $\det(\boldsymbol{Z}_{\mathrm{con}} + \boldsymbol{Z}_{\mathrm{g}})$ to 0, which suggests that the system is marginally stable and tends to be unstable. In other way, $1/\det(\boldsymbol{Z}_{\mathrm{con}} + \boldsymbol{Z}_{\mathrm{g}})$ appears to be infinite at a certain frequency, which suggests that the system is marginally stable and tends to be unstable.

A case study, which is a PLL-based converter system as described in Section 4.4.3, is adopted for impedance stability criterion analysis via the Bode plot. It is the same case study, which was used for impedance stability analysis based on the Generalized Nyquist Criterion in Section 5.2.3.1. Two cases at different natural frequencies of the PLL are applied for a solidly stable and an almost marginally stable scenario, respectively, as indicated in Figure 5.6a, including:

1. when $\omega_{\mathrm{PLL}}^{\mathrm{n}} = 100 \times 2\pi$, the system is solidly stable;
2. when $\omega_{\mathrm{PLL}}^{\mathrm{n}} = 180 \times 2\pi$, the system is almost marginally stable.

The Bode plot is applied to the impedance factor $1/\det(\boldsymbol{Z}_{\mathrm{con}} + \boldsymbol{Z}_{\mathrm{g}})$ as shown in Figure 5.6b. The results show that:

1. when $\omega_{\mathrm{PLL}}^{\mathrm{n}} = 180 \times 2\pi$, the gain sharply increases at 39 Hz, which indicates the system is close to marginal stability;
2. when $\omega_{\mathrm{PLL}}^{\mathrm{n}} = 100 \times 2\pi$, the gain remains low across the frequency range, indicating that the system is solidly stable.

Note that the same case has been applied in the transfer-function-based stability analysis method and the state-space stability analysis method in both Sections 4.4.2 and 5.1.3. To avoid repetition, the simulation validation will not be presented.

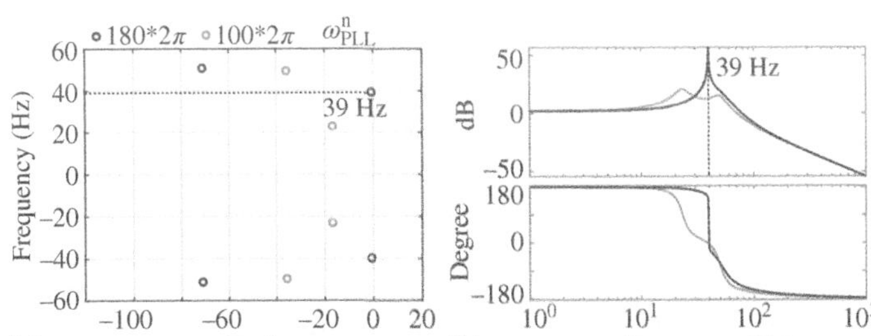

Figure 5.6 Impedance Stability Analysis Using Bode Plot

5.2.4 *dq*-sequence Impedance

The *dq* impedance forms the foundation for stability analysis and can be equivalently transformed into different forms to provide various views on system stability, such as sequence impedance [3–5], $\alpha\beta$ impedance [5], and impedance in the phasor domain [6]. Actually, all these impedance forms share the same stability accuracy.

The so-called sequence impedance refers to impedance in the *dq* frame and does not represent the traditional positive and negative sequences of an AC system. This distinction is explained in detail in Sections 7.1 and 7.2. Therefore, it is renamed as the *dq*-sequence impedance to distinguish it from the common AC system positive and negative sequence.

An unbalanced signal in a *dq* frame can be separated into balanced positive- and negative-frequency (PNF) as shown in Figure 5.7. This imbalance is mainly caused by controls, such as PLL-based or power-droop-based controls, which will be further explained in Section 7.5. Accordingly, the *dq* impedance representation is aligned with the unbalanced signals. If the separation process of the unbalanced signals is applied to the *dq* impedance, *dq*-sequence impedance will be formed, providing a new approach to understand impedance stability via the PNF components and analyzing the degree of imbalance in a system. It should be noted that both impedance representations are derived in the *dq* frame.

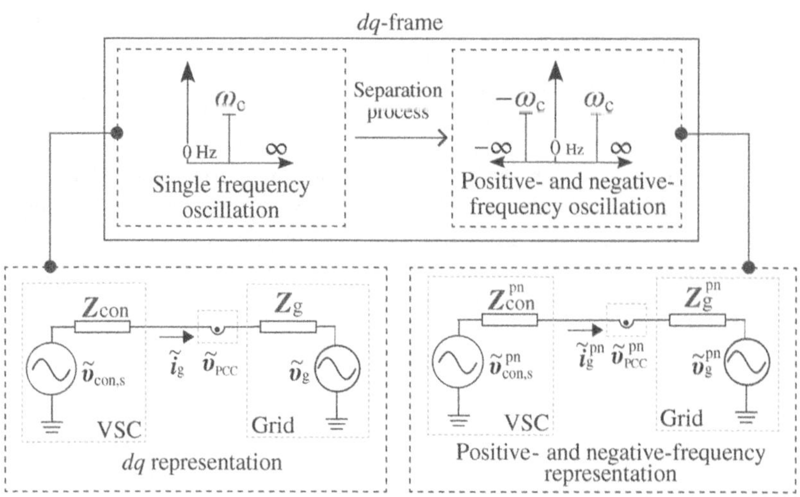

Figure 5.7 *dq*-impedance-represented system and *dq*-sequence-impedance-represented system.

As shown in Figure 5.3, if taking converter equivalent *dq* impedance as an example, the following relations are presented as below:

$$\tilde{\boldsymbol{v}}_{\text{con,s}} - \tilde{\boldsymbol{v}}_{\text{PCC}} = \boldsymbol{Z}_{\text{con}}\tilde{\boldsymbol{i}}_{\text{g}} \tag{5.37}$$

The separation process was derived as shown in Equation 7.20, using $\tilde{\boldsymbol{v}}_{\text{PCC}}$ as an example:

$$\underbrace{\begin{bmatrix} \tilde{v}^{d}_{\text{PCC}} \\ \tilde{v}^{q}_{\text{PCC}} \end{bmatrix}}_{\tilde{\boldsymbol{v}}_{\text{PCC}}} = \underbrace{\begin{bmatrix} 1 & 1 \\ -\text{j} & \text{j} \end{bmatrix}}_{\boldsymbol{T}_{\text{pn}-dq}} \underbrace{\begin{bmatrix} \tilde{v}^{\text{p}}_{\text{PCC}} \\ \tilde{v}^{\text{n}}_{\text{PCC}} \end{bmatrix}}_{\tilde{\boldsymbol{v}}^{\text{pn}}_{\text{PCC}}} \tag{5.38}$$

where the superscripts ^{p} and ^{n} represent the PNF components, respectively.

Note that matrix elements like $\tilde{v}^{d}_{\text{con}}$ were previously presented in vector form as $\tilde{\boldsymbol{v}}^{d}_{\text{con}}$ in Section 7.1.2.2. In this section, they are shown in non-bold symbol as $\tilde{v}^{d}_{\text{con}}$, because the bold symbol $\tilde{\boldsymbol{v}}_{\text{con}}$ indicates the matrix form.

Applying Equation 5.37, as shown below:

$$\boldsymbol{T}_{\text{pn}-dq}\tilde{\boldsymbol{v}}^{\text{pn}}_{\text{con,s}} - \boldsymbol{T}_{\text{pn}-dq}\tilde{\boldsymbol{v}}^{\text{pn}}_{\text{PCC}} = \boldsymbol{Z}_{\text{con}}\boldsymbol{T}_{\text{pn}-dq}\tilde{\boldsymbol{i}}^{\text{pn}}_{\text{g}} \tag{5.39}$$

Rearranging yields:

$$\tilde{\boldsymbol{v}}^{\text{pn}}_{\text{con,s}} - \tilde{\boldsymbol{v}}^{\text{pn}}_{\text{PCC}} = \underbrace{\boldsymbol{T}^{-1}_{\text{pn}-dq}\boldsymbol{Z}_{\text{con}}\boldsymbol{T}_{\text{pn}-dq}}_{\boldsymbol{Z}^{\text{pn}}_{\text{con}}}\tilde{\boldsymbol{i}}^{\text{pn}}_{\text{g}} \tag{5.40}$$

To summarize in a general form, the transformation from a *dq* impedance to *dq*-sequence impedance is shown below:

$$\boldsymbol{Z}^{\text{pn}} = \boldsymbol{T}^{-1}_{\text{pn}-dq}\boldsymbol{Z}^{dq}\boldsymbol{T}_{\text{pn}-dq} \tag{5.41}$$

$$\boldsymbol{Z}^{dq} = \boldsymbol{T}_{\text{pn}-dq}\boldsymbol{Z}^{\text{pn}}\boldsymbol{T}^{-1}_{\text{pn}-dq} \tag{5.42}$$

The detailed expression of the *dq*-sequence impedance is shown as follows by expanding Equation 5.41:

$$\boldsymbol{Z}^{\text{pn}} = 0.5\begin{bmatrix} Z^{dd} + Z^{qq} - \text{j}Z^{dq} + \text{j}Z^{qd} & Z^{dd} - Z^{qq} + \text{j}Z^{dq} + \text{j}Z^{qd} \\ Z^{dd} - Z^{qq} - \text{j}Z^{dq} - \text{j}Z^{qd} & Z^{dd} + Z^{qq} + \text{j}Z^{dq} - \text{j}Z^{qd} \end{bmatrix} \tag{5.43}$$

5.2.4.1 Stability Criteria for *dq*-sequence Impedance

Actually, the *dq*-sequence impedance is equivalently transformed from the *dq* impedance, and this transformation does not impact the system poles. This means that both *dq*-sequence impedance and *dq* impedance are equivalent for stability analysis. The mathematical proof is provided below.

As shown in Figure 5.3, both impedance representations have the following relations:

$$\begin{aligned}
\widetilde{\boldsymbol{v}}_{\mathrm{con,s}} - \widetilde{\boldsymbol{v}}_{\mathrm{g}} &= (\boldsymbol{Z}_{\mathrm{con}} + \boldsymbol{Z}_{\mathrm{g}})\widetilde{\boldsymbol{i}}_{\mathrm{g}} \\
\widetilde{\boldsymbol{v}}^{\mathrm{pn}}_{\mathrm{con,s}} - \widetilde{\boldsymbol{v}}^{\mathrm{pn}}_{\mathrm{g}} &= (\boldsymbol{Z}^{\mathrm{pn}}_{\mathrm{con}} + \boldsymbol{Z}^{\mathrm{pn}}_{\mathrm{g}})\widetilde{\boldsymbol{i}}^{\;\mathrm{pn}}_{\mathrm{g}}
\end{aligned} \tag{5.44}$$

The dq impedance stability of a converter system is determined by the determinant of the impedance matrix $1/\det(\boldsymbol{Z}\mathrm{con} + \boldsymbol{Z}\mathrm{g})$, as described in Section 5.2.3.2.

To find the relation between $\det(\boldsymbol{Z}\mathrm{con} + \boldsymbol{Z}\mathrm{g})$ and $\det(\boldsymbol{Z}^{\mathrm{pn}}_{\mathrm{con}})$, applying the transformation from a dq impedance to dq-sequence impedance yields:

$$\begin{aligned}
\det(\boldsymbol{Z}^{\mathrm{pn}}_{\mathrm{con}} + \boldsymbol{Z}^{\mathrm{pn}}_{\mathrm{g}}) &= \det(\boldsymbol{T}^{-1}_{\mathrm{pn}-dq}(\boldsymbol{Z}_{\mathrm{con}} + \boldsymbol{Z}_{\mathrm{g}})\boldsymbol{T}_{\mathrm{pn}-dq}) \\
&= \det(\boldsymbol{T}^{-1}_{\mathrm{pn}-dq}) \det(\boldsymbol{Z}_{\mathrm{con}} + \boldsymbol{Z}_{\mathrm{g}}) \det(\boldsymbol{T}_{\mathrm{pn}-dq}) \\
&= (-0.5\mathrm{j}) \det(\boldsymbol{Z}_{\mathrm{con}} + \boldsymbol{Z}_{\mathrm{g}})(2\mathrm{j})
\end{aligned} \tag{5.45}$$

Rearranging Equation 5.45 yields:

$$\det(\boldsymbol{Z}^{\mathrm{pn}}_{\mathrm{con}} + \boldsymbol{Z}^{\mathrm{pn}}_{\mathrm{g}}) = \det(\boldsymbol{Z}_{\mathrm{con}} + \boldsymbol{Z}_{\mathrm{g}}) \tag{5.46}$$

It is found that $\det(\boldsymbol{Z}\mathrm{con} + \boldsymbol{Z}\mathrm{g})$ is exactly equal to $\det(\boldsymbol{Z}^{\mathrm{pn}}_{\mathrm{con}} + \boldsymbol{Z}\mathrm{g}^{\mathrm{pn}})$. Therefore, the stability criterion for a system represented by dq-sequence impedance is the same as that for a system represented by dq impedance, which is given below:

If any poles of $1/\det(\boldsymbol{Z}^{\mathrm{pn}}_{\mathrm{con}} + \boldsymbol{Z}^{\mathrm{pn}}_{\mathrm{g}})$ cross to the RHP, it indicates that the system is unstable. The Bode plot can also be applied; for any poles located on the imaginary axis of the pole map, $1/\det(\boldsymbol{Z}^{\mathrm{pn}}_{\mathrm{con}} + \boldsymbol{Z}^{\mathrm{pn}}_{\mathrm{g}})$ appears as infinity at a certain frequency in the Bode plot, which suggests that the system is marginally stable and tends to be unstable.

5.2.5 $\alpha\beta$ Impedance

$\alpha\beta$ impedance refers to the impedance represented in the $\alpha\beta$ frame. It is almost same to the dq-sequence impedance. As the transformation between a dq-sequence signal and $\alpha\beta$ signal is almost a unit matrix, as explained in Section 7.1.2.2, or as shown in Figure 5.8, dq–$\alpha\beta$ transformation only shift the oscillation frequency from $\pm w_c$ to $\omega_0 - \omega_c$ and $\omega_0 + \omega_c$, which has been presented by Equation 7.25. Accordingly, Equation 7.25 can be re-presented as below:

$$\begin{array}{c|c}
0 < \omega_c < \omega_0 & \omega_c > \omega_0 \\
\underbrace{\widetilde{\boldsymbol{v}}^{\,\mathrm{p}}}_{\text{at } \omega_c} = \underbrace{\widetilde{\boldsymbol{v}}_1}_{\text{at } \omega_0+\omega_c} & \underbrace{\widetilde{\boldsymbol{v}}^{\,\mathrm{p}}}_{\text{at } \omega_c} = \underbrace{\widetilde{\boldsymbol{v}}_1}_{\text{at } \omega_0+\omega_c} \\
\underbrace{\widetilde{\boldsymbol{v}}^{\,\mathrm{n}}}_{\text{at } -\omega_c} = \underbrace{\overline{\widetilde{\boldsymbol{v}}_2}}_{\text{at } \omega_0-\omega_c} & \underbrace{\widetilde{\boldsymbol{v}}^{\,\mathrm{n}}}_{\text{at } -\omega_c} = \underbrace{\widetilde{\boldsymbol{v}}_2}_{\text{at } \omega_0-\omega_c}
\end{array} \tag{5.47}$$

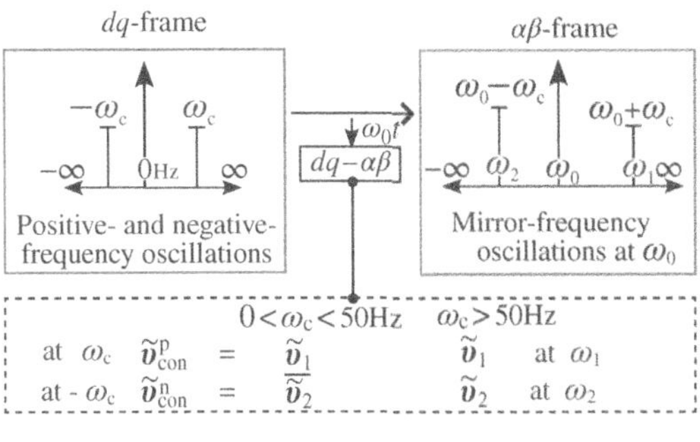

Figure 5.8 Relation between *dq*-sequence signals and *αβ* signals.

where $\bar{\tilde{\boldsymbol{v}}}$ indicates the complex conjugate.

Note that matrix elements like $\tilde{v}^d$ were previously presented in vector form as $\tilde{\boldsymbol{v}}^d$ in Section 7.1.2.2; in this section, they are shown in non-bold symbol as $\tilde{v}^d$, because the bold symbol $\tilde{\boldsymbol{v}}$ indicates the matrix form.

The *dq*-sequence impedance on the left side of the converter system, as shown in Figure 5.3, is presented below:

$$-\begin{bmatrix} \tilde{v}_{\mathrm{PCC}}^{\mathrm{p}} \\ \tilde{v}_{\mathrm{PCC}}^{\mathrm{n}} \end{bmatrix} = \begin{bmatrix} Z_{\mathrm{con}}^{\mathrm{pp}} & Z_{\mathrm{con}}^{\mathrm{pn}} \\ Z_{\mathrm{con}}^{\mathrm{np}} & Z_{\mathrm{con}}^{\mathrm{nn}} \end{bmatrix} \begin{bmatrix} \tilde{i}_{\mathrm{g}}^{\mathrm{p}} \\ \tilde{i}_{\mathrm{g}}^{\mathrm{n}} \end{bmatrix} \tag{5.48}$$

Accordingly, replacing $[\tilde{v}_{\mathrm{con}}^{\mathrm{p}}; \tilde{v}_{\mathrm{con}}^{\mathrm{n}}]$ and $[\tilde{i}_{\mathrm{g}}^{\mathrm{p}}; \tilde{i}_{\mathrm{g}}^{\mathrm{n}}]$ in Equation 5.48 with $\alpha\beta$ components yields:

If $\omega_c > \omega_0$:

$$-\begin{bmatrix} \tilde{v}_{\mathrm{PCC}}^{1} \\ \tilde{v}_{\mathrm{PCC}}^{2} \end{bmatrix} = \underbrace{\begin{bmatrix} Z_{\mathrm{con}}^{\mathrm{pp}} & Z_{\mathrm{con}}^{\mathrm{pn}} \\ Z_{\mathrm{con}}^{\mathrm{np}} & Z_{\mathrm{con}}^{\mathrm{nn}} \end{bmatrix}}_{Z_{\mathrm{con}}^{\alpha\beta}} \begin{bmatrix} \tilde{i}_{\mathrm{g}}^{1} \\ \tilde{i}_{\mathrm{g}}^{2} \end{bmatrix} \tag{5.49}$$

If $0 < \omega_c < \omega_0$:

$$-\begin{bmatrix} \tilde{v}_{\mathrm{PCC}}^{1} \\ \overline{\tilde{v}_{\mathrm{PCC}}^{2}} \end{bmatrix} = \underbrace{\begin{bmatrix} Z_{\mathrm{con}}^{\mathrm{pp}} & Z_{\mathrm{con}}^{\mathrm{pn}} \\ Z_{\mathrm{con}}^{\mathrm{np}} & Z_{\mathrm{con}}^{\mathrm{nn}} \end{bmatrix}}_{Z_{\mathrm{con}}^{\alpha\beta}} \begin{bmatrix} \tilde{i}_{\mathrm{g}}^{1} \\ \overline{\tilde{i}_{\mathrm{g}}^{2}} \end{bmatrix} \tag{5.50}$$

To keep $\boldsymbol{Z}_{\mathrm{con}}^{\alpha\beta}$ and $\boldsymbol{Z}_{\mathrm{con}}^{\mathrm{pn}}$ identical, it should be noted that $\tilde{v}_{\mathrm{con}}^{2}$ and $\tilde{i}_{\mathrm{g}}^{2}$ have different relationships with $\tilde{v}_{\mathrm{con}}^{\mathrm{n}}$ and $\tilde{i}_{\mathrm{g}}^{2}$ in different frequency intervals, at $0 < \omega_c < \omega_0$ and $\omega_c > \omega_0$.

Therefore, an $\alpha\beta$ impedance defined by Equations 5.49 and 5.50 is identical to the *dq*-sequence impedance across the entire frequency range.

5.3 Comparison Among Stability Analysis Methods

All the presented stability analysis methods are based on the same small-signal model but in different representations. All these methods can identify the stability of a converter-integrated power system, as demonstrated before, including:

- state-space stability analysis method, the model is presented as Equation 5.51a;
- transfer-function-based stability analysis method, the model is presented as Equation 5.51b;
- impedance stability analysis method based on Generalized Nyquist Criterion, the model is presented as Equation 5.51d;
- impedance stability analysis method based on the criterion of Bode plot, the model is presented as Equation 5.51c;
- measurement stability analysis method based on impedance stability criterion, this method shares the same criterion as the above impedance stability analysis method.

However, their different representations lead to various assumptions for stability analysis. If the system does not align with the assumptions of the method, the stability analysis could lose its effectiveness. Therefore, the comparison of assumptions among these stability analysis methods will be uniformly presented and compared in this section to provide a clearer understanding of these methods and their application cases.

For a unified comparison, the models for these methods are re-presented below:

$$\underbrace{(\boldsymbol{I}s - \boldsymbol{A})\boldsymbol{C}^{-1}}_{M_1}\tilde{\boldsymbol{i}}_{\mathrm{g}} = \underbrace{(\boldsymbol{B}_1\tilde{\boldsymbol{u}} + \boldsymbol{B}_2\tilde{\boldsymbol{v}}_{\mathrm{g}})}_{M_2} \tag{5.51a}$$

$$\underbrace{(\boldsymbol{Z}_{\mathrm{f}} + \boldsymbol{Z}_{\mathrm{g}} - \boldsymbol{G}^{i}_{\mathrm{col}} - \boldsymbol{G}^{v}_{\mathrm{col}}\boldsymbol{Z}_{\mathrm{g}})}_{M_1}\tilde{\boldsymbol{i}}_{\mathrm{g}} = \underbrace{\boldsymbol{G}_{\mathrm{ref}}\tilde{\boldsymbol{u}} - (\boldsymbol{I} - \boldsymbol{G}^{v}_{\mathrm{col}})\tilde{\boldsymbol{v}}_{\mathrm{g}}}_{M_2} \tag{5.51b}$$

$$\underbrace{(\boldsymbol{Z}_{\mathrm{con}} + \boldsymbol{Z}_{\mathrm{g}})}_{M_1}\tilde{\boldsymbol{i}}_{\mathrm{g}} = \underbrace{(\tilde{\boldsymbol{v}}_{\mathrm{con,s}} - \tilde{\boldsymbol{v}}_{\mathrm{g}})}_{M_2} \tag{5.51c}$$

$$\underbrace{(\boldsymbol{I} + \boldsymbol{Z}_{\mathrm{con}}^{-1}\boldsymbol{Z}_{\mathrm{g}})}_{M_1}\tilde{\boldsymbol{i}}_{\mathrm{g}} = \underbrace{\boldsymbol{I} + \boldsymbol{Z}_{\mathrm{con}}^{-1}\boldsymbol{Z}_{\mathrm{g}}[\boldsymbol{Z}_{\mathrm{con}}^{-1}(\tilde{\boldsymbol{v}}_{\mathrm{con,s}} - \tilde{\boldsymbol{v}}_{\mathrm{g}})]}_{M_2} \tag{5.51d}$$

where $\boldsymbol{Z}_{\mathrm{con}} = (\boldsymbol{I} - \boldsymbol{G}^{v}_{\mathrm{col}})^{-1}(\boldsymbol{Z}_{\mathrm{f}} - \boldsymbol{G}^{i}_{\mathrm{col}})$ and $\tilde{\boldsymbol{v}}_{\mathrm{con,s}} = (\boldsymbol{I} - \boldsymbol{G}^{v}_{\mathrm{col}})^{-1}\boldsymbol{G}_{\mathrm{ref}}\tilde{\boldsymbol{u}}$.

As presented in Equation 5.51, for each stability analysis method, $\boldsymbol{M}_1$ is the matrix for stability analysis, and $\boldsymbol{M}_2$ is the matrix assumed to have no RHP poles. It was clearly demonstrated in the previous sections that all $\boldsymbol{M}_1$ matrices in Equation 5.51 can represent the system poles. A detailed relationship among these stability analysis methods is summarized in Figure 5.9. The state-space method, as shown by Equation 5.51a, provides the mathematical foundation for stability analysis, while the other equations used by various stability analysis methods are simply rearrangements of this Equation 5.51a. Due to the rearrangement of equations to meet the requirements of the other stability analysis methods, $\boldsymbol{M}_2$ becomes more complicated and assumptions for no RHP poles have to be applied to additional elements in $\boldsymbol{M}_2$.

Therefore, application cases for different stability analysis methods require more rigorous consideration to ensure they fit the assumptions. It is found that the impedance stability analysis method based on the Generalized Nyquist Criterion has the most complex $\boldsymbol{M}_2$; its application cases need to be carefully considered. If an impedance stability analysis method has to be adopted, the method based on the criterion of the Bode plot is recommended due to the simpler $\boldsymbol{M}_2$.

The impedance measurement method shares the same criterion with impedance stability analysis method.

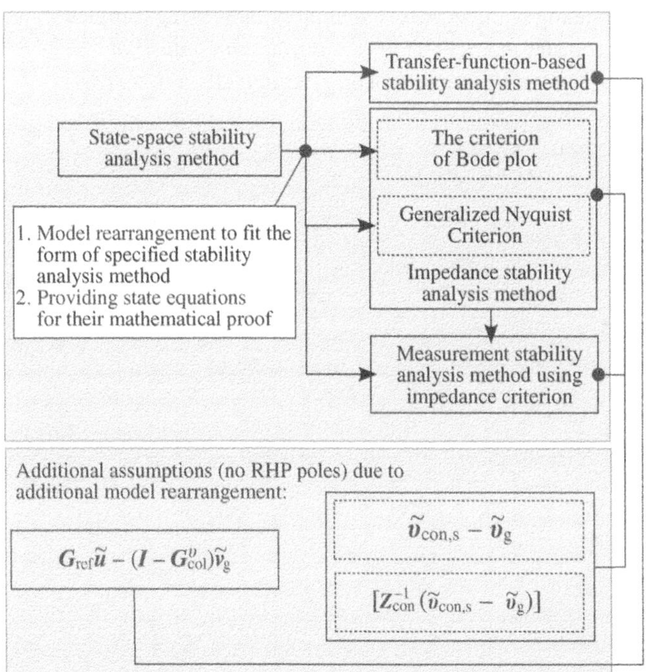

Figure 5.9 Relationship among various stability analysis methods.

References

1 Kundur Prabha. *Power System Stability and Control*. McGraw-Hill Professional, 1994. ISBN 007035958X.

2 A.G.J. MacFarlane and I. Postlethwaite. *A Complex Variable Approach to the Analysis of Linear Multivariable Feedback Systems*. Springer-Verlag, Berlin Heidelberg, New York, 1979.

3 Shahil Shah and Leila Parsa. Impedance modeling of three-phase voltage source converters in dq, sequence, and phasor domains. *IEEE Transactions on Energy Conversion*, 32(3): 1139–1150, 2017. https://doi.org/10.1109/TEC.2017.2698202.

4 Jian Sun. Small-signal methods for AC distributed power systems – a review. *IEEE Transactions on Power Electronics*, 24(11): 2545–2554, 2009. https://doi.org/10.1109/TPEL.2009.2029859.

5 Xiongfei Wang, Lennart Harnefors, and Frede Blaabjerg. Unified impedance model of grid-connected voltage-source converters. *IEEE Transactions on Power Electronics*, 33(2): 1775–1787, 2018. https://doi.org/10.1109/TPEL.2017.2684906.

6 Lennart Harnefors. Modeling of three-phase dynamic systems using complex transfer functions and transfer matrices. *IEEE Transactions on Industrial Electronics*, 54(4): 2239–2248, 2007. https://doi.org/10.1109/TIE.2007.894769.

6

Impedance Interaction Analysis for Converter-integrated Power Systems

Jun Liang[1,*] *and Chuanyue Li*[2]

[1] *School of Engineering, Cardiff University, Cardiff, UK*
[2] *School of Electric Power Engineering, South China University of Technology, Guangzhou, China*

*Corresponding author: LiangJ1@cardiff.ac.uk

6.1 Impedance Interaction Analysis

For a DC system, its impedance representation is a single-dimensional system, and the impedance stability criterion follows the relation below:

$$\tilde{i} = \frac{\tilde{v}_1 - \tilde{v}_2}{Z_1 + Z_2} \tag{6.1}$$

$Z_1 + Z_2$ is the factor that contains the system characteristic equation and will be applied for the impedance stability analysis. The impedance interaction analysis is to further simplify the stability analysis without identifying the RHP poles of $Z_1 + Z_2$, providing a simpler understanding or criterion of system stability for engineers. Therefore, analyzing the characteristic of $Z_1 + Z_2$ is essential for understanding the impedance interaction.

6.1.1 Negative Resistor Criterion in an Equivalent *RLC* Circuit

The negative resistor criterion, as noted by Middlebrook [1], is considered to be a representative and convenient method for analyzing system stability. If one component in a system exhibits negative resistor, it could lead to system-wide oscillation or non-convergence. A simple case is presented in Figure 6.2 to demonstrate the mechanism by which a negative resistor can lead to system instability.

Stability Analysis of Converter-Rich Power Grids, First Edition. Jun Liang and Chuanyue Li.
© 2026 The Institute of Electrical and Electronics Engineers, Inc. Published 2026 by
John Wiley & Sons, Inc. All rights reserved.

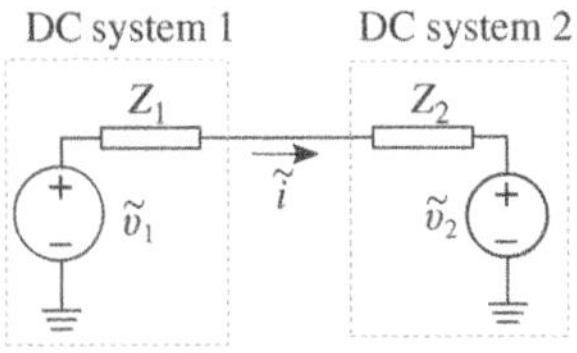

Figure 6.1 Stability analysis via impedance representation.

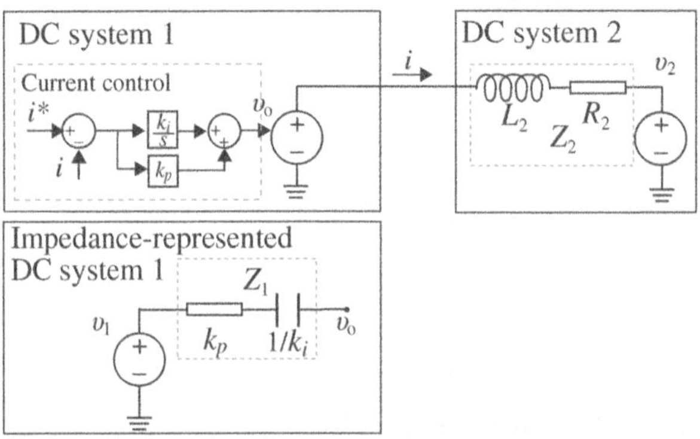

Figure 6.2 *RLC* circuit with negative impedance insertion.

As shown in Figure 6.2, DC system 1 is a current-controlled DC voltage source, and its formulation is given below:

$$(\tilde{i}^* - \tilde{i})\left(k_\mathrm{p} + \frac{k_\mathrm{i}}{s}\right) = \tilde{v}_\mathrm{o} \tag{6.2}$$

Rearranging into the form of an impedance representation, as shown in Figure 6.1, yields:

$$\underbrace{\left(k_\mathrm{p} + \frac{k_\mathrm{i}}{s}\right)\tilde{i}^*}_{\tilde{v}_1} = \tilde{v}_\mathrm{o} + \left(k_\mathrm{p} + \frac{k_\mathrm{i}}{s}\right)\tilde{i} \tag{6.3}$$

Let $R_1 = k_p$ and $C_1 = 1/k_\mathrm{i}$, which yields:

$$\underbrace{\left(k_\mathrm{p} + \frac{k_\mathrm{i}}{s}\right)\tilde{i}^*}_{\tilde{v}_1} = \tilde{v}_\mathrm{o} + \underbrace{\left(R_1 + \frac{1}{C_1 s}\right)}_{Z_1}\tilde{i} \tag{6.4}$$

Equation 6.4 represents the impedance of DC system 1, as depicted in Figure 6.2. The impedance of DC system 2 is formulated as follows:

$$\tilde{v}_\mathrm{o} = (L_2 s + R_2)\tilde{i} + \tilde{v}_2 \tag{6.5}$$

The impedance representation of the entire system is derived by substituting $\widetilde{v}_o$ in Equation 6.4 with Equation 6.5:

$$\widetilde{v}_1 - \widetilde{v}_2 = (\underbrace{R_1 + \frac{1}{C_1 s}}_{Z_1} + \underbrace{L_2 s + R_2}_{Z_2})\widetilde{i} \tag{6.6}$$

Further rearranging Equation 6.6 yields:

$$\widetilde{i} = \frac{s/L_2(\widetilde{v}_1 - \widetilde{v}_2)}{\left(s^2 + \frac{(R_1+R_2)}{L_2}s + \frac{1}{L_2C_1}\right)} \tag{6.7}$$

Therefore, the system poles are the zeros of $Z_1 + Z_2$, which are represented as follows:

$$\lambda_1, \lambda_2 = \frac{-\frac{R_1+R_2}{L_2} \pm \sqrt{\left(\frac{R_1+R_2}{L_2}\right)^2 - \frac{4}{L_2C_1}}}{2} \tag{6.8}$$

Typically, $\sqrt{\left(\frac{R_1+R_2}{L_2}\right)^2 - \frac{4}{L_2C_1}} < 0$, which corresponds to the imaginary part of the poles. Therefore, if $\frac{-(R_1+R_2)}{L_2} > 0$, it leads to RHP poles. Assuming $L_2 > 0$, the term $-(R_1 + R_2)$ determines the position of the poles.

If $R_1 + R_2 < 0$, the poles will cross into the RHP. This is the mechanism of the negative resistor criterion. When either R_1 or R_2 is negative, the summed resistance could be negative. As a result, a negative resistor in either system 1 or system 2 may cause the system poles to approach $x = 0$ or move into the RHP, which could induce oscillation at the frequency $\sqrt{\left(\frac{R_1+R_2}{L_2}\right)^2 - \frac{4}{L_2C_1}}$, or lead to non-convergence.

6.1.2 Stability Impact of Negative Inductors and Capacitors in an *RLC* Circuit

Besides the negative resistor, in the *RLC* circuit, any negative inductor or capacitor will also cause the *RLC* circuit to become unstable. Therefore, a generalized negative *RLC* criterion is introduced below.

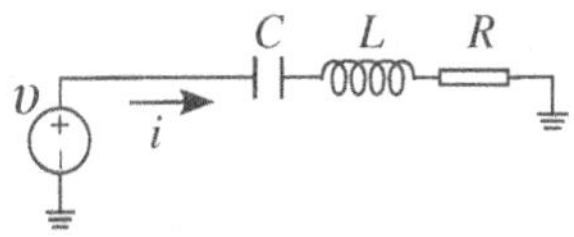

Figure 6.3 *RLC* circuit.

The system in Figure 6.2 is generalized as an *RLC* circuit in DC system, as shown in Figure 6.3, and used to explain the characteristic of impedance and its transfer function is given below:

$$i = \frac{1}{\left(\frac{1}{Cs} + Ls + R\right)} v = \frac{s/L}{s^2 + \underbrace{(R/L)}_{2\zeta\omega_n} s + \underbrace{1/(LC)}_{\omega_n^2}} v \tag{6.9}$$

Equation 6.9 can be a typical second-order system; the damping ratio ζ and natural frequency (ω_n) have been noted. R can be considered as the damping ratio, if $1/LC$ is constant. The resonant frequency can be approximated as $\sqrt{1/LC}$ if R is not too large. This is basically how passive components impact on the circuit stability. For more accurate analysis, please refer to the second-order system, which will not be explained here.

For any voltage deviation, the impedance transfer function in Equation 6.9 has following poles:

$$\lambda_1, \lambda_2 = \frac{-\frac{R}{L} \pm \sqrt{\left(\frac{R}{L}\right)^2 - \frac{4}{LC}}}{2} \tag{6.10}$$

If R, L, and C are all positive, λ_1 and λ_2 are in the left half-plane, and the current in the *RLC* circuit converges. However, if the negative impedance manifests, including a negative inductor, capacitor, or resistor, the pole λ_1 or λ_2, or both of them, could reach the right half-plane, which leads to divergence or instability; therefore, the negative impedance criterion is summarized below:

1. negative inductance result in $-R/L > 0$ and $\sqrt{\left(\frac{R}{L}\right)^2 - \frac{4}{LC}}$ being a real value, which causes the system to purely diverge;
2. negative capacitance may result in $-R/L + \sqrt{\left(\frac{R}{L}\right)^2 - \frac{4}{LC}} > 0$, which causes the system to purely diverge;
3. negative resistance results in $-R/L > 0$. When $\sqrt{\left(\frac{R}{L}\right)^2 - \frac{4}{LC}}$ is an imaginary value, the system oscillates and fails to converge; when $\sqrt{\left(\frac{R}{L}\right)^2 - \frac{4}{LC}}$ is a real value, the system purely diverges.

The above mathematically explains how negative impedance causes instabilities. In the real world, passive components do not induce negative impedance, while improper control design of active devices or control interaction could cause negative impedance and lead the system to instability. Meanwhile, the negative impedance effect may only manifest in a certain frequency range rather than a pure negative impedance crossing the whole frequency domain.

This impedance criterion basically refers to characteristics of how negative components, including resistor, inductor, and capacitor, impact on the *RLC* circuit stability.

6.1.3 Effective Frequency Range of a Negative Resistor, Inductor, and Capacitor in an *RLC* Circuit

While a system is typically more complex than a pure *RLC* circuit, the manifestation of negative impedance, including a negative resistor, inductor, or capacitor, may occur only within a certain frequency range.

To address the above situation, a follow-up case of the *RLC* circuit with filters to isolate the negative impedance within a certain range, as shown in Figure 6.4, is conducted to demonstrate the effective frequency range of negative impedance on stability.

A controllable voltage source ($\widetilde{v}$) with negative impedance insertion is added in the *RLC* circuit. Filters, including low-pass (LP) filter, high-pass (HP) filter, and band-pass (BP) filter, are used to limit the negative impedance effect at a certain frequency range to present how the negative impedance destabilizes the system:

$$v = \left(\frac{1}{Cs} + Ls + R\right) \times i + \underbrace{\Delta v}_{-L_{\text{v}}s, -R_{\text{v}}, -1/(C_{\text{v}}s)} \times G_{\text{fil}} \tag{6.11}$$

where subscript $_{\text{v}}$ means virtual impedance and G_{fil} denotes the transfer function of applied filters.

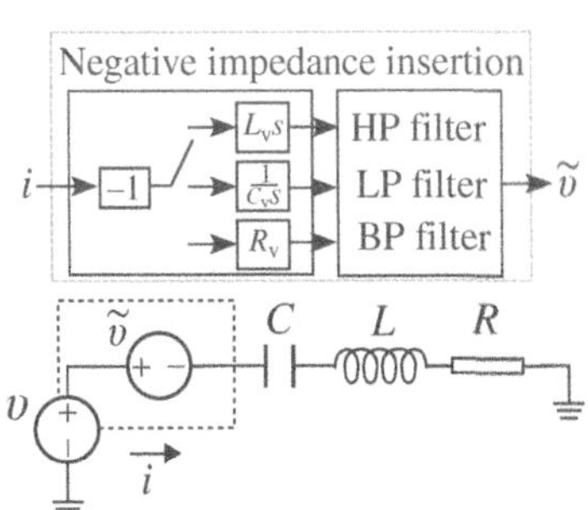

Figure 6.4 *RLC* circuit with negative impedance insertion.

Transfer functions G_{fil} of low-pass filter, high-pass filter, and band-pass filter are given below:

$$\frac{\omega_c}{s+\omega_c}$$
$$\frac{s}{s+\omega_c} \tag{6.12}$$
$$\frac{sQ/\omega_{\text{cen}}}{s^2+sQ/\omega_{\text{cen}}+\omega_{\text{cen}}^2}$$

where ω_c is the cut-off frequency, ω_{cen} is the center frequency of the band-pass filter, and Q is the quality factor.

The insertion of a negative impedance, including R, L, and C, in an *RLC* circuit is demonstrated to show how the effective frequency range of the negative impedance impacts the circuit's stability. A case study of an *RLC* circuit, as shown in Figure 6.4, is adopted using the following parameters: $v = 20$ V, $R = 5\ \Omega$, $L = 0.05$ H, $C = 0.001$ F; resonance frequency $\omega_r = 1/\sqrt{LC}$; cut-off frequency of high-pass and low-pass filters $\omega_c = \omega_r$; and ω_{cen} of band-pass filter is set to ω_r.

A negative resistor is selected for the simulation case demonstration, as shown in Figure 6.5. When a negative resistor of $-1.1R$ is inserted at 1 s, the system's overall resistance becomes negative $R + Rv(= -1.1R)$, which leads to current instability, as shown in Figure 6.5a. The transfer function from Equation 6.11 is also validated through the simulation results. This is the case where the circuit's negative impedance spans the entire frequency range. If the circuit's negative impedance is limited to a certain frequency range, it may not destabilize the system. For example, a low-pass filter, where $\omega_c = 1/\sqrt{LC}$, is applied to $-R_v$ to keep the negative impedance only within a low frequency range. It is found that the system returns to a stable condition. As shown in Figure 6.5b, after adding the low-pass filter to

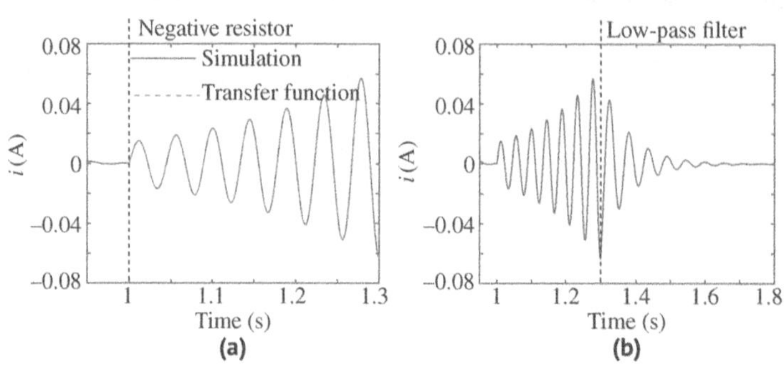

Figure 6.5 *RLC* circuit with negative resistor insertion.

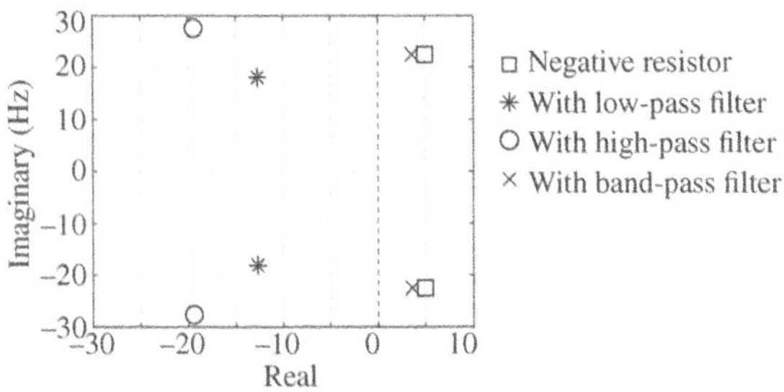

Figure 6.6 *RLC* circuit with negative resistor insertion at different frequency ranges.

the negative resistor at 1.3 s, the current stabilizes. This demonstrates that the negative resistance has an effective frequency range that can destabilize an *RLC* circuit. Otherwise, if the negative resistance operates in a different frequency range, it will not cause instability.

To further analyze the effectiveness of negative resistance, the frequency is divided into three parts: low frequency range, mid frequency range, and high frequency range, using a low-pass filter, band-pass filter, and high-pass filter, respectively.

The resonance frequency $\omega_r = 1/\sqrt{LC}$ of the *RLC* circuit is adopted as the cut-off frequency and center frequency of these filters. The results are shown as a pole map in Figure 6.6. It is found that only the negative impedance appearing at the circuit's resonant frequency ω_r will cause the system to become unstable, exhibiting similar destabilization as negative impedance without filters. Otherwise, if the negative resistance appears below or above ω_r, it will not lead to instability.

Although the analysis in Section 6.1.2 indicated that (1) negative inductance and capacitance cause only divergence and (2) negative resistance causes both oscillation and divergence, if the negative impedance appears in different frequency ranges, its stabilization effectiveness varies. In certain frequency ranges, these negative impedances will not cause instability, as demonstrated with the negative resistance above.

Therefore, the same analysis is applied to negative inductance and capacitance to identify their effective frequency ranges for destabilization. The results, showing how negative impedance impacts the *RLC* circuit's stability across various frequency ranges, are summarized in Table 6.1 and visualized in Figure 6.7, with further details explained below:

- when an *RLC* circuit manifests negative resistance at resonant frequency, the circuit will lose stability. Otherwise, the negative resistance at the frequency range below or the resonant frequency will not cause the instability.

- when an *RLC* circuit manifests negative inductance at high frequency range, the circuit will lose stability. Otherwise, the negative inductance at the low frequency range will not suffer instability.
- when an *RLC* circuit manifests negative capacitance at low frequency range, the circuit will lose stability. Otherwise, the negative capacitance at the high frequency range will not induce instability.

Table 6.1 Negative impedance's impact on an *RLC* circuit across various frequency ranges.

	Value	Oscillation	Convergence	Stabilization
Negative resistance	$-1.1R$	Yes at $\sqrt{\left(\frac{R}{L}\right)^2 - \frac{4}{LC}}/2$	No	Low-pass filter or high-pass filter
Negative inductance	$-1.1L$	No	No	Low-pass filter
Negative capacitance	$-0.9C$	No	No	High-pass filter

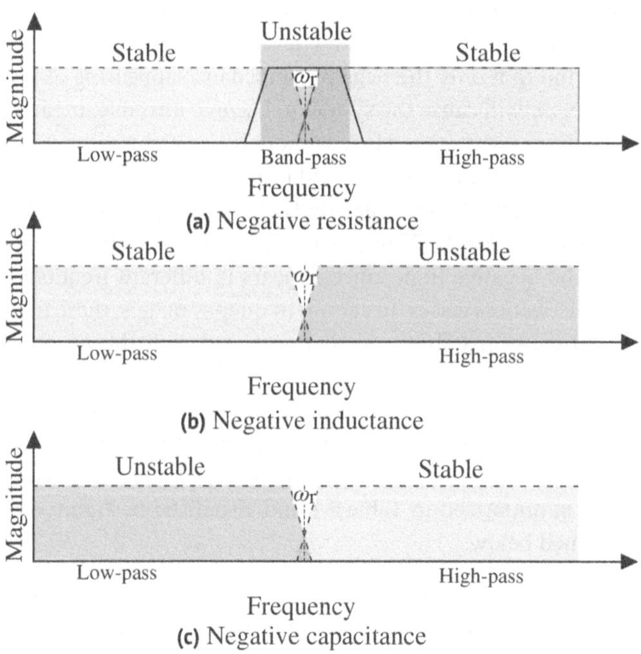

Figure 6.7 Effective frequency range for negative impedance in an *RLC* circuit.

It should be noticed that the unstable frequency range will increase if the negative impedance increases. The values of negative impedance applied in the case are shown in Table 6.1 and set as 110% of *RLC* impedance.

6.2 Impedance Interaction in a High-order System

As mentioned in the above section, the negative impedance criterion is developed for *R*, *L*, or *C* individually. They cause instability in the *RLC* circuit only if the overall impedance $Z_1 + Z_2$ appears as a negative *R*, *L*, or *C* within their specific frequency range. In more complicated and higher-order systems, the system characteristics are a combination of *RLC* components rather than being purely determined by individual *R*, *L*, *C* elements, making the analysis exceptionally complex. As a result, an additional impedance interaction analysis method is deemed necessary to present the system's stability in a clearer and simpler way, which is introduced below. Based on the characteristics of an impedance-representation system, $Z_1 + Z_2$ is the factor that contains the system characteristic equation and can always be reformed as follows:

$$Z_1 + Z_2 = G_{\text{imp}}(s)(s - p_1) \tag{6.13}$$

where p_1 is a pole of the system, $p_1 = \sigma_1 + j\omega_1$; $G_{\text{imp}}(s)$ represents the remaining system transfer function.

When the system is marginally stable, p_1 lies on $x = 0$, which indicates $\sigma = 0$. When $s = j\omega_1$, resulting in:

$$Z_1 + Z_2 = G_{\text{imp}}(s)\underbrace{(s - j\omega_1)}_{=0 \text{ at } s=j\omega_1} = 0 \tag{6.14}$$

Therefore, the gain of the system will be infinite at ω_1 in the Bode plot, which serves as the stability criterion via the Bode plot, as detailed in Section 3.3.

6.2.1 Impedance Interaction Criterion

Equation 6.14 serves as the stability criterion, indicating that the system is marginally stable if $Z_1 + Z_2 = 0$ at ω_1.

Rewriting Z_1 and Z_2 in vector form at ω_1 gives:

$$\begin{aligned} &Z_{1,\omega_1}^{\text{mag}} \angle \theta_{1,\omega_1} \\ &Z_{2,\omega_1}^{\text{mag}} \angle \theta_{2,\omega_2} \end{aligned} \tag{6.15}$$

Combining Equations 6.14 and 6.15 specifies the impedance interaction mechanism that can indicate marginal system stability, as shown below:

$$-Z_{1,\omega_1}^{mag}\angle\theta_{1,\omega_1} = Z_{2,\omega_1}^{mag}\angle\theta_{2,\omega_2} \tag{6.16}$$

Based on Equation 6.16, the impedance interaction criterion 1 is developed as follows:

Impedance interaction criterion 1: *if Z_1 and Z_2 of an impedance-representation DC system, as shown in Figure 6.8 with current reference direction D1, satisfy the condition in Equation 6.16 at a frequency ω_1, where $Z_{1,\omega_1}^{mag} = Z_{2,\omega_1}^{mag}$ and $\theta_{1,\omega_1} - \theta_{2,\omega_2} = \pi$, this indicates the system is marginally stable.*

Conditions of Use: since the mechanism of this stability analysis is based on any pole with $\sigma = 0$, it can clearly specify whether the system contains a pole close to or at $x = 0$. Therefore, if a system is stable, this method can identify how close the pole is to $x = 0$ and at what oscillation frequency.

An example, detailed in the following case study in Section 6.2.3, presents the relevant case related to "impedance interaction criterion 1" to provide a demonstration. Keeping Z_2 fixed and changing Z_1 to modify the overall system poles, two cases are considered: Case A, where all poles are in the LHP, and Case B, where there is a pole on the line $x = 0$. The Bode plot of both Cases A and B is shown in Figure 6.9.

In Case A, it is found that $Z_{1,\omega_1}^{mag} = Z_{2,\omega_1}^{mag}$ at point A1 at 14 Hz. Furthermore, the phase difference between A2 and A3 is 180°. This indicates that there is a pole of the system close to $x = 0$, and the system is marginally stable or nearly unstable.

In Case B, it is found that $Z_{1,\omega_1}^{mag} = Z_{2,\omega_1}^{mag}$ at point B1 at 12 Hz, while the phase difference between B2 and B3 is 195.5°, which is not close to 180°. This indicates that all the system poles are not close to $x = 0$, and the system is stable.

More information and detailed analysis about this case study can be found in Section 6.2.3.

6.2.2 Modified Impedance Interaction Criterion

Impedance interaction criterion 1 is already sufficient for stability analysis via the impedance interaction of Z_1 and Z_2. However, the phase difference of 180° on

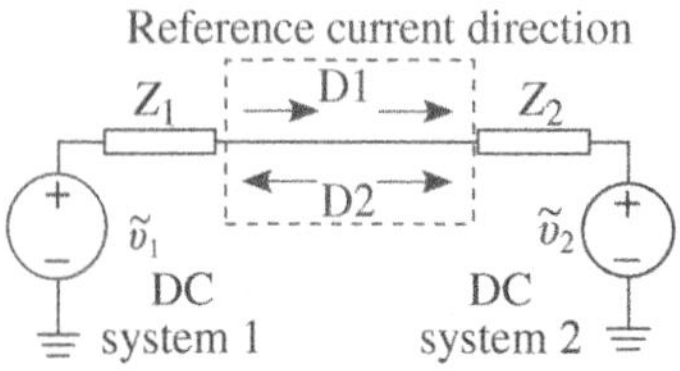

Figure 6.8 Impedance-representation system with the definition of the reference current direction.

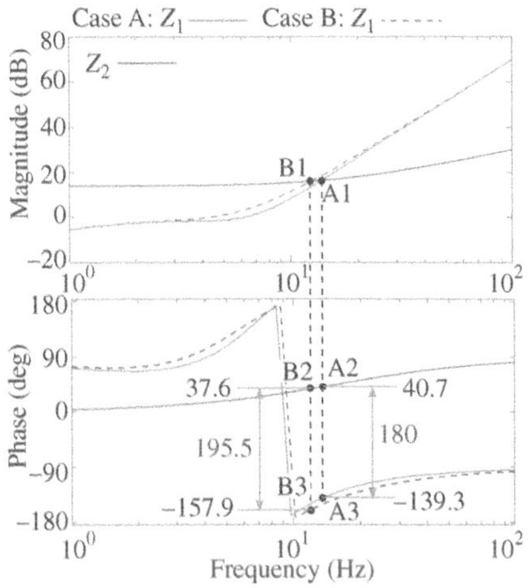

Figure 6.9 Stability analysis via impedance interaction criterion 1.

the Bode plot still needs to be calculated as shown in Figure 6.9, such as between A2–A3 and B2–B3. To represent the phase difference of 180° more simply and clearly, a modified impedance interaction criterion has been developed.

If the reference current direction is changed in the form of D2, as shown in Figure 6.8, the impedance interaction changes from $Z_1 + Z_2$ to $-Z_1 + Z_2$, leading to a change in the stability judgment of Equation 6.16 to the following form:

$$Z_{1,\omega_1}^{\text{mag}} \angle\theta_{1,\omega_1} = Z_{2,\omega_1}^{\text{mag}} \angle\theta_{2,\omega_2} \tag{6.17}$$

Based on Equation 6.17, a modified impedance interaction criterion is developed as follows:

Impedance interaction criterion 2: *if Z_1 and Z_2 of an impedance-represented DC system, as shown in Figure 6.8 with current reference direction D2, satisfy the condition in Equation 6.17 at a frequency ω_1, where $Z_{1,\omega_1}^{\text{mag}} = Z_{2,\omega_1}^{\text{mag}}$ and $\theta_{1,\omega_1} = \theta_{2,\omega_2}$, this indicates the system is marginally stable.* The same case presented in Figure 6.9 is redrawn based on impedance interaction criterion 2, as shown in Figure 6.10. The stability analysis for calculating the 180° phase difference is avoided. Instead, if the magnitude crossover point and phase crossover point occur at the same frequency, this indicates that the system is marginally stable or nearly unstable. For example, in Case A, points A1 and A2 in the Bode plot in Figure 6.10 provide a clearer analysis than in Figure 6.9.

More information and detailed analysis about this case study can be found in Section 6.2.3.

6.2.3 Case Study for Impedance Interaction Analysis

A relatively complex fifth-order system, as shown in Figure 6.10, is adopted for stability analysis via impedance interaction. Its parameters are provided in Table 6.2. DC system 1 is a controllable DC voltage source with DC voltage control and inner current control. DC system 2 is a simple RL circuit. If the reference current direction is adopted as D1 in Figure 6.8, the equivalent impedance Z_1 of DC system 1 is derived as follows:

$$\underbrace{v_{\text{PCC}}^*}_{v_1} = \underbrace{\frac{R_1 + L_1 s + k_{\text{p}}^{\text{cur}} + k_{\text{i}}^{\text{cur}}/s}{(k_{\text{p}}^{\text{cur}} + k_{\text{i}}^{\text{cur}}/s)(k_{\text{p}}^{\text{vol}} + k_{\text{i}}^{\text{vol}}/s)}[C_1 s(R_{\text{g}} + L_{\text{g}} s) + 1]}_{Z_1} i + v_{\text{PCC}} \tag{6.18}$$

The impedance of DC system 2 is given as:

$$v_{\text{PCC}} = \underbrace{(R_2 + L_2 s)}_{Z_2} i + v_2 \tag{6.19}$$

The system's impedance representation is as follows:

$$v_{\text{PCC}}^* - v_2 = (Z_1 + Z_2)i \tag{6.20}$$

Two cases are selected as follows:

- Case A with $k_{\text{i}}^{\text{cur}} = 80$
- Case B with $k_{\text{i}}^{\text{cur}} = 50$

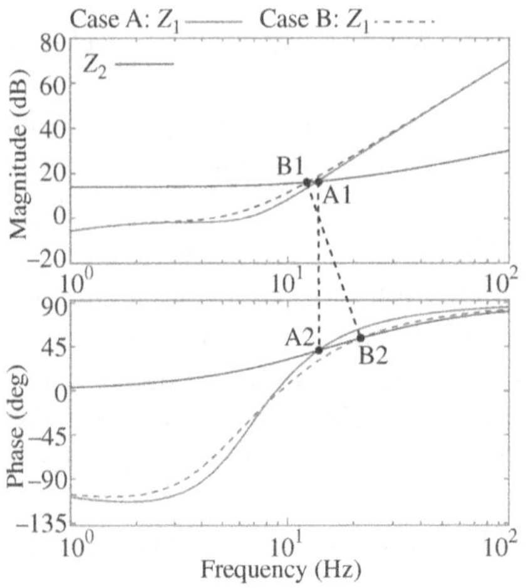

Figure 6.10 Stability analysis via impedance interaction criterion 2.

Table 6.2 Parameters of the DC system in Figure 6.11.

Symbol	Parameter	Value
L_1	inductor	0.05 H
R_1	resistor	1 Ω
C_1	capacitor	0.005 F
L_2	inductor	0.05 H
R_2	resistor	1 Ω
k_i^{vol}	voltage control PI controller	10
k_p^{vol}	voltage control PI controller	1
k_i^{cur}	current control PI controller	Case A: 50; Case B: 80
k_p^{cur}	current control PI controller	1
v_{PCC}^*	voltage control reference	20 V
v_2	DC voltage	0

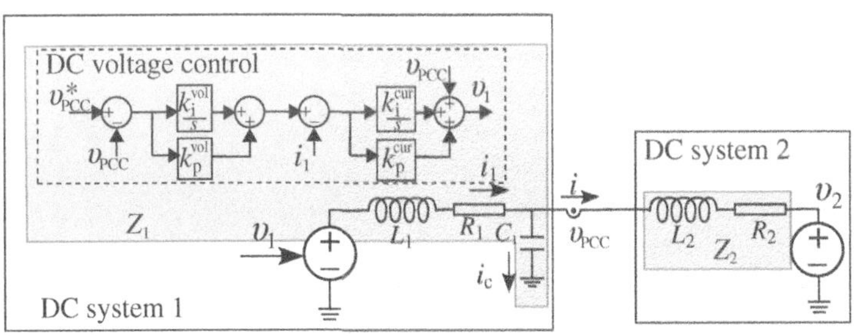

Figure 6.11 Case study of impedance interaction analysis: a system with a voltage-controlled DC voltage source.

Case A is a marginally stable case, as one pair of its poles is at $x = 0$, while Case B is stable, as its poles are relatively far from $x = 0$. After stability analysis of $Z_1 + Z_2$, the system poles are presented into Figure 6.12. The pole's y-axis indicates the system oscillation is 14 Hz.

The Bode plot of $Z_1 + Z_2$, as shown in Figure 6.13, further confirms the stability analysis results from the pole map. When there are poles on the line $x = 0$, the Bode plot will correspondingly show infinite gain.

A simulation validation of Case A is presented in Figure 6.14. After 1 s, k_i^{cur} is changed from 50 to 80. The simulation results confirm the above stability analysis using the pole map and Bode plot. The current i is marginally stable at an

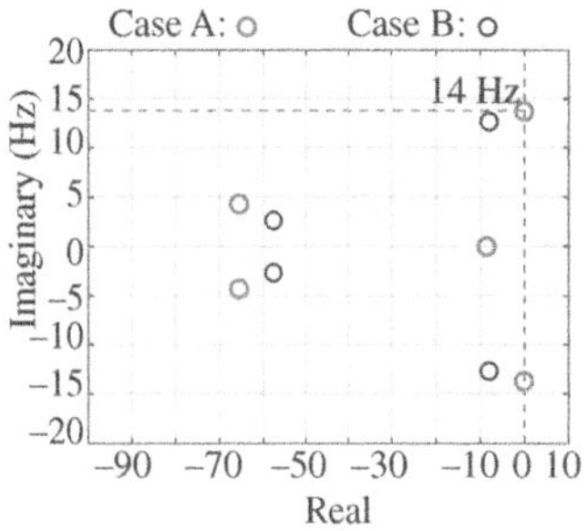

Figure 6.12 Pole analysis the DC system via $Z_1 + Z_2$.

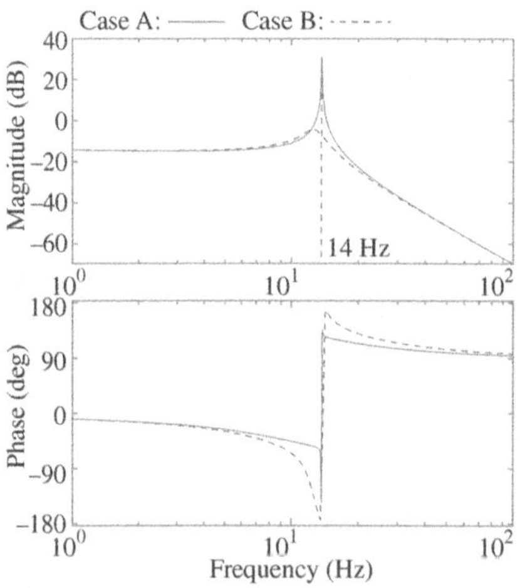

Figure 6.13 Bode plot analysis for the DC system via $Z_1 + Z_2$.

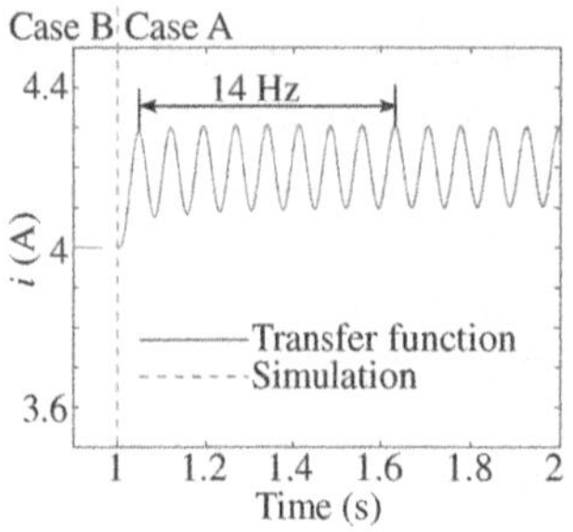

Figure 6.14 Simulation result of i with a step change in v^*_{PCC} to 21 V at 1 s.

oscillation frequency of 14 Hz. The transfer function of i is also derived based on the system's impedance representation in Equation 6.20, as follows:

$$i = (Z_1 + Z_2)^{-1}(v_{\text{PCC}}^* - v_2) \tag{6.21}$$

An additional validation of the system's impedance representation is shown in Figure 6.14, where Equation 6.21 matches the system's simulation accurately.

The impedance interaction analysis, presenting the same results as the above stability analysis, has been conducted and repositioned in Sections 6.2.1 and 6.2.2. These sections detail the impedance interaction criterion, making it more convenient for readers to follow.

6.2.4 Why Is the Impedance in the Form of a 2 × 2 Matrix?

In general, a passive impedance in a DC or AC system is typically a single-dimensional component. However, when a converter control system is introduced, the impedance is represented as a matrix. This change occurs because the converter control system regulates both the magnitude and phase of a sinusoidal waveform, leading to a two-dimensional objective when performing frequency decoupling to convert the sinusoidal waveform into a linear time-invariant system. A more detailed explanation is provided below.

To simplify the analysis, a three-phase system is transformed into its $\alpha - \beta$ representation using the Clarke transformation. For the converter control adopting magnitude and phase control, the small-signal model can be expressed as:

$$\begin{aligned} &(v_1^{\alpha,\text{mag}} + \widetilde{v}_1^{\alpha,\text{mag}}) \cos(\omega_0 t + \theta_1 + \widetilde{\theta}_1) \\ &(v_1^{\beta,\text{mag}} + \widetilde{v}_1^{\beta,\text{mag}}) \sin(\omega_0 t + \theta_1 + \widetilde{\theta}_1) \end{aligned} \tag{6.22}$$

It should be noted that a single-phase system can also be represented using the above equations. One of these equations represents the single-phase system, while the other serves as an auxiliary term for further analysis.

Expanding Equation 6.22 yields:

$$\begin{aligned} &v_1^{\alpha,\text{mag}} \cos(\omega_0 t + \theta_1) - v_1^{\alpha,\text{mag}} \sin(\omega_0 t + \theta_1)\widetilde{\theta}_1 + \cos(\omega_0 t + \theta_1)\widetilde{v}_1^{\alpha,\text{mag}} \\ &v_1^{\beta,\text{mag}} \sin(\omega_0 t + \theta_1) + v_1^{\beta,\text{mag}} \cos(\omega_0 t + \theta_1)\widetilde{\theta}_1 + \sin(\omega_0 t + \theta_1)\widetilde{v}_1^{\beta,\text{mag}} \end{aligned} \tag{6.23}$$

where the terms $-\sin(\omega_0 t + \theta_1)\widetilde{v}_1^{\alpha,\text{mag}}\widetilde{\theta}_1 \approx 0$ and $\cos(\omega_0 t + \theta_1)\widetilde{v}_1^{\beta,\text{mag}}\widetilde{\theta}_1 \approx 0$ are therefore ignored.

The transfer function is defined only for a linear time-invariant system, which serves as the basis for stability analysis. However, the terms $\sin(\omega_0 t + \theta_1)$ and $\cos(\omega_0 t + \theta_1)$ in Equation 6.23 are not time invariant, as shown below:

$$\cos(\omega_0 t) \neq \cos(\omega_0 (t + t_0)) \tag{6.24}$$

Therefore, to convert the system into a linear time-invariant system, the following relations are applied using the vector form, taking the first equation in Equation 6.24 as an example:

$$\boldsymbol{v}_1^{\alpha\beta} = v_1^{\mathrm{mag}} \cos(\omega_0 t + \theta_1) + \mathrm{j} v_1^{\mathrm{mag}} \sin(\omega_0 t + \theta_1) = v_1^{\mathrm{mag}} = \mathrm{e}^{\omega_0 t\mathrm{j}} \mathrm{e}^{\theta_1 \mathrm{j}} \tag{6.25}$$

where the system operates under balance conditions, leading to $v_1^{\alpha,\mathrm{mag}} = v_1^{\beta,\mathrm{mag}} = v_1^{\mathrm{mag}}$.

To decouple the frequency component $\mathrm{e}^{\omega_0 t\mathrm{j}}$, the following transformation is applied:

$$\mathrm{e}^{-\omega_0 t\mathrm{j}} \boldsymbol{v}_1^{\alpha\beta} = \mathrm{e}^{-\omega_0 t\mathrm{j}} v_1^{\mathrm{mag}} \mathrm{e}^{\omega_0 t\mathrm{j}} \mathrm{e}^{\theta_1 \mathrm{j}} = \underbrace{v_1^{\mathrm{mag}} \mathrm{e}^{\theta_1 \mathrm{j}}}_{v_1^{dq}} \tag{6.26}$$

This frequency decoupling process by multiplying with $\mathrm{e}^{-\omega_0 t\mathrm{j}}$ is also known as the *abc–dq* transformation; therefore, the superscript *dq* is used.

For Equation 6.23, the *abc–dq* transformation is essential to convert the system into a linear time-invariant form as represented below:

$$\boldsymbol{v}_1^{dq} + \mathrm{j}\boldsymbol{v}_1^{dq}\widetilde{\theta} + \underbrace{\widetilde{\boldsymbol{v}}_1^{dq}}_{\mathrm{e}^{\mathrm{j}\theta_1}(\widetilde{v}_1^{\alpha,\mathrm{mag}} + \mathrm{j}\widetilde{v}_1^{\beta,\mathrm{mag}})} \tag{6.27}$$

As the voltage becomes two-dimensional variable, similarly, the current is also a two-dimensional variable; therefore, the impedance is a 2×2 matrix:

$$\begin{bmatrix} v^d \\ v^q \end{bmatrix} = \begin{bmatrix} Z^{dd} & Z^{dq} \\ Z^{qd} & Z^{qq} \end{bmatrix} \begin{bmatrix} i^d \\ i^q \end{bmatrix} \tag{6.28}$$

Taking an inductor as an example, the impedance is transformed using the *abc–dq* transformation via $\mathrm{e}^{-\mathrm{j}\omega_0 t}$, which requires:

$$\mathrm{e}^{-\mathrm{j}\omega_0 t} \widetilde{\boldsymbol{v}}^{dq} \mathrm{e}^{\omega_0 t\mathrm{j}} = \mathrm{e}^{-\mathrm{j}\omega_0 t} L \frac{\mathrm{d}\widetilde{\boldsymbol{i}}^{dq} \mathrm{e}^{\omega_0 t\mathrm{j}}}{\mathrm{d}t} \tag{6.29}$$

Rearranging the above equation yields:

$$\widetilde{\boldsymbol{v}}^{dq} = (Ls + L\omega_0 \mathrm{j})\widetilde{\boldsymbol{i}}^{dq} \tag{6.30}$$

In Equation 6.30, the vector $(Ls + L\omega_0\mathrm{j})$ is a 2×2 matrix. Rewriting it yields:

$$\begin{bmatrix} v^d \\ v^q \end{bmatrix} = \begin{bmatrix} Ls & -L\omega_0 \\ L\omega_0 & Ls \end{bmatrix} \begin{bmatrix} i^d \\ i^q \end{bmatrix} \tag{6.31}$$

Reference

1 Robert David Middlebrook. Input filter considerations in design and application of switching regulators. In *IEEE Industry Applications Society Annual Meeting*, pages 91–107, 1976.

7

Mirror-frequency Oscillation Due to Converters

Jun Liang[1,*] *and Chuanyue Li*[2]

[1] *School of Engineering, Cardiff University, Cardiff, UK*
[2] *School of Electric Power Engineering, South China University of Technology, Guangzhou, China*

*Corresponding author: LiangJ1@cardiff.ac.uk

7.1 Mirror-frequency Effect of Converter Systems

Mirror-frequency effect (MFE) refers to any oscillation at single frequency ω_c in the control frame being transferred into the power system, resulting in a pair of mirror-frequency oscillations, as shown below:

$$\begin{aligned} \omega_{1,2} &= \omega_0 \pm \omega_c \\ \omega_1 &= \omega_0 + \omega_c, \omega_2 = \omega_0 - \omega_c \end{aligned} \tag{7.1}$$

ω_1 is the normal oscillation frequency in the power system, ω_2 is the mirrored oscillation frequency in the power system, and ω_0 is the rated operating frequency of the power system.

A converter voltage signal v_{con} receives commands from the control system to regulate the magnitude and phase of the output voltage of a voltage source converter (VSC) in order to achieve the desired control objectives. Any oscillation in the magnitude or phase of v_{con} may induce MFE. The detailed mechanism of MFE in both single-phase and three-phase control is presented below.

Stability Analysis of Converter-Rich Power Grids, First Edition. Jun Liang and Chuanyue Li.
© 2026 The Institute of Electrical and Electronics Engineers, Inc. Published 2026 by John Wiley & Sons, Inc. All rights reserved.

7.1.1 Mirror-frequency Effect in a Single-phase Control

For oscillations in the magnitude and phase of a converter voltage, as shown in Figure 7.1, it can be expressed as follows:

$$(v_{\text{con}}^{\text{mag}} + \tilde{v}_{\text{con}}^{\text{mag}})\cos(\omega_0 t + \tilde{\theta}_{\text{con}}) \tag{7.2}$$

Expanding and simplifying Equation 7.2 yields:

$$v_{\text{con}}^{\text{mag}}\cos(\omega_0 t) \underbrace{- v_{\text{con}}^{\text{mag}}\tilde{\theta}_{\text{con}}\sin(\omega_0 t)}_{\text{A2}} + \underbrace{\tilde{v}_{\text{con}}^{\text{mag}}\cos(\omega_0 t)}_{\text{A1}} \tag{7.3}$$

Parts A1 and A2 of Equation 7.3 are responsible for the MFE. Taking Part A1 as an example, let $\tilde{v}_{\text{con}}^{\text{mag}} = v_{\text{con}}^{\text{mag},\Delta}\cos(\omega_c t)$ and expanding A1 yields:

$$\begin{aligned}\tilde{v}_{\text{con}}^{\text{mag}}\cos(\omega_0 t) &= v_{\text{con}}^{\text{mag},\Delta}\cos(\omega_c t)\cos(\omega_0 t)\\ &= 0.5v_{\text{con}}^{\text{mag},\Delta}(\cos((\omega_0 - \omega_c)t) + \cos((\omega_0 + \omega_c)t))\end{aligned} \tag{7.4}$$

Equation 7.4 clearly shows that any oscillation at ω_c in the magnitude control of a converter voltage is transferred to the power system as a pair of mirror frequencies at $\omega_0 + \omega_c$ and $\omega_0 - \omega_c$.

Similarly, any oscillation in phase control, such as $\tilde{\theta}_{\text{con}} = \theta_{\text{con}}^{\Delta}\cos(\omega_c t)$, can also induce mirror-frequency oscillations (Figure 7.2). Its expression is Part A2 of Equation 7.3, which is expanded below:

$$\begin{aligned}-v_{\text{con}}^{\text{mag}}\tilde{\theta}_{\text{con}}\sin(\omega_0 t) &= -v_{\text{con}}^{\text{mag}}\theta_{\text{con}}^{\Delta}\cos(\omega_c t)\sin(\omega_0 t)\\ &= -0.5v_{\text{con}}^{\text{mag},\Delta}\theta_{\text{con}}^{\Delta}(\sin((\omega_0 - \omega_c)t) + \sin((\omega_0 + \omega_c)t))\end{aligned} \tag{7.5}$$

Case studies insert an oscillation $0.1\cos(20 * 2\pi)$ into the magnitude or phase of the converter voltage control. The results are presented in Figure 7.3. Mirror-frequency oscillations in both cases match the mathematical expressions in Equations 7.4 and 7.5.

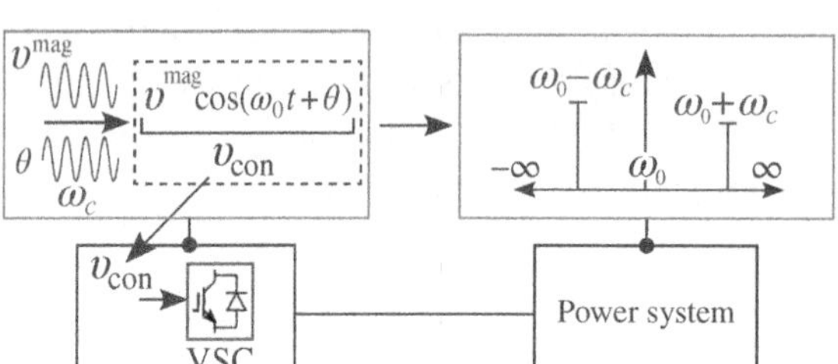

Figure 7.1 Mirror-frequency effect caused by a single-phase control system of a converter.

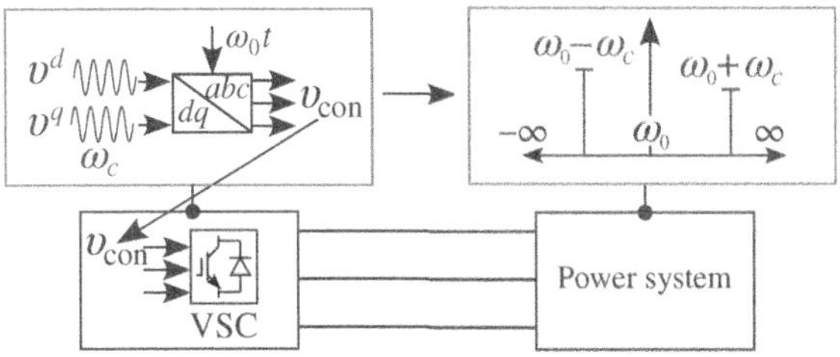

Figure 7.2 Mirror-frequency effect caused by a three-phase control system of a converter.

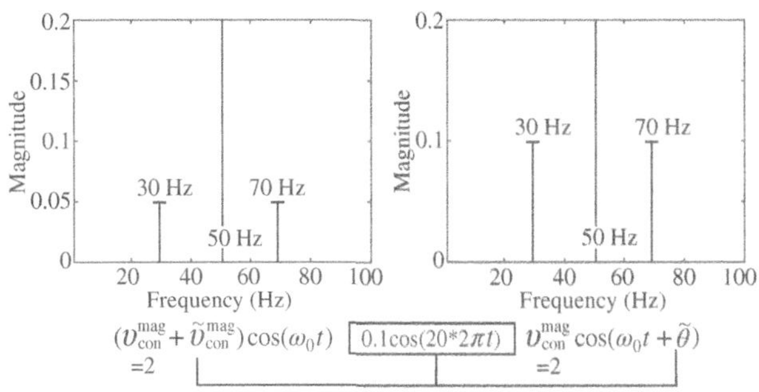

(a) The oscillation in magnitude control **(b)** The oscillation in phase control

Figure 7.3 Examples of mirror-frequency oscillation caused by a single-phase control system of a converter.

7.1.2 Mirror Frequency in a Three-phase System

While in a three-phase power system the converter is generally controlled in the *dq*-frame, the MFE also exists due to *dq–abc* transformation for converter *abc* voltage, as shown in Figure 7.2.

For a *dq*-frame control, an oscillation at ω_c in the d and q axes results in $\tilde{v}_{con}^d$ and $\tilde{v}_{con}^q$. Let $\tilde{v}_{con}^d = v_{con}^{d,\Delta}\cos(\omega_c t + \theta_{con}^d)$ and $\tilde{v}_{con}^q = v_{con}^{q,\Delta}\sin(\omega_c t + \theta_{con}^q)$. Such single-frequency oscillation can induce the MFE, and its transformation from the control frame to the power system, ultimately manifesting as mirror-frequency oscillations, is shown in Figure 7.4. **The transformation of mirror-frequency oscillations** is summarized in the two steps below:

1. separation process: although $\tilde{v}_{con}^d$ and $\tilde{v}_{con}^q$ oscillate at a single frequency ω_c, they can be separated into two components oscillating at positive ω_c and negative ω_c, as shown in Figure 7.4;
2. *dq–αβ* transformation: these positive and negative oscillations undergo the *dq–αβ* transformation, resulting in mirror-frequency oscillations in the power system at $\omega_0 + \omega_c$ and $\omega_0 - \omega_c$.

The mathematical mechanism is provided below.

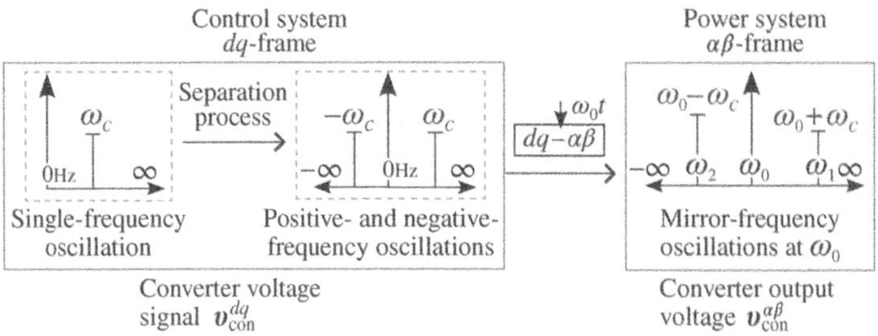

Figure 7.4 Transformation mechanism of mirror-frequency oscillations from the control frame to the power system.

7.1.2.1 Brief Understanding of Transformation Mechanism of Mirror-frequency Oscillation

As the mathematical analysis of the mirror-frequency oscillation transformation is relatively complex, a simplified transformation is provided here with the following assumptions to offer a complete and concise understanding of the MFE. A detailed mathematical explanation will be given in Section 7.1.2.2.

Assumption for simplification: the phases of both $\widetilde{v}_{\text{con}}^{d}$ and $\widetilde{v}_{\text{con}}^{q}$ are ignored, leading to $\widetilde{v}_{\text{con}}^{d} = v_{\text{con}}^{d,\Delta}\cos(\omega_c t)$ and $\widetilde{v}_{\text{con}}^{q} = v_{\text{con}}^{q,\Delta}\sin(\omega_c t)$.

Any dq quantities of $\widetilde{v}_{\text{con}}^{d}$ and $\widetilde{v}_{\text{con}}^{q}$ can be split into two components as follows:

$$\begin{aligned}\widetilde{v}_{\text{con}}^{d} + \text{j}\widetilde{v}_{\text{con}}^{q} &= v_{\text{con}}^{d,\Delta}\cos(\omega_c t) + \text{j}v_{\text{con}}^{q,\Delta}\sin(\omega_c t)\\ &= v_{\text{con}}^{\text{p},\Delta}(\cos(\omega_c t) + \text{j}\sin(\omega_c t)) + v_{\text{con}}^{\text{n},\Delta}(\cos(\omega_c t) - \text{j}\sin(\omega_c t))\end{aligned} \tag{7.6}$$

Where the superscripts ^{p} and ^{n} represent the positive frequency component and the negative frequency component, respectively.

$v_{\text{con}}^{\text{p},\Delta}$ and $v_{\text{con}}^{\text{n},\Delta}$ are calculated as follows:

$$\begin{aligned}v_{\text{con}}^{\text{p},\Delta} &= \frac{v_{\text{con}}^{d,\Delta} + v_{\text{con}}^{q,\Delta}}{2} \quad \text{at} \quad \omega_c\\ v_{\text{con}}^{\text{n},\Delta} &= \frac{v_{\text{con}}^{d,\Delta} - v_{\text{con}}^{q,\Delta}}{2} \quad \text{at} \quad -\omega_c\end{aligned} \tag{7.7}$$

In exponential form, the signal is represented as $\text{e}^{\text{j}\omega t}$, rewriting Equation 7.6 yields:

$$\widetilde{v}_{\text{con}}^{d} + \text{j}\widetilde{v}_{\text{con}}^{q} = v_{\text{con}}^{\text{p},\Delta}\text{e}^{\omega_c t\text{j}} + v_{\text{con}}^{\text{n},\Delta}\text{e}^{-\omega_c t\text{j}} \tag{7.8}$$

Equations 7.6 and 7.8 explain the process of splitting an oscillation at ω_c into positive- and negative-frequency (PNF) components at ω_c and $-\omega_c$, as shown in Figure 7.4.

To transfer the control signal to a converter voltage in the power system, the dq–$\alpha\beta$ transformation is applied using $e^{\omega_0 tj}$. Therefore, Equation 7.8 is transformed as:

$$\begin{aligned} e^{\omega_0 tj}(\tilde{v}_{\text{con}}^{d} + j\tilde{v}_{\text{con}}^{q}) &= v_{\text{con}}^{\text{p},\Delta} e^{(\omega_0+\omega_c)tj} + v_{\text{con}}^{\text{n},\Delta} e^{(\omega_0-\omega_c)tj} \\ &= v_1^{\Delta} e^{(\omega_0+\omega_c)tj} + v_2^{\Delta} e^{(\omega_0-\omega_c)tj} \end{aligned} \tag{7.9}$$

where v_1^{Δ} and v_2^{Δ} are the values after the dq–$\alpha\beta$ transformation in the power system.

Transformation of mirror-frequency oscillations: the two steps of the transformation in Figure 7.4, including the separation process and the dq–$\alpha\beta$ transformation, are mathematically explained by Equations 7.8 and 7.9. The relationship between a single-frequency oscillation in the control frame and the mirror-frequency oscillations in the power system is described below:

$$\begin{aligned} v_1^{\Delta} &= \frac{v_{\text{con}}^{d,\Delta} + v_{\text{con}}^{q,\Delta}}{2} \quad \text{at} \quad \omega_1 \\ v_2^{\Delta} &= \frac{v_{\text{con}}^{d,\Delta} - v_{\text{con}}^{q,\Delta}}{2} \quad \text{at} \quad \omega_2 \end{aligned} \tag{7.10}$$

Equation 7.9 can also be represented in the matrix form, as below:

$$\begin{aligned} &\begin{bmatrix} \cos(\omega_0 t) & -\sin(\omega_0 t) \\ \sin(\omega_0 t) & \cos(\omega_0 t) \end{bmatrix} \begin{bmatrix} \tilde{v}_{\text{con}}^{d} \\ \tilde{v}_{\text{con}}^{q} \end{bmatrix} \\ &= \begin{bmatrix} v_1^{\Delta} \cos(\omega_0 + \omega_c)t \\ v_1^{\Delta} \sin(\omega_0 + \omega_c)t \end{bmatrix} + \begin{bmatrix} v_2^{\Delta} \cos(\omega_0 - \omega_c)t \\ v_2^{\Delta} \sin(\omega_0 - \omega_c)t \end{bmatrix} \end{aligned} \tag{7.11}$$

7.1.2.2 Detailed Understanding of the Mirror-frequency Effect

A detailed representation of a pair of oscillations, $\tilde{v}_{\text{con}}^{d} = v_{\text{con}}^{d,\Delta} \cos(\omega_c t + \theta^d)$ and $\tilde{v}_{\text{con}}^{q} = v_{\text{con}}^{q,\Delta} \sin(\omega_c t + \theta^q)$, is considered in this section to provide a detailed and comprehensive understanding.

Any dq vector of $\tilde{v}_{\text{con}}^{d}$ and $\tilde{v}_{\text{con}}^{q}$ can be split into the following forms of components, as shown below:

$$\begin{aligned} \begin{bmatrix} \tilde{v}_{\text{con}}^{d} \\ \tilde{v}_{\text{con}}^{q} \end{bmatrix} &= 0.5\left(v_{\text{con}}^{d,\Delta} \begin{bmatrix} \cos(\omega_c t + \theta^d) \\ \sin(\omega_c t + \theta^d) \end{bmatrix} + v_{\text{con}}^{q,\Delta} \begin{bmatrix} \cos(\omega_c t + \theta^q) \\ \sin(\omega_c t + \theta^q) \end{bmatrix} \right) \\ &+ 0.5\left(v_{\text{con}}^{d,\Delta} \begin{bmatrix} \cos(\omega_c t + \theta^d) \\ -\sin(\omega_c t + \theta^d) \end{bmatrix} - v_{\text{con}}^{q,\Delta} \begin{bmatrix} \cos(\omega_c t + \theta^q) \\ -\sin(\omega_c t + \theta^q) \end{bmatrix} \right) \end{aligned} \tag{7.12}$$

Expanding Equation 7.12 yields:

$$\begin{bmatrix} \tilde{v}_{\mathrm{con}}^{d} \\ \tilde{v}_{\mathrm{con}}^{q} \end{bmatrix} = \begin{bmatrix} v_{\mathrm{con}}^{\mathrm{p},\Delta} \cos\theta^{\mathrm{p}} \cos\omega_c t - v_{\mathrm{con}}^{\mathrm{p},\Delta} \sin\theta^{\mathrm{p}} \sin\omega_c t \\ v_{\mathrm{con}}^{\mathrm{p},\Delta} \cos\theta^{\mathrm{p}} \sin\omega_c t + v_{\mathrm{con}}^{\mathrm{p},\Delta} \sin\theta^{\mathrm{p}} \cos\omega_c t \end{bmatrix} + \begin{bmatrix} v_{\mathrm{con}}^{\mathrm{n},\Delta} \cos\omega_c t - v_{\mathrm{con}}^{\mathrm{n},\Delta} \sin\theta^{\mathrm{n}} \sin\omega_c t \\ -(v_{\mathrm{con}}^{\mathrm{n},\Delta} \sin\omega_c t + v_{\mathrm{con}}^{\mathrm{n},\Delta} \sin\theta^{\mathrm{n}} \cos\omega_c t) \end{bmatrix} \tag{7.13}$$

where $v_{\mathrm{con}}^{\mathrm{p},\Delta}$ and $v_{\mathrm{con}}^{\mathrm{n},\Delta}$ are proposed to follow the following relations to obtain a concise expression of Equation 7.13:

$$\begin{cases} v_{\mathrm{con}}^{\mathrm{p},\Delta} \cos(\theta^{\mathrm{p}}) = 0.5(v_{\mathrm{con}}^{d,\Delta} \cos\theta_{\mathrm{con}}^{d} + v_{\mathrm{con}}^{q,\Delta} \cos\theta_{\mathrm{con}}^{q}) \\ v_{\mathrm{con}}^{\mathrm{p},\Delta} \sin(\theta^{\mathrm{p}}) = 0.5(v_{\mathrm{con}}^{d,\Delta} \sin\theta_{\mathrm{con}}^{d} + v_{\mathrm{con}}^{q,\Delta} \sin\theta_{\mathrm{con}}^{q}) \\ v_{\mathrm{con}}^{\mathrm{n},\Delta} \cos(\theta^{\mathrm{n}}) = 0.5(v_{\mathrm{con}}^{d,\Delta} \cos\theta_{\mathrm{con}}^{d} - v_{\mathrm{con}}^{q,\Delta} \cos\theta_{\mathrm{con}}^{q}) \\ v_{\mathrm{con}}^{\mathrm{n},\Delta} \sin(\theta^{\mathrm{n}}) = 0.5(v_{\mathrm{con}}^{d,\Delta} \sin\theta_{\mathrm{con}}^{d} - v_{\mathrm{con}}^{q,\Delta} \sin\theta_{\mathrm{con}}^{q}) \end{cases} \tag{7.14}$$

Combining Equation 7.13 gives:

$$\begin{bmatrix} \tilde{v}_{\mathrm{con}}^{d} \\ \tilde{v}_{\mathrm{con}}^{q} \end{bmatrix} = \begin{bmatrix} v_{\mathrm{con}}^{\mathrm{p},\Delta} \cos(\omega_c t + \theta^{\mathrm{p}}) \\ v_{\mathrm{con}}^{\mathrm{p},\Delta} \sin(\omega_c t + \theta^{\mathrm{p}}) \end{bmatrix} + \begin{bmatrix} v_{\mathrm{con}}^{\mathrm{n},\Delta} \cos(\omega_c t + \theta^{\mathrm{n}}) \\ -v_{\mathrm{con}}^{\mathrm{n},\Delta} \sin(\omega_c t + \theta^{\mathrm{n}}) \end{bmatrix} \tag{7.15}$$

To transfer the control signal to a converter voltage in the power system, the *dq*–*αβ* transformation is applied using $e^{\omega_0 t j}$. Therefore, Equation 7.14 is transformed as:

$$\begin{bmatrix} \cos(\omega_0 t) & -\sin(\omega_0 t) \\ \sin(\omega_0 t) & \cos(\omega_0 t) \end{bmatrix} \begin{bmatrix} \tilde{v}_{\mathrm{con}}^{d} \\ \tilde{v}_{\mathrm{con}}^{q} \end{bmatrix} = \begin{bmatrix} v_1^{\Delta} \cos(\omega_0 t + \omega_c t + \theta^{\mathrm{p}}) \\ v_1^{\Delta} \sin(\omega_0 t + \omega_c t + \theta^{\mathrm{p}}) \end{bmatrix} + \begin{bmatrix} v_2^{\Delta} \cos(\omega_0 t - \omega_c t - \theta^{\mathrm{n}}) \\ v_2^{\Delta} \sin(\omega_0 t - \omega_c t - \theta^{\mathrm{n}}) \end{bmatrix} \tag{7.16}$$

where $v_1^{\Delta} = v_{\mathrm{con}}^{\mathrm{p},\Delta}$ and $v_2^{\Delta} = v_{\mathrm{con}}^{\mathrm{n},\Delta}$.

Transformation of mirror-frequency oscillations in the analytical solution representation: the two steps of the transformation in Figure 7.4, including the separation process and the *dq*–*αβ* transformation, are mathematically explained by Equations 7.15 and 7.16. The relationship between a single-frequency oscillation in the control frame and the mirror-frequency oscillations in the power system is described by Equation 7.14.

However, the above representation of the mirror-frequency oscillation transformation is still not concise enough due to the complexity of Equation 7.15. Therefore, an equivalent vector representation of the transformation is developed to meet the conciseness requirements.

First, define a new vector representation $\boldsymbol{v}$ of $v\cos(\omega t+\theta)$ at ω as:

$$v\cos(\omega t+\theta)\rightarrow v(\cos\theta+\mathrm{j}\sin\theta)=\boldsymbol{v} \tag{7.17}$$

Accordingly, $v\sin(\omega t+\theta)$ at ω is represented as $-\mathrm{j}\boldsymbol{v}$, as derived below:

$$\begin{aligned}&v\sin(\omega t+\theta)=v\cos\left(\omega t+\theta-\frac{\pi}{2}\right)\\&\rightarrow v\left(\cos\left(\theta-\frac{\pi}{2}\right)+\mathrm{j}\sin\left(\theta-\frac{\pi}{2}\right)\right)=-\mathrm{j}\boldsymbol{v}\end{aligned} \tag{7.18}$$

Based on the above definition of vector representation, Equation 7.15 can be rewritten as:

$$\begin{bmatrix}\tilde{\boldsymbol{v}}_{\text{con}}^{d}\\ \tilde{\boldsymbol{v}}_{\text{con}}^{q}\end{bmatrix}=\tilde{\boldsymbol{v}}_{\text{con}}^{\text{p}}\begin{bmatrix}1\\-\mathrm{j}\end{bmatrix}+\tilde{\boldsymbol{v}}_{\text{con}}^{\text{n}}\begin{bmatrix}1\\ \mathrm{j}\end{bmatrix} \tag{7.19}$$

where $\tilde{\boldsymbol{v}}_{\text{con}}^{d}=v_{\text{con}}^{d,\Delta}(\cos\theta_{\text{con}}^{d}+\mathrm{j}\sin\theta_{\text{con}}^{d})$, $\tilde{\boldsymbol{v}}_{\text{con}}^{q}=v_{\text{con}}^{q,\Delta}(\cos\theta_{\text{con}}^{q}+\mathrm{j}\sin\theta_{\text{con}}^{q})$, $\tilde{\boldsymbol{v}}_{\text{con}}^{\text{n}}=v_{\text{con}}^{\text{p},\Delta}(\cos\theta^{\text{p}}+\mathrm{j}\sin\theta^{\text{p}})$, and $\tilde{\boldsymbol{v}}_{\text{con}}^{\text{n}}=v_{\text{con}}^{\text{n},\Delta}(\cos\theta^{\text{n}}+\mathrm{j}\sin\theta^{\text{n}})$.

Equation 7.19 can be rewritten as:

$$\begin{bmatrix}\tilde{\boldsymbol{v}}_{\text{con}}^{d}\\ \tilde{\boldsymbol{v}}_{\text{con}}^{q}\end{bmatrix}=\begin{bmatrix}1&1\\-\mathrm{j}&\mathrm{j}\end{bmatrix}\begin{bmatrix}\tilde{\boldsymbol{v}}_{\text{con}}^{\text{p}}\\ \tilde{\boldsymbol{v}}_{\text{con}}^{\text{n}}\end{bmatrix} \tag{7.20}$$

$\tilde{\boldsymbol{v}}_{\text{con}}^{\text{p}}$ and $\tilde{\boldsymbol{v}}_{\text{con}}^{\text{n}}$ can be calculated by inversing Equation 7.20, as follows:

$$\begin{bmatrix}\tilde{\boldsymbol{v}}_{\text{con}}^{\text{p}}\\ \tilde{\boldsymbol{v}}_{\text{con}}^{\text{n}}\end{bmatrix}=0.5\begin{bmatrix}1&\mathrm{j}\\1&-\mathrm{j}\end{bmatrix}\begin{bmatrix}\tilde{\boldsymbol{v}}_{\text{con}}^{d}\\ \tilde{\boldsymbol{v}}_{\text{con}}^{q}\end{bmatrix}\begin{matrix}\text{at }\omega_c\\ \text{at }-\omega_c\end{matrix} \tag{7.21}$$

If we define the vector-form mirror-frequency oscillations in the power system after the dq–$\alpha\beta$ transformation as $\boldsymbol{v}_1$ and $\boldsymbol{v}_2$, then based on the relation in Equation 7.16, the dq–$\alpha\beta$ transformation can be mathematically presented as:

$$\begin{aligned}&\begin{bmatrix}v_{\text{con}}^{\text{p},\Delta}\cos(\omega_c t+\theta^{\text{p}})\\ v_{\text{con}}^{\text{p},\Delta}\sin(\omega_c t+\theta^{\text{p}})\end{bmatrix}+\begin{bmatrix}v_{\text{con}}^{\text{n},\Delta}\cos(\omega_c t+\theta^{\text{n}})\\ -v_{\text{con}}^{\text{n},\Delta}\sin(\omega_c t+\theta^{\text{n}})\end{bmatrix}\xrightarrow{dq-\alpha\beta}\\&\begin{bmatrix}v^{1,\Delta}\cos(\omega_0 t+\omega_c t+\theta^{\text{p}})\\ v^{1,\Delta}\sin(\omega_0 t+\omega_c t+\theta^{\text{p}})\end{bmatrix}+\begin{bmatrix}v^{2,\Delta}\cos(\omega_0 t-\omega_c t-\theta^{\text{n}})\\ v^{2,\Delta}\sin(\omega_0 t-\omega_c t-\theta^{\text{n}})\end{bmatrix}\end{aligned} \tag{7.22}$$

where $v_1^{\Delta}=v_{\text{con}}^{\text{p},\Delta}$ and $v_2^{\Delta}=v_{\text{con}}^{\text{n},\Delta}$.

Represented in vector form as:
If $\omega_c > \omega_0$:

$$\underbrace{\widetilde{\boldsymbol{v}}^{\mathrm{p}}_{\mathrm{con}}\begin{bmatrix}1\\-\mathrm{j}\end{bmatrix}+\widetilde{\boldsymbol{v}}^{\mathrm{n}}_{\mathrm{con}}\begin{bmatrix}1\\\mathrm{j}\end{bmatrix}}_{\text{at } \omega_c}=\underbrace{\widetilde{\boldsymbol{v}}^{1}\begin{bmatrix}1\\-\mathrm{j}\end{bmatrix}}_{\text{at } \omega_1}+\underbrace{\widetilde{\boldsymbol{v}}^{2}\begin{bmatrix}1\\\mathrm{j}\end{bmatrix}}_{\text{at } \omega_2} \tag{7.23}$$

If $0 < \omega_c < \omega_0$:

$$\underbrace{\widetilde{\boldsymbol{v}}^{\mathrm{p}}_{\mathrm{con}}\begin{bmatrix}1\\-\mathrm{j}\end{bmatrix}+\widetilde{\boldsymbol{v}}^{\mathrm{n}}_{\mathrm{con}}\begin{bmatrix}1\\\mathrm{j}\end{bmatrix}}_{\text{at } \omega_c}=\underbrace{\widetilde{\boldsymbol{v}}^{1}\begin{bmatrix}1\\-\mathrm{j}\end{bmatrix}}_{\text{at } \omega_1}+\underbrace{\overline{\widetilde{\boldsymbol{v}}^{2}}\begin{bmatrix}1\\\mathrm{j}\end{bmatrix}}_{\text{at } \omega_2} \tag{7.24}$$

where $\overline{\boldsymbol{v}}$ indicates the complex conjugate.

Equations 7.23 and 7.24 can be written in a more concise form:

$$\begin{array}{lll} 0 < \omega_c < \omega_0 & \omega_c > \omega_0 & \\ \widetilde{\boldsymbol{v}}_1 = \widetilde{\boldsymbol{v}}^{\mathrm{p}}_{\mathrm{con}} & \widetilde{\boldsymbol{v}}_1 = \widetilde{\boldsymbol{v}}^{\mathrm{p}}_{\mathrm{con}} & \text{at} \quad \omega_1 \\ \widetilde{\boldsymbol{v}}_2 = \overline{\widetilde{\boldsymbol{v}}^{\mathrm{n}}_{\mathrm{con}}} & \widetilde{\boldsymbol{v}}_2 = \widetilde{\boldsymbol{v}}^{\mathrm{n}}_{\mathrm{con}} & \text{at} \quad \omega_2 \end{array} \tag{7.25}$$

Transformation of mirror-frequency oscillations in the vector representation: the separation process is clearly and simply explained by Equation 7.21. The dq–$\alpha\beta$ transformation is presented as Equation 7.25. These equations in the vector form provide a very simple explanation of the entire transformation process.

Additionally, the vector representation corresponds to the complex Fourier transformation, which is an essential method for measuring oscillations at a specific frequency ω. Therefore, any measured oscillation can be seamlessly applied to the above vector-form equations.

An example of how a single-frequency oscillation in the control is transferred into the power system as mirror-frequency oscillations is demonstrated by inserting a single-frequency oscillation into v^d_{con} and v^q_{con} of the system in Figure 7.2, respectively, as shown below:

$$\begin{aligned} \widetilde{v}^d_{\mathrm{con}} &= 0.1\cos(20 * 2\pi + \pi/3) \\ \widetilde{v}^q_{\mathrm{con}} &= 0.2\sin(20 * 2\pi - \pi/4) \end{aligned} \tag{7.26}$$

The Fourier analysis results are shown in Figure 7.5. Their values match the calculated values using the developed Equations 7.21 and 7.25.

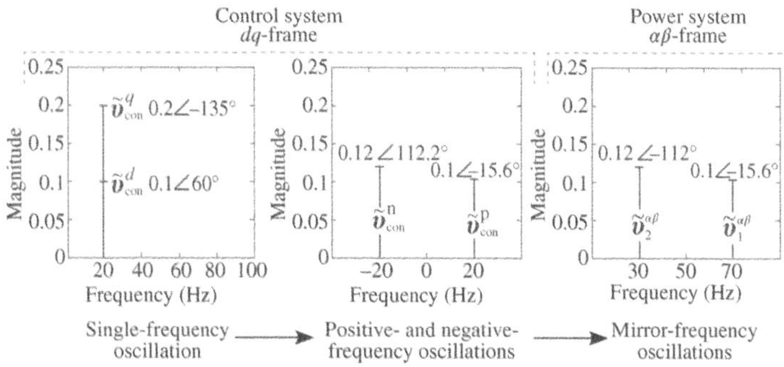

Figure 7.5 An oscillation $\widetilde{v}_{con}$ at ω_c is transferred from the control frame to the power system via the *dq–abc* transformation in vector representation.

7.2 The Difference Between Mirror-frequency Oscillations and Positive/Negative Sequence Oscillations

Positive and negative sequence oscillations are similar in form to the previously mentioned PNF oscillations in the MFE, which may cause confusion among system engineers. In fact, they are distinct phenomena but share similar form. Therefore, the similarities and differences will be clarified in the following section.

The expression of the positive and negative sequence voltages is presented below:

$$\begin{array}{ll} \text{Positive sequence} & \text{Negative sequence} \\ \widetilde{v}_a^+ = v^{+,\Delta}\cos(\omega t + \theta^+) & \widetilde{v}_a^- = v^{-,\Delta}\cos(\omega t + \theta^-) \\ \widetilde{v}_b^+ = v^{+,\Delta}\cos\left(\omega t + \theta^+ - \dfrac{2\pi}{3}\right) & \widetilde{v}_b^- = v^{-,\Delta}\cos\left(\omega t + \theta^- + \dfrac{2\pi}{3}\right) \\ \widetilde{v}_c^+ = v^{+,\Delta}\cos\left(\omega t + \theta^+ - \dfrac{4\pi}{3}\right) & \widetilde{v}_c^- = v^{-,\Delta}\cos\left(\omega t + \theta^- + \dfrac{4\pi}{3}\right) \end{array} \tag{7.27}$$

where the superscripts $^+$ and $^-$ represent the positive and negative sequences, respectively.

A voltage containing both positive and negative sequence components can be represented as:

$$\begin{aligned} \widetilde{v}_a &= \widetilde{v}_a^+ + \widetilde{v}_a^- \\ \widetilde{v}_b &= \widetilde{v}_b^+ + \widetilde{v}_b^- \\ \widetilde{v}_c &= \widetilde{v}_c^+ + \widetilde{v}_c^- \end{aligned} \tag{7.28}$$

Transferring Equation 7.28 into the $\alpha\beta$-frame yields:

$$\begin{bmatrix} \widetilde{v}^\alpha \\ \widetilde{v}^\beta \end{bmatrix} = \begin{bmatrix} v^{+,\Delta}\cos(\omega t + \theta^+) \\ v^{+,\Delta}\sin(\omega t + \theta^+) \end{bmatrix} + \begin{bmatrix} v^{-,\Delta}\cos(\omega t + \theta^-) \\ -v^{-,\Delta}\sin(\omega t + \theta^-) \end{bmatrix} \tag{7.29}$$

Rewriting Equation 7.29 in vector form, as defined in Equations 7.17 and 7.18, at ω yields:

$$\begin{bmatrix} \widetilde{\boldsymbol{v}}^{\alpha} \\ \widetilde{\boldsymbol{v}}^{\beta} \end{bmatrix} = \widetilde{\boldsymbol{v}}^{+} \begin{bmatrix} 1 \\ -\mathrm{j} \end{bmatrix} + \widetilde{\boldsymbol{v}}^{-} \begin{bmatrix} 1 \\ \mathrm{j} \end{bmatrix} \tag{7.30}$$

where $\widetilde{\boldsymbol{v}}^{+} = v^{+,\Delta}(\cos\theta^{+} + \mathrm{j}\sin\theta^{+})$ and $\widetilde{\boldsymbol{v}}^{-} = v^{-,\Delta}(\cos\theta^{-} + \mathrm{j}\sin\theta^{-})$.

Applying [1 j; 1 −j] to Equation 7.30 yields:

$$\begin{bmatrix} 1 & \mathrm{j} \\ 1 & -\mathrm{j} \end{bmatrix} \begin{bmatrix} \widetilde{\boldsymbol{v}}^{\alpha} \\ \widetilde{\boldsymbol{v}}^{\beta} \end{bmatrix} = \begin{bmatrix} 1 & \mathrm{j} \\ 1 & -\mathrm{j} \end{bmatrix} \left(\widetilde{\boldsymbol{v}}^{+} \begin{bmatrix} 1 \\ -\mathrm{j} \end{bmatrix} + \widetilde{\boldsymbol{v}}^{-} \begin{bmatrix} 1 \\ \mathrm{j} \end{bmatrix} \right) \tag{7.31}$$

Rearranging Equation 7.31 yields:

$$\begin{bmatrix} \widetilde{\boldsymbol{v}}^{+} \\ \widetilde{\boldsymbol{v}}^{-} \end{bmatrix} = 0.5 \begin{bmatrix} 1 & \mathrm{j} \\ 1 & -\mathrm{j} \end{bmatrix} \begin{bmatrix} \widetilde{\boldsymbol{v}}^{\alpha} \\ \widetilde{\boldsymbol{v}}^{\beta} \end{bmatrix} \tag{7.32}$$

An alternative method to derive the expression of Equation 7.32 is also provided. The three-phase voltage can be represented as the vector form $[\widetilde{\boldsymbol{v}}_a; \widetilde{\boldsymbol{v}}_b; \widetilde{\boldsymbol{v}}_c]$. The positive or negative sequence components can then be calculated based on the following relation:

$$\begin{bmatrix} \widetilde{\boldsymbol{v}}^{+} \\ \widetilde{\boldsymbol{v}}^{-} \end{bmatrix} = \frac{1}{3} \begin{bmatrix} 1 & \boldsymbol{a} & \boldsymbol{a}^2 \\ 1 & \boldsymbol{a}^2 & \boldsymbol{a} \end{bmatrix} \begin{bmatrix} \widetilde{\boldsymbol{v}}_a \\ \widetilde{\boldsymbol{v}}_b \\ \widetilde{\boldsymbol{v}}_c \end{bmatrix} \tag{7.33}$$

where $\boldsymbol{a} = 1\angle\frac{2\pi}{3} = \cos\frac{2\pi}{3} + \mathrm{j}\sin\frac{2\pi}{3}$.

Substituting $[\widetilde{\boldsymbol{v}}_a; \widetilde{\boldsymbol{v}}_b; \widetilde{\boldsymbol{v}}_c]$ with $\alpha\beta$ representation yields:

$$\begin{bmatrix} \widetilde{\boldsymbol{v}}^{+} \\ \widetilde{\boldsymbol{v}}^{-} \end{bmatrix} = \frac{1}{3} \begin{bmatrix} 1 & \boldsymbol{a} & \boldsymbol{a}^2 \\ 1 & \boldsymbol{a}^2 & \boldsymbol{a} \end{bmatrix} \begin{bmatrix} 1 & 0 \\ -0.5 & \frac{\sqrt{3}}{2} \\ -0.5 & -\frac{\sqrt{3}}{2} \end{bmatrix} \begin{bmatrix} \widetilde{\boldsymbol{v}}^{\alpha} \\ \widetilde{\boldsymbol{v}}^{\beta} \end{bmatrix} \tag{7.34}$$

Combining and organizing Equation 7.33 yields:

$$\begin{bmatrix} \widetilde{\boldsymbol{v}}^{+} \\ \widetilde{\boldsymbol{v}}^{-} \end{bmatrix} = 0.5 \begin{bmatrix} 1 & \mathrm{j} \\ 1 & -\mathrm{j} \end{bmatrix} \begin{bmatrix} \widetilde{\boldsymbol{v}}^{\alpha} \\ \widetilde{\boldsymbol{v}}^{\beta} \end{bmatrix} \begin{matrix} \text{at} & \omega \\ \text{at} & -\omega \end{matrix} \tag{7.35}$$

By comparing the MFE's separation process in Equation 7.21, the positive and negative representations in Equation 7.35 share a similar mathematical form. However, they are in different frames: Equation 7.21 is in the *dq*-frame, while

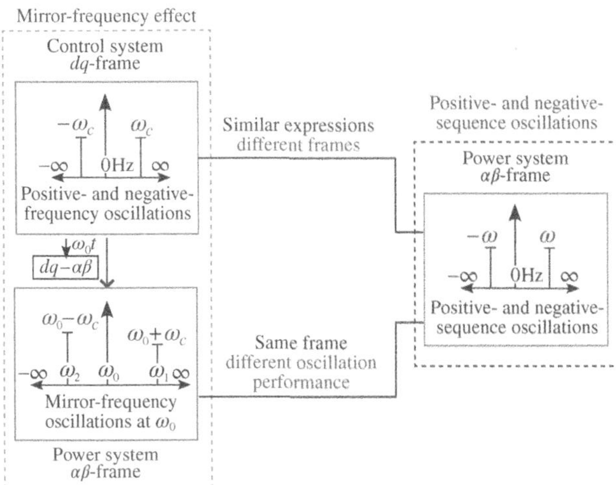

Figure 7.6 Similarities and differences between mirror-frequency oscillations and positive/negative oscillations.

Equation 7.35 is in the $\alpha\beta$-frame, which indicates that, despite their similar mathematical form, they represent entirely different oscillation performance, as shown in Figure 7.6. The frequency performance of mirror-frequency oscillations relating to positive/negative sequence are summarized below at the situation of an oscillation ω in the power system:

- when $\omega_c < \omega_0$, this leads to ω_1 and $\omega_2 > 0$, meaning both mirror-frequency oscillations are in the positive sequence;
- when $\omega_c > \omega_0$, this leads to $\omega_1 > 0$ and $\omega_2 < 0$, where the normal oscillation at ω_1 is in the positive sequence, and the mirrored oscillation at ω_2 is in the negative sequence.

7.3 The Performance of Mirror-frequency Oscillations Across Different Electrical Quantities

Mirror-frequency oscillations exhibit different frequency characteristics; for example, in a dq-frame, they appear as unbalanced single-frequency oscillations. However, in the power system, instantaneous voltages contain mirror-frequency oscillations. In summary, any quantity requiring separation from the fundamental frequency will exhibit a single-frequency oscillation denoted as ω_c, while instantaneous quantities such as voltage and current will show mirror-frequency oscillations at $\omega_1 = \omega_0 + \omega_c$ and $\omega_2 = \omega_0 - \omega_c$. A summary of different electrical quantities are provided below:

- frequency-separated electrical quantities: dq values, root mean square (RMS) values, and apparent power-**single-frequency oscillation**;
- instantaneous electrical quantities: voltage and current mirror-frequency **oscillation**.

As the mirror-frequency oscillation of instantaneous values has been explained in detail in Section 7.1, this section focuses primarily on a mathematical explanation of why frequency-separated electrical quantities exhibit single-frequency oscillation. Assuming that mirror-frequency oscillations appear in the instantaneous voltage and current, represented as below:

$$\begin{aligned} &v^{\mathrm{mag}}\mathrm{e}^{\mathrm{j}\omega_0 t}+\widetilde{v}_1^{\mathrm{mag}}\mathrm{e}^{\mathrm{j}\omega_1 t}+\widetilde{v}_2^{\mathrm{mag}}\mathrm{e}^{\mathrm{j}\omega_2 t}\\ &i^{\mathrm{mag}}\mathrm{e}^{\mathrm{j}\omega_0 t}+\widetilde{i}_1^{\mathrm{mag}}\mathrm{e}^{\mathrm{j}\omega_1 t}+\widetilde{i}_2^{\mathrm{mag}}\mathrm{e}^{\mathrm{j}\omega_2 t}\end{aligned} \tag{7.36}$$

For dq values, a $\alpha\beta$–dq transformation is required, taking instantaneous voltage as an example:

$$\begin{aligned} v^{dq}+\widetilde{v}^{dq} &= (v^{\mathrm{mag}}\mathrm{e}^{\mathrm{j}\omega_0 t}+\widetilde{v}_1^{\mathrm{mag}}\mathrm{e}^{\mathrm{j}\omega_1 t}+\widetilde{v}_2^{\mathrm{mag}}\mathrm{e}^{\mathrm{j}\omega_2 t})\times \mathrm{e}^{-\mathrm{j}\omega_0 t}\\ &= (v^{\mathrm{mag}}+\widetilde{v}_1^{\mathrm{mag}}\mathrm{e}^{\mathrm{j}\omega_c t}+\widetilde{v}_2^{\mathrm{mag}}\mathrm{e}^{-\mathrm{j}\omega_c t})\end{aligned} \tag{7.37}$$

It is evident from Equation 7.37 that it contains a single-frequency oscillation in an unbalanced manner.

The apparent power can be calculated as follows:

$$\begin{aligned} S+\widetilde{S} = &(v^{\mathrm{mag}}\mathrm{e}^{\mathrm{j}\omega_0 t}+\widetilde{v}_1^{\mathrm{mag}}\mathrm{e}^{\mathrm{j}\omega_1 t}+\widetilde{v}_2^{\mathrm{mag}}\mathrm{e}^{\mathrm{j}\omega_2 t})\\ &\times \mathrm{conj}(i^{\mathrm{mag}}\mathrm{e}^{\mathrm{j}\omega_0 t}+\widetilde{i}_1^{\mathrm{mag}}\mathrm{e}^{\mathrm{j}\omega_1 t}+\widetilde{i}_2^{\mathrm{mag}}\mathrm{e}^{\mathrm{j}\omega_2 t})\end{aligned} \tag{7.38}$$

Eliminating S and ignoring the term $\widetilde{v}^{\mathrm{mag}}\widetilde{i}^{\mathrm{mag}}$ yields:

$$\begin{aligned} \widetilde{S} = &(\widetilde{v}_1^{\mathrm{mag}}\mathrm{e}^{\mathrm{j}\omega_1 t}+\widetilde{v}_2^{\mathrm{mag}}\mathrm{e}^{\mathrm{j}\omega_2 t})i^{\mathrm{mag}}\mathrm{e}^{-\mathrm{j}\omega_0 t}\\ &+(\widetilde{i}_1^{\mathrm{mag}}\mathrm{e}^{-\mathrm{j}\omega_1 t}+\widetilde{i}_2^{\mathrm{mag}}\mathrm{e}^{-\mathrm{j}\omega_2 t})v^{\mathrm{mag}}\mathrm{e}^{\mathrm{j}\omega_0 t}\end{aligned} \tag{7.39}$$

Rearranging and organizing Equation 7.39 yields:

$$\begin{aligned} \widetilde{S} = &i^{\mathrm{mag}}\widetilde{v}_1^{\mathrm{mag}}\mathrm{e}^{\mathrm{j}\omega_c t}+i^{\mathrm{mag}}\widetilde{v}_2^{\mathrm{mag}}\mathrm{e}^{\mathrm{j}(-\omega_c)t}\\ &+v^{\mathrm{mag}}\widetilde{i}_1^{\mathrm{mag}}\mathrm{e}^{\mathrm{j}(-\omega_c)t}+v^{\mathrm{mag}}\widetilde{i}_2^{\mathrm{mag}}\mathrm{e}^{\mathrm{j}\omega_c t}\end{aligned} \tag{7.40}$$

It is evident from Equation 7.40 that S contains a single-frequency oscillation in an unbalanced manner, which means both active and reactive power will contain this single-frequency oscillation.

The RMS value of an instantaneous voltage can be calculated as follows:

$$\mathrm{RMS} = \sqrt{\frac{1}{T}\int_0^T x(t)^2\,dt} \tag{7.41}$$

As the operation $\sqrt{\frac{1}{T}\int_0^T x(t)^2 dt}$ does not alter the frequency of $x(t)^2$, a simplified explanation is provided below, considering only $x(t)^2$:

$$\begin{aligned}\underbrace{(\cos(\omega_0 t) + \cos(\omega_1 t))^2}_{x(t)} &= \cos(\omega_0 t)\cos(\omega_1 t) \\ &= 0.5(\cos(\omega_0 - \omega)t + \cos(\omega_0 + \omega)t)\end{aligned} \tag{7.42}$$

No matter whether ω is ω_1 or ω_2, Equation 7.42 contains a component of single-frequency oscillations.

7.4 Understanding Balanced and Unbalanced Control Systems and Their Role in the Mirror-frequency Effect

To explain the cause of MFE clearly, the following definition is developed:

- precondition: all definitions are developed in the control frame (*dq*-frame) to align with the analysis presented in the sections above;
- an unbalanced signal refers to a signal, as represented below, where $v^{d,\text{mag}} \neq v^{q,\text{mag}}$ or $\theta^d \neq \theta^q$:

$$\begin{bmatrix} v^{d,\text{mag}}\cos(\omega t + \theta^d) \\ v^{q,\text{mag}}\sin(\omega t + \theta^q) \end{bmatrix} \tag{7.43}$$

- a balanced signal refers to a signal that follows one of the following forms: positive frequency component:

$$\begin{bmatrix} v^{\text{mag}}\cos(\omega t + \theta^{\text{p}}) \\ v^{\text{mag}}\sin(\omega t + \theta^{\text{p}}) \end{bmatrix} \xrightarrow{\text{vector}} v^{\text{mag}}(\cos\theta^{\text{p}} + \text{j}\sin\theta^{\text{p}})\begin{bmatrix} 1 \\ -\text{j} \end{bmatrix} \tag{7.44}$$

  negative frequency component:

$$\begin{bmatrix} v^{\text{mag}}\cos(\omega t + \theta^{\text{n}}) \\ -v^{\text{mag}}\sin(\omega t + \theta^{\text{n}}) \end{bmatrix} \xrightarrow{\text{vector}} v^{\text{mag}}(\cos\theta^{\text{n}} + \text{j}\sin\theta^{\text{n}})\begin{bmatrix} 1 \\ \text{j} \end{bmatrix} \tag{7.45}$$

- an unbalanced control system refers to a system where, if a balanced signal is input, its output signal becomes unbalanced.

The cause of the MFE is the unbalanced converter control, which transforms a balanced oscillation into an unbalanced converter voltage signal that contains PNF components, further leading to mirror-frequency oscillation in the converter output voltage due to the *dq*–$\alpha\beta$ transformation.

The clear definition of an unbalanced control system is specified and quantified below.

Vector representations of signals are used, and a control system can be represented as a 2 × 2 transfer function matrix $\boldsymbol{G}_{\text{cs}}$, as shown below:

$$\begin{bmatrix} G_{\text{cs}}^{dd} & G_{\text{cs}}^{dq} \\ G_{\text{cs}}^{qd} & G_{\text{cs}}^{qq} \end{bmatrix} \begin{bmatrix} \tilde{\boldsymbol{x}}^{d} \\ \tilde{\boldsymbol{x}}^{q} \end{bmatrix} = \begin{bmatrix} \tilde{\boldsymbol{v}}_{\text{con}}^{d} \\ \tilde{\boldsymbol{v}}_{\text{con}}^{q} \end{bmatrix} \tag{7.46}$$

where $\tilde{\boldsymbol{x}}^{d}$ and $\tilde{\boldsymbol{x}}^{q}$ represent the imputed oscillation signals at frequency ω.

Applying the corresponding PNF components to Equation 7.46 using Equation 7.20 yields the following:

$$\begin{bmatrix} G_{\text{cs}}^{dd} & G_{\text{cs}}^{dq} \\ G_{\text{cs}}^{qd} & G_{\text{cs}}^{qq} \end{bmatrix} \begin{bmatrix} 1 & 1 \\ -\text{j} & \text{j} \end{bmatrix} \begin{bmatrix} \tilde{\boldsymbol{x}}^{\text{p}} \\ \tilde{\boldsymbol{x}}^{\text{n}} \end{bmatrix} = \begin{bmatrix} 1 & 1 \\ -\text{j} & \text{j} \end{bmatrix} \begin{bmatrix} \tilde{\boldsymbol{v}}_{\text{con}}^{\text{p}} \\ \tilde{\boldsymbol{v}}_{\text{con}}^{\text{n}} \end{bmatrix} \tag{7.47}$$

Rearranging yields:

$$\underbrace{\begin{bmatrix} 1 & 1 \\ -\text{j} & \text{j} \end{bmatrix}^{-1} \begin{bmatrix} G_{\text{cs}}^{dd} & G_{\text{cs}}^{dq} \\ G_{\text{cs}}^{qd} & G_{\text{cs}}^{qq} \end{bmatrix} \begin{bmatrix} 1 & 1 \\ -\text{j} & \text{j} \end{bmatrix}}_{G_{\text{cs}}^{\text{pn}}} \begin{bmatrix} \tilde{\boldsymbol{x}}^{\text{p}} \\ \tilde{\boldsymbol{x}}^{\text{n}} \end{bmatrix} = \begin{bmatrix} \tilde{\boldsymbol{v}}_{\text{con}}^{\text{p}} \\ \tilde{\boldsymbol{v}}_{\text{con}}^{\text{n}} \end{bmatrix} \tag{7.48}$$

Based on Equation 7.48, the control system matrix is transformed into the form of the positive- and negative- (PN) frequency axis as follows:

$$\boldsymbol{G}_{\text{cs}}^{\text{pn}} = \begin{bmatrix} G_{\text{cs}}^{\text{pp}} & G_{\text{cs}}^{\text{pn}} \\ G_{\text{cs}}^{\text{np}} & G_{\text{cs}}^{\text{nn}} \end{bmatrix} \tag{7.49}$$

$$= 0.5 \begin{bmatrix} G_{\text{cs}}^{dd} + G_{\text{cs}}^{qq} - \text{j}G_{\text{cs}}^{dq} + \text{j}G_{\text{cs}}^{qd} & G_{\text{cs}}^{dd} - G_{\text{cs}}^{qq} + \text{j}G_{\text{cs}}^{dq} + \text{j}G_{\text{cs}}^{qd} \\ G_{\text{cs}}^{dd} - G_{\text{cs}}^{qq} - \text{j}G_{\text{cs}}^{dq} - \text{j}G_{\text{cs}}^{qd} & G_{\text{cs}}^{dd} + G_{\text{cs}}^{qq} + \text{j}G_{\text{cs}}^{dq} - \text{j}G_{\text{cs}}^{qd} \end{bmatrix} \tag{7.50}$$

For a balanced input signal, such as positive frequency component or negative frequency component, Equation 7.48 is rewritten as follows:

$$\begin{bmatrix} \tilde{\boldsymbol{v}}_{\text{con}}^{\text{p}} \\ \tilde{\boldsymbol{v}}_{\text{con}}^{\text{n}} \end{bmatrix} = \begin{bmatrix} G_{\text{cs}}^{\text{pp}} & G_{\text{cs}}^{\text{pn}} \\ G_{\text{cs}}^{\text{np}} & G_{\text{cs}}^{\text{nn}} \end{bmatrix} \begin{bmatrix} \tilde{\boldsymbol{x}}^{\text{p}} \\ 0 \end{bmatrix} = \begin{bmatrix} G_{\text{cs}}^{\text{pp}} \tilde{\boldsymbol{x}}^{\text{p}} \\ G_{\text{cs}}^{\text{np}} \tilde{\boldsymbol{x}}^{\text{p}} \end{bmatrix} \tag{7.51a}$$

$$\begin{bmatrix} \tilde{\boldsymbol{v}}_{\text{con}}^{\text{p}} \\ \tilde{\boldsymbol{v}}_{\text{con}}^{\text{n}} \end{bmatrix} = \begin{bmatrix} G_{\text{cs}}^{\text{pp}} & G_{\text{cs}}^{\text{pn}} \\ G_{\text{cs}}^{\text{np}} & G_{\text{cs}}^{\text{nn}} \end{bmatrix} \begin{bmatrix} 0 \\ \tilde{\boldsymbol{x}}^{\text{n}} \end{bmatrix} = \begin{bmatrix} G_{\text{cs}}^{\text{pn}} \tilde{\boldsymbol{x}}^{\text{n}} \\ G_{\text{cs}}^{\text{nn}} \tilde{\boldsymbol{x}}^{\text{n}} \end{bmatrix} \tag{7.51b}$$

For a balanced control system, it is required that if the input is a positive frequency component, the output signal must also be a positive frequency component, and the same applies to the negative frequency component. Therefore, based on Equation 7.51, a balanced control system requires both $G_{\text{cs}}^{\text{pn}} = 0$ in Equation 7.51a and $G_{\text{cs}}^{\text{np}} = 0$ in Equation 7.51b.

Otherwise, the control system is unbalanced, as it will transfer a balanced input signal into both PNF components in the output.

Based on the above analysis, the balanced and unbalanced system is quantified as:

- if the control system is balanced, it requires $G_{\text{cs}}^{\text{pn}} = 0$ and $G_{\text{cs}}^{\text{np}} = 0$, which further implies $G_{\text{cs}}^{dd} = G_{\text{cs}}^{qq}$ and $G_{\text{cs}}^{dq} = -G_{\text{cs}}^{qd}$;
- if the control system is unbalanced, it requires $G_{\text{cs}}^{\text{pn}} \neq 0$ or $G_{\text{cs}}^{\text{np}} \neq 0$, which further implies $G_{\text{cs}}^{dd} \neq G_{\text{cs}}^{qq}$ or $G_{\text{cs}}^{dq} \neq -G_{\text{cs}}^{qd}$.

A case study using a converter system, as shown in Figure 7.7 with parameters listed in Table 4.1 with $k_{\text{i}} = 76.4$ $k_{\text{p}} = 0.377$, is provided to demonstrate how balanced and unbalanced control systems interact with an oscillation. Only current control is applied to the converter, leading to the following relations:

$$\begin{bmatrix} \tilde{v}_{\text{con}}^{\text{c},d} \\ \\ \tilde{v}_{\text{con}}^{\text{c},q} \end{bmatrix} = \underbrace{\begin{bmatrix} -\left(k_{\text{p}}^{\text{cur}} + \dfrac{k_{\text{i}}^{\text{cur}}}{s}\right) & -L_{\text{f}}\omega_0 \\ L_{\text{f}}\omega_0 & -\left(k_{\text{p}}^{\text{cur}} + \dfrac{k_{\text{i}}^{\text{cur}}}{s}\right) \end{bmatrix}}_{G_{\text{cs}}} \begin{bmatrix} \tilde{i}_{\text{g}}^{\text{c},d} \\ \\ \tilde{i}_{\text{g}}^{\text{c},q} \end{bmatrix} \tag{7.52}$$

$\boldsymbol{G}_{\text{cs}}$ is a balanced control system based on the definition of a balanced control system. Three cases are developed as:

Case 1: a balanced control system, keeping $\boldsymbol{G}_{\text{cs}}$;
Case 2: an unbalanced control system with $G_{\text{cs}}^{dd} \neq G_{\text{cs}}^{qq}$, where G_{cs}^{dd} is set to $G_{\text{cs}+1}^{dd}$;
Case 3: an unbalanced control system with $G_{\text{cs}}^{dq} \neq -G_{\text{cs}}^{qd}$, where G_{cs}^{dq} is set to $G_{\text{cs}+1}^{dq}$.

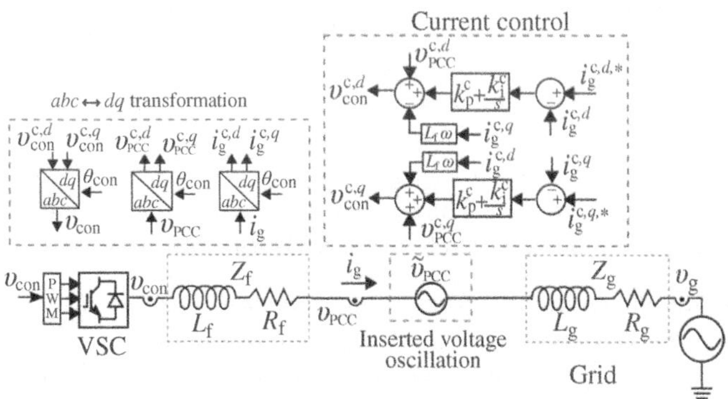

Figure 7.7 Current control of a converter.

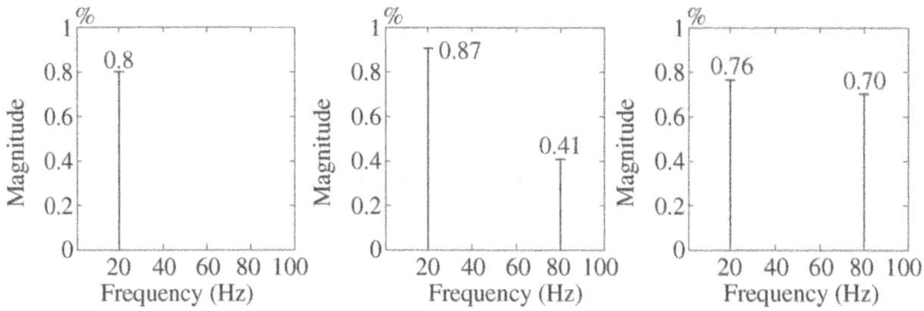

(a) Case 1 (b) Case 2 (c) Case 3

Figure 7.8 Examples of oscillation frequency distribution, in a mirror-frequency manner, for balanced and unbalanced control systems.

A balanced oscillation of $\widetilde{\boldsymbol{v}}_{\text{PCC}}(=0.01[1;-\text{j}]$ p.u. in the $\alpha\beta$-frame at 20 Hz) is injected in series at the point of common coupling (PCC). The dq-frame oscillation frequency is $\omega_c = 30$ Hz. If a mirror-frequency oscillation occurs, the oscillation frequencies in the power system will be $\pm\omega_c + 50 = 20, 80$ Hz. These three cases are applied to the converter system, and the Fourier analysis results of v_{con} are shown in Figure 7.8.

In Case 1, a balanced control system is used, resulting in a 20 Hz oscillation in v_{con}. However, when the other two unbalanced cases are applied, mirror-frequency oscillations occur, as shown in Figure 7.8b, c. This case study validates the definition of an unbalanced control system and demonstrates how it leads to mirror-frequency oscillations.

7.5 Why Is Converter Control an Unbalanced System?

A balanced control system in a dq-frame requires a symmetrical control structure and identical control parameters, such as current control, shown in Figure 7.9, which has been explained in Section 7.4. Conversely, in a converter control system, phase alignment is essential, using methods such as phase-locked loop (PLL) or P–ω droop control, to synchronize with the grid. These phase alignment controls are unbalanced, which leads to an overall unbalanced performance of the converter system. As a result, the converter control transfers a balanced oscillation into mirror-frequency oscillations within the power system.

To illustrate the difference between balanced control and unbalanced control, typical cases are presented in Figure 7.9, and summarized as below:

- balanced control: taking current control as an example, it is a typical symmetrical control structure in the dq-frame. If the control parameters of the PI controllers in both the d- and q-axes are identical, the system satisfies the definition of balanced control;
- unbalanced control:

1. a symmetrical control structure where the control parameters are not identical;
2. an asymmetrical control structure, such as *PV* control, PLL, and *P*–*ω* droop control, as shown in Figure 7.9.

The PLL or *P*–*ω* droop control generates the phase θ for *dq*–*abc* and *abc*–*dq* transformations, which are fundamental for *dq*-frame converter control. Therefore, using the PLL and *P*–*ω* droop control as examples demonstrates how to identify unbalanced control.

The model for *dq*–*abc* transformation, incorporating the dynamics of PLL or *P*–*ω* droop control, is developed in Section 4.3. For instance, the *dq*–*abc* transformation of $\tilde{\boldsymbol{v}}_{\mathrm{con}}$ is considered, as shown in Equation 4.31 for the PLL and Equation 4.39 for *P*–*ω* droop control, both of which are re-presented below in a bold matrix form:

PLL:

$$\tilde{\boldsymbol{v}}_{\mathrm{con}} = -\boldsymbol{G}^{v}_{\mathrm{con}}\tilde{\boldsymbol{v}}_{\mathrm{PCC}} + \tilde{\boldsymbol{v}}^{\mathrm{c}}_{\mathrm{con}} \tag{7.53}$$

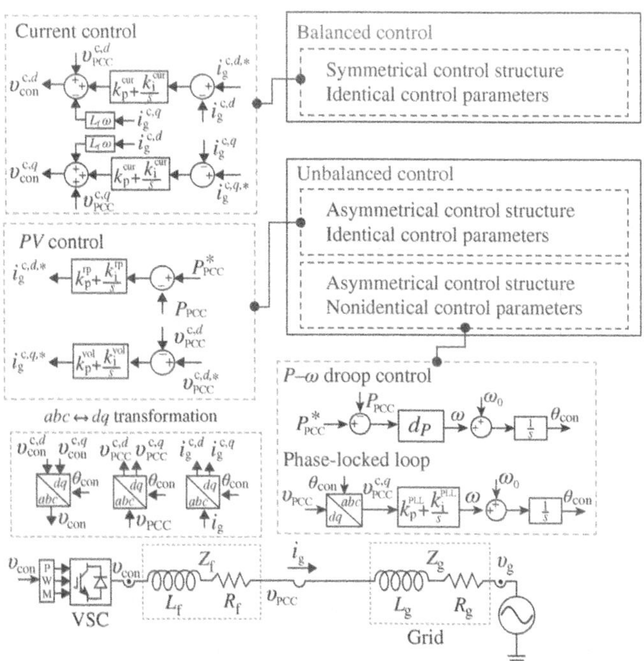

Figure 7.9 Characteristics of Unbalanced and Balanced Control Systems.

P–ω droop control:

$$\tilde{\boldsymbol{v}}_{\text{con}} = -\boldsymbol{H}^{i}_{\text{con}}\tilde{\boldsymbol{i}}_{\text{g}} - \boldsymbol{H}^{v}_{\text{con}}\tilde{\boldsymbol{v}}_{\text{PCC}} + \tilde{\boldsymbol{v}}^{\text{c}}_{\text{con}} \tag{7.54}$$

Based on the definition of an unbalanced control system in Section 7.4, it was found that:

- the PLL is synchronized with the grid frequency via v^{q}_{PCC}, which does not represent a symmetrical design. Its dynamics, modeled by $\boldsymbol{G}^{v}_{\text{con}}$ in the *dq*–*abc* transformation, fall into the definition of unbalanced control where $G^{dd}_{\text{con}} \neq G^{qq}_{\text{con}}$ or $G^{dq}_{\text{con}} \neq -G^{qd}_{\text{con}}$;
- P–ω droop control is synchronized with the grid frequency via P, which also does not represent a symmetrical design. Its dynamics, modeled by $\boldsymbol{H}^{i}_{\text{con}}$ and $\boldsymbol{H}^{v}_{\text{con}}$ in the *dq*–*abc* transformation, fall into the definition of unbalanced control where $H^{dd}_{\text{con}} \neq H^{qq}_{\text{con}}$ or $H^{dq}_{\text{con}} \neq -H^{qd}_{\text{con}}$.

It is concluded that any converter control based on PLL or P–ω control is always unbalanced and will inevitably lead to an MFE.

8

Impedance Measurement Techniques for Power System Stability Analysis

Jun Liang[1,*] *and Chuanyue Li*[2]

[1] *School of Engineering, Cardiff University, Cardiff, UK*
[2] *School of Electric Power Engineering, South China University of Technology, Guangzhou, China*

*Corresponding author: LiangJ1@cardiff.ac.uk

8.1 Stability Analysis Method Based on Measurement

In a black-box system, the system's parameters are unknown, making it impossible to build a small-signal model for stability analysis using transfer function-based, state-space, or impedance stability analysis methods. However, measurements taken from outside the black-box system provide a way to form the small-signal model of the system externally. Since the impedance stability criterion is well established, measuring impedance is a widely adopted method, which is detailedly described in the following sections.

8.1.1 Mechanism of Impedance Measurement

As described in Section 5.2.1 and Figure 8.1, a converter-integrated power system can be represented as a Thevenin equivalent circuit, forming the following relationships:

$$\tilde{\boldsymbol{v}}_{\mathrm{con,s}} - \tilde{\boldsymbol{v}}_{\mathrm{PCC}} = \boldsymbol{Z}_{\mathrm{con}}\tilde{\boldsymbol{i}}_{\mathrm{g}} \tag{8.1a}$$

$$\tilde{\boldsymbol{v}}_{\mathrm{PCC}} - \tilde{\boldsymbol{v}}_{\mathrm{g}} = \boldsymbol{Z}_{\mathrm{g}}\tilde{\boldsymbol{i}}_{\mathrm{g}} \tag{8.1b}$$

Here $\boldsymbol{v}_{\mathrm{con,s}}$ and $\boldsymbol{v}_{\mathrm{g}}$ are normally the control references for the system; for example, $\tilde{\boldsymbol{v}}_{\mathrm{con,s}} = (\boldsymbol{G}^{v}_{\mathrm{col}} - \boldsymbol{I})^{-1}\boldsymbol{G}_{\mathrm{ref}}\tilde{\boldsymbol{u}}$, as described by Equation 5.21. If these control references

Stability Analysis of Converter-Rich Power Grids, First Edition. Jun Liang and Chuanyue Li.
© 2026 The Institute of Electrical and Electronics Engineers, Inc. Published 2026 by John Wiley & Sons, Inc. All rights reserved.

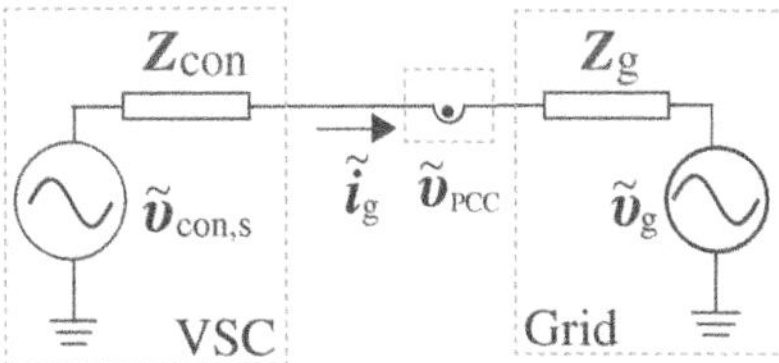

Figure 8.1 The Thevenin equivalent circuit of a converter system.

are kept constant, this leads to $\tilde{\boldsymbol{v}}_{\text{con,s}} = 0$ and $\tilde{\boldsymbol{v}}_{\text{g}} = 0$. Therefore, Equation 8.1 can be rewritten as:

$$-\tilde{\boldsymbol{v}}_{\text{PCC}} = \boldsymbol{Z}_{\text{con}}\tilde{\boldsymbol{i}}_{\text{g}} \tag{8.2a}$$

$$\tilde{\boldsymbol{v}}_{\text{PCC}} = \boldsymbol{Z}_{\text{g}}\tilde{\boldsymbol{i}}_{\text{g}} \tag{8.2b}$$

Further rearranging yields:

$$-\tilde{\boldsymbol{v}}_{\text{PCC}}\tilde{\boldsymbol{i}}_{\text{g}}^{-1} = \boldsymbol{Z}_{\text{con}} \tag{8.3a}$$

$$\tilde{\boldsymbol{v}}_{\text{PCC}}\tilde{\boldsymbol{i}}_{\text{g}}^{-1} = \boldsymbol{Z}_{\text{g}} \tag{8.3b}$$

Equation 8.3 describes the mechanism of impedance measurement, as follows: *When a small disturbance occurs in a system, if the terminal small-disturbance voltage and current at the converter or the grid are measured, such as $\tilde{\boldsymbol{v}}_{\text{PCC}}$ and $\tilde{\boldsymbol{i}}_{\text{g}}$, then the corresponding impedance $\boldsymbol{Z}_{\text{con}}$ or $\boldsymbol{Z}_{\text{g}}$ can be calculated using Equation 8.3.*

The requirements of an impedance measurement are as follows:

1. the voltage source of a Thevenin equivalent circuit, such as $\boldsymbol{v}_{\text{con,s}}$ and $\boldsymbol{v}_{\text{g}}$, remains unchanged during the disturbance;
2. the disturbance is small enough that does not alter system states, such as changing the current or voltage level, as the system's small-signal impedance includes the steady-state values; for example, the linearization of the *abc–dq* transformation in voltage source converter (VSC) control, as described in Equation 4.19, where steady-state values $[v_{\text{con}}^{d}; v_{\text{con}}^{q}]$ are included.

8.2 Impedance Measurement Methods

8.2.1 DC Impedance Measurement

To clearly demonstrate how to measure impedance, a simple direct current (DC) impedance measurement is presented in this section. In the case of a DC system, to induce a controllable small disturbance, a series voltage source or parallel current source can be applied to inject a small-disturbance voltage $\tilde{v}_{\text{inj},n}$ or current $\tilde{i}_{\text{inj},n}$, as shown in Figure 8.2. The subscript n indicates that the injection is at a certain frequency $\omega_n \in [0, \infty]$.

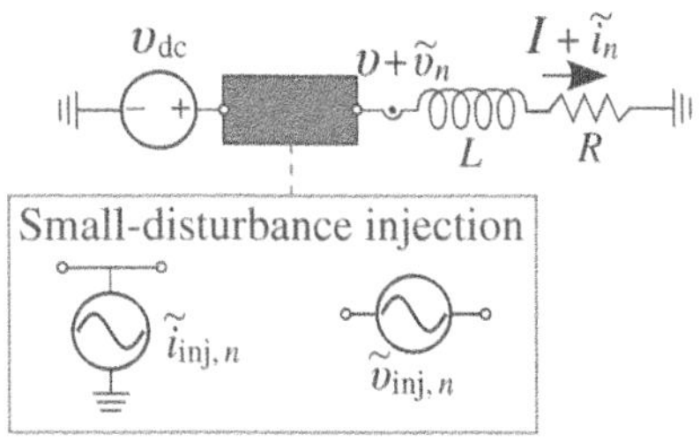

Figure 8.2 DC impedance measurement.

The transfer function of a DC impedance is presented as below:

$$\widetilde{v} = (Ls + R)\widetilde{i} \tag{8.4}$$

Equation 8.5 can also be expressed in the discrete frequency domain:

$$\underbrace{\widetilde{v}_{mag}e^{(\omega_n t+\theta_1)}}_{\widetilde{v}_n} = (L\omega_n j + R)\underbrace{\widetilde{i}_{mag}e^{(\omega_n t+\theta_2)}}_{\widetilde{i}_n} \tag{8.5}$$

where $\omega_n \in [0, \infty]$ and $\widetilde{v}_{mag}$ and $\widetilde{i}_{mag}$ are the magnitudes.

If a small disturbance is applied at ω_1, $V + \widetilde{v}_1$ and $I + \widetilde{i}_1$ are measured by sensors. By applying a Fourier transform at ω_1, Equation 8.5 can be expressed as:

$$\widetilde{v}_{mag}e^{(\theta_1)} = (L\omega_1 j + R)\widetilde{i}_{mag}e^{(\theta_2)} \tag{8.6}$$

Therefore, the impedance at ω_1 can be calculated as:

$$\frac{\widetilde{v}_{mag}e^{\theta_1}}{\widetilde{i}_{mag}e^{\theta_2}} = (L\omega_1 j + R) \tag{8.7}$$

If the disturbances are applied N times with $\omega_n \in [0, \infty]$, the full impedance can be measured. An example is shown in Figure 8.3. The impedance $Ls + R$ is presented in a Bode plot, and the measured impedances at $\omega_n = [10\ 30\ 70] \times 2\pi$ are matched to $Ls + R$ accurately.

8.2.2 AC System Impedance Measurement

To transform the alternating current (AC) system into a linear time-invariant system, the *abc–dq* transformation is adopted, and the corresponding impedance is represented as a 2×2 dq impedance matrix. The system equivalent impedances, as illustrated in Figure 8.4, $\boldsymbol{Z}_{con}$ and $\boldsymbol{Z}_g$, consist of four unknowns: Z^{dd}, Z^{dq}, Z^{qd}, and Z^{qq}. Compared to DC impedance measurement, the method remains the same, but instead of measuring a single impedance value, the 2×2 impedance matrix is measured. To perform AC impedance measurement, a parallel-connected current

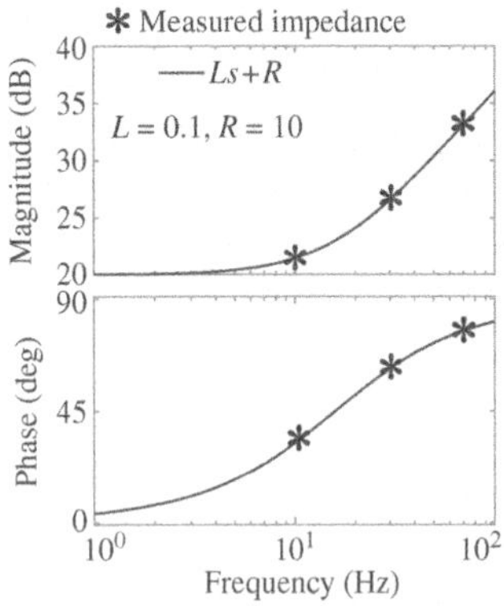

Figure 8.3 Example of impedance measurements for $Ls + R$.

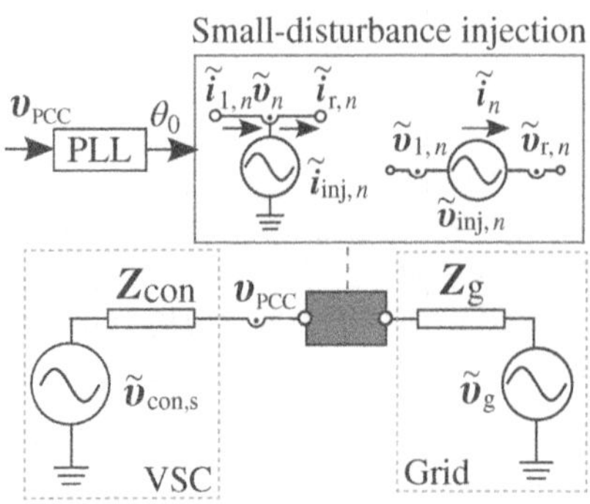

Figure 8.4 AC impedance measurement.

source or a series-connected voltage source is introduced to create a controllable small disturbance at desired frequencies. Measurements are taken over the full frequency range [0, ∞] Hz, and at least two measurements are required at each frequency ω_n to calculate the impedance matrix. The density of ω_n can vary depending on the trade-off between measurement efficiency and the desired precision of the impedance.

For the current-injection method, the measurement uses a parallel current source to measure the impedance at frequency ω_n, and two current disturbances are injected, respectively, as follows:

$$\begin{aligned} \tilde{\boldsymbol{i}}_{\text{inj},n,1} &= [\tilde{i}^d_{\text{inj},n,1}; \tilde{i}^q_{\text{inj},n,1}] \\ \tilde{\boldsymbol{i}}_{\text{inj},n,2} &= [\tilde{i}^d_{\text{inj},n,2}; \tilde{i}^q_{\text{inj},n,2}] \end{aligned} \tag{8.8}$$

where $\tilde{\boldsymbol{i}}_{\text{inj},n,1}$ and $\tilde{\boldsymbol{i}}_{\text{inj},n,2}$ are linearly independent.

The corresponding measurement includes two ACs on both sides of the connection point and the voltage at the connection point, as follows:

$$\begin{aligned} \tilde{\boldsymbol{i}}_{\mathrm{l},n,1} &= [\tilde{i}^d_{\mathrm{l},n,1}; \tilde{i}^q_{\mathrm{l},n,1}] & \tilde{\boldsymbol{i}}_{\mathrm{l},n,2} &= [\tilde{i}^d_{\mathrm{l},n,2}; \tilde{i}^q_{\mathrm{l},n,2}] \\ \tilde{\boldsymbol{i}}_{\mathrm{r},n,1} &= [\tilde{i}^d_{\mathrm{r},n,1}; \tilde{i}^q_{\mathrm{r},n,1}] & \tilde{\boldsymbol{i}}_{\mathrm{r},n,2} &= [\tilde{i}^d_{\mathrm{r},n,2}; \tilde{i}^q_{\mathrm{r},n,2}] \\ \tilde{\boldsymbol{v}}_{n,1} &= [\tilde{v}^d_{n,1}; \tilde{v}^q_{n,1}] & \tilde{\boldsymbol{v}}_{n,2} &= [\tilde{v}^d_{n,2}; \tilde{v}^q_{n,2}] \end{aligned} \tag{8.9}$$

Based on the mechanism of impedance measurement described in Section 8.1.1, $\boldsymbol{Z}_{\mathrm{con}}$ and $\boldsymbol{Z}_{\mathrm{g}}$ at ω_n can be calculated as follows:

$$\begin{aligned} \boldsymbol{Z}_{\mathrm{con},n} &= [-\tilde{\boldsymbol{v}}_{n,1} \; -\tilde{\boldsymbol{v}}_{n,2}][\tilde{\boldsymbol{i}}_{\mathrm{l},n,1} \; \tilde{\boldsymbol{i}}_{\mathrm{l},n,2}]^{-1} \\ \boldsymbol{Z}_{\mathrm{g},n} &= [\tilde{\boldsymbol{v}}_{n,1} \; \tilde{\boldsymbol{v}}_{n,2}][\tilde{\boldsymbol{i}}_{\mathrm{r},n,1} \; \tilde{\boldsymbol{i}}_{\mathrm{r},n,2}]^{-1} \end{aligned} \tag{8.10}$$

Expanding Equation 8.10 yields:

$$Z^{dd}_{\mathrm{con},n} = \frac{-\tilde{i}^q_{\mathrm{l},n,1}\tilde{v}^d_{n,2} + \tilde{i}^q_{\mathrm{l},n,2}\tilde{v}^d_{n,1}}{-\tilde{i}^d_{\mathrm{l},n,1}\tilde{i}^q_{\mathrm{l},n,2} + \tilde{i}^d_{\mathrm{l},n,2}\tilde{i}^q_{\mathrm{l},n,1}} \tag{8.11a}$$

$$Z^{dq}_{\mathrm{con},n} = \frac{\tilde{i}^d_{\mathrm{l},n,1}\tilde{v}^d_{n,2} - \tilde{i}^d_{\mathrm{l},n,2}\tilde{v}^d_{n,1}}{-\tilde{i}^d_{\mathrm{l},n,1}\tilde{i}^q_{\mathrm{l},n,2} + \tilde{i}^d_{\mathrm{l},n,2}\tilde{i}^q_{\mathrm{l},n,1}} \tag{8.11b}$$

$$Z^{qd}_{\mathrm{con},n} = \frac{-\tilde{i}^q_{\mathrm{l},n,1}\tilde{v}^q_{n,2} + \tilde{i}^q_{\mathrm{l},n,2}\tilde{v}^q_{n,1}}{-\tilde{i}^d_{\mathrm{l},n,1}\tilde{i}^q_{\mathrm{l},n,2} + \tilde{i}^d_{\mathrm{l},n,2}\tilde{i}^q_{\mathrm{l},n,1}} \tag{8.11c}$$

$$Z^{qq}_{\mathrm{con},n} = \frac{\tilde{i}^d_{\mathrm{l},n,1}\tilde{v}^q_{n,2} - \tilde{i}^d_{\mathrm{l},n,2}\tilde{v}^q_{n,1}}{-\tilde{i}^d_{\mathrm{l},n,1}\tilde{i}^q_{\mathrm{l},n,2} + \tilde{i}^d_{\mathrm{l},n,2}\tilde{i}^q_{\mathrm{l},n,1}} \tag{8.11d}$$

$$Z^{dd}_{\mathrm{g},n} = \frac{-\tilde{i}^q_{\mathrm{r},n,1}\tilde{v}^d_{n,2} + \tilde{i}^q_{\mathrm{r},n,2}\tilde{v}^d_{n,1}}{\tilde{i}^d_{\mathrm{r},n,1}\tilde{i}^q_{\mathrm{r},n,2} - \tilde{i}^d_{\mathrm{r},n,2}\tilde{i}^q_{\mathrm{r},n,1}} \tag{8.12a}$$

$$Z^{dq}_{\mathrm{g},n} = \frac{\tilde{i}^d_{\mathrm{r},n,1}\tilde{v}^d_{n,2} - \tilde{i}^d_{\mathrm{r},n,2}\tilde{v}^d_{n,1}}{\tilde{i}^d_{\mathrm{r},n,1}\tilde{i}^q_{\mathrm{r},n,2} - \tilde{i}^d_{\mathrm{r},n,2}\tilde{i}^q_{\mathrm{r},n,1}} \tag{8.12b}$$

$$Z^{qd}_{\mathrm{g},n} = \frac{-\tilde{i}^q_{\mathrm{r},n,1}\tilde{v}^q_{n,2} + \tilde{i}^q_{\mathrm{r},n,2}\tilde{v}^q_{n,1}}{\tilde{i}^d_{\mathrm{r},n,1}\tilde{i}^q_{\mathrm{r},n,2} - \tilde{i}^d_{\mathrm{r},n,2}\tilde{i}^q_{\mathrm{r},n,1}} \tag{8.12c}$$

$$Z^{qq}_{\mathrm{g},n} = \frac{\tilde{i}^d_{\mathrm{r},n,1}\tilde{v}^q_{n,2} - \tilde{i}^d_{\mathrm{r},n,2}\tilde{v}^q_{n,1}}{\tilde{i}^d_{\mathrm{r},n,1}\tilde{i}^q_{\mathrm{r},n,2} - \tilde{i}^d_{\mathrm{r},n,2}\tilde{i}^q_{\mathrm{r},n,1}} \tag{8.12d}$$

For the voltage-injection method, the measurement uses a series voltage source to measure the impedance at frequency ω_n, and two voltage disturbances are injected, respectively, as below:

$$\begin{aligned} \tilde{\boldsymbol{v}}_{\mathrm{inj},n,1} &= [\tilde{v}^d_{\mathrm{inj},n,1}; \tilde{v}^q_{\mathrm{inj},n,1}] \\ \tilde{\boldsymbol{v}}_{\mathrm{inj},n,2} &= [\tilde{v}^d_{\mathrm{inj},n,2}; \tilde{v}^q_{\mathrm{inj},n,2}] \end{aligned} \tag{8.13}$$

where $\tilde{\boldsymbol{v}}_{\mathrm{inj},n,1}$ and $\tilde{\boldsymbol{v}}_{\mathrm{inj},n,2}$ are linearly independent.

AC voltages on both sides of the voltage source and the current are measured as following forms:

$$\begin{aligned}
\widetilde{\boldsymbol{v}}_{\mathrm{l},n,1} &= [\widetilde{v}^{d}_{\mathrm{l},n,1}; \widetilde{v}^{q}_{\mathrm{l},n,1}] & \widetilde{\boldsymbol{v}}_{\mathrm{l},n,2} &= [\widetilde{v}^{d}_{\mathrm{l},n,2}; \widetilde{v}^{q}_{\mathrm{l},n,2}] \\
\widetilde{\boldsymbol{v}}_{\mathrm{r},n,1} &= [\widetilde{v}^{d}_{\mathrm{r},n,1}; \widetilde{v}^{q}_{\mathrm{r},n,1}] & \widetilde{\boldsymbol{v}}_{\mathrm{r},n,2} &= [\widetilde{v}^{d}_{\mathrm{r},n,2}; \widetilde{v}^{q}_{\mathrm{r},n,2}] \\
\widetilde{\boldsymbol{i}}_{n,1} &= [\widetilde{i}^{d}_{n,1}; \widetilde{i}^{q}_{n,1}] & \widetilde{\boldsymbol{i}}_{n,2} &= [\widetilde{i}^{d}_{n,2}; \widetilde{i}^{q}_{n,2}]
\end{aligned} \tag{8.14}$$

Accordingly, $\boldsymbol{Z}_{\mathrm{con}}$ and $\boldsymbol{Z}_{\mathrm{g}}$ at ω_n can be calculated as:

$$\begin{aligned}
\boldsymbol{Z}_{\mathrm{con},n} &= [-\widetilde{\boldsymbol{v}}_{\mathrm{l},n,1} \; -\widetilde{\boldsymbol{v}}_{\mathrm{l},n,2}][\widetilde{\boldsymbol{i}}_{n,1} \; \widetilde{\boldsymbol{i}}_{n,2}]^{-1} \\
\boldsymbol{Z}_{\mathrm{g},n} &= [\widetilde{\boldsymbol{v}}_{\mathrm{r},n,1} \; \widetilde{\boldsymbol{v}}_{\mathrm{r},n,2}][\widetilde{\boldsymbol{i}}_{n,1} \; \widetilde{\boldsymbol{i}}_{n,2}]^{-1}
\end{aligned} \tag{8.15}$$

After extensive impedance measurement for $\omega_n \in [0, \infty]$, the entire system impedance can be measured using either the current-injection method or the voltage-injection method, based on Equations 8.10 and 8.15.

The requirements of the AC impedance measurement are given below:

- *the sources in Thevenin or Norton equivalent circuits, such as* $\mathbf{v}_{\mathrm{con,s}}$ *and* $\mathbf{v}_{\mathrm{g}}$*, remain unchanged during the disturbance;*
- *the injected disturbance cannot alter system states, as the system's small-signal impedance includes the steady-state values;*
- *during the measurement, any control of the injected devices, such as the parallel current source or series voltage source, should be abandoned to avoid impacting the measurement accuracy.*

A case study, based on a phase-locked loop (PLL)-based converter system as described in Section 4.4.3, is adopted to demonstrate the impedance measurement. It is the same case study that was used for impedance stability analysis in Section 5.2.3.1. The PLL is tuned based on $\omega^{n}_{\mathrm{PLL}} = 180 \times 2\pi$. The derived converter impedance $\boldsymbol{Z}_{\mathrm{con}}$ is represented as a Bode plot, shown as the solid line in Figure 8.5. The impedance measurement is carried out at $\omega_n = [10\ 150\ 300] \times 2\pi$. It is validated that the proposed impedance measurement method can accurately represent the system impedance.

8.3 Impact of Noise on Measurement Accuracy and Elimination Methods

In a power system, noise oscillations are always present, which disrupt impedance measurements and reduce accuracy. This issue occurs because impedance measurement relies on the assumption that no sources other than the injection device are present at the measurement frequency. However, when noise acts as a source of oscillation at this same frequency, it leads to reduced measurement accuracy.

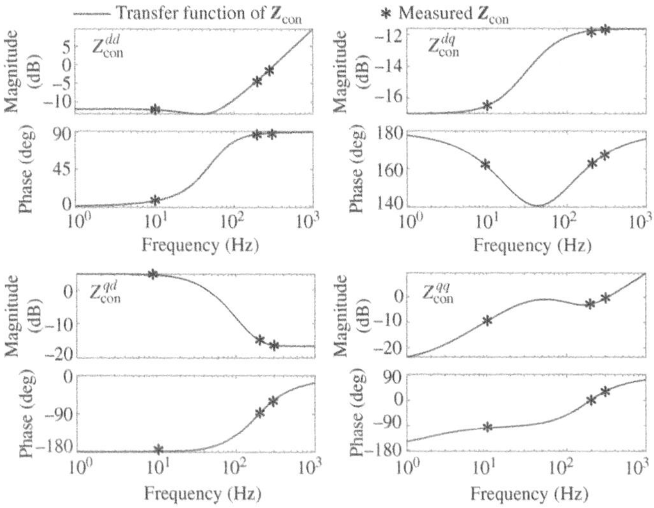

Figure 8.5 Measured converter impedances.

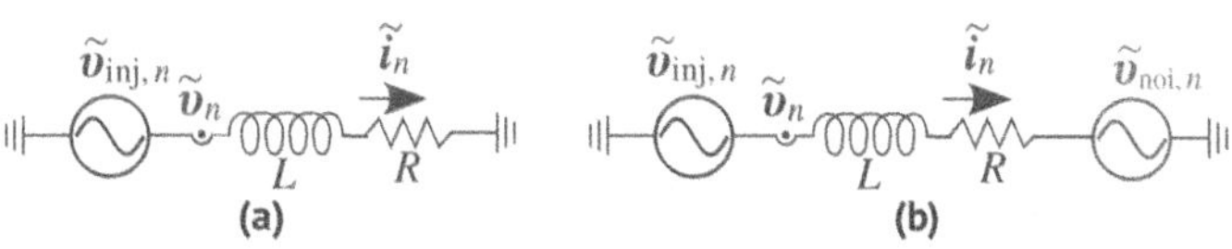

Figure 8.6 Impact of noise on the impedance measurements.

8.3.1 Impact of Noise on DC Impedance Measurements and Elimination Methods

As shown in Figure 8.6a, which illustrates a typical impedance measurement process, a small voltage is injected at frequency ω_n. With no other noise present at this frequency, the impedance $Z(\omega_n \mathrm{j})$ then is calculated using the measurement below:

$$\frac{\tilde{v}_n}{\tilde{i}_n} = Z(\omega_n \mathrm{j}) \tag{8.16}$$

where $Z(\omega_n \mathrm{j}) = \omega_n \mathrm{j} L + R$.

However, if noise exists at the same frequency and is treated as a voltage source, $\tilde{v}_n^{\mathrm{noi}}$, then Equation 8.16 can be represented as:

$$\frac{\tilde{v}_n}{\tilde{i}_n} = \frac{\tilde{v}_{\mathrm{noi},n} + Z(\omega_n \mathrm{j})\tilde{i}_n}{\tilde{i}_n} = Z(\omega_n \mathrm{j}) + \underbrace{\frac{\tilde{v}_{\mathrm{noi},n}}{\tilde{i}_n}}_{\text{error}} \tag{8.17}$$

If the impedance calculation method in Equation 8.16 is followed, the measured impedance will contain an error, as indicated in Equation 8.17.

To mitigate the impact of noise, an additional measurement is proposed to help eliminate its effects, which is named as noise-elimination measurement, as presented below:

$$\begin{aligned} \widetilde{v}_{n,1} &= Z\widetilde{i}_{n,1} + \widetilde{v}_{\mathrm{noi},n} \\ \widetilde{v}_{n,2} &= Z\widetilde{i}_{n,2} + \widetilde{v}_{\mathrm{noi},n} \end{aligned} \tag{8.18}$$

where the subscripts 1 and 2 indicate the first and second measurements, respectively.

Therefore, the noise is eliminated, and the impedance can be calculated as:

$$Z = \frac{\widetilde{v}_{n,1} - \widetilde{v}_{n,2}}{\widetilde{i}_{n,1} - \widetilde{i}_{n,2}} \tag{8.19}$$

It should be noted that the two injections are not necessarily linearly independent.

In the second measurement, it is not essential to inject a small signal; measuring only the noise at the injection point is also effective. If further simplification is desired, noise can be identified via the fast Fourier transform (FFT) across a wide range of frequencies in a single measurement, eliminating the need for the second measurement at each frequency. This approach assumes that noise remains constant during each measurement.

The same impedance measurement case from Section 8.2.1 is used here to illustrate how noise affects measurement accuracy and how noise-elimination techniques can mitigate this impact. Three impedance values at [10 30 70] Hz are selected for the measurement.

A controllable voltage source with $0.01V_g[\cos(10 \times 2\pi t + \pi/3) + \cos(30 \times 2\pi t + \pi/3) + \cos(70 \times 2\pi t + \pi/3)]$ is added to the circuit to simulate same frequency noise during measurement. The measurement results are summarized in Table 8.1 and illustrated in Figure 8.7. It is found that noise introduces errors in each normal measurement. However, the noise-elimination measurement mitigates the noise impact, providing accurate impedance measurements.

Table 8.1 Measured DC impedances Ω.

Frequency	Z	Noise-elimination measurement	Normal measurement
10 Hz	10 + 5.96j	10 + 5.96j	9.48 + 8.45j
30 Hz	10 + 18.74j	10 + 18.74j	6.85 + 22.14j
70 Hz	10 + 43.98j	10 + 43.98j	1.65 + 49.13j

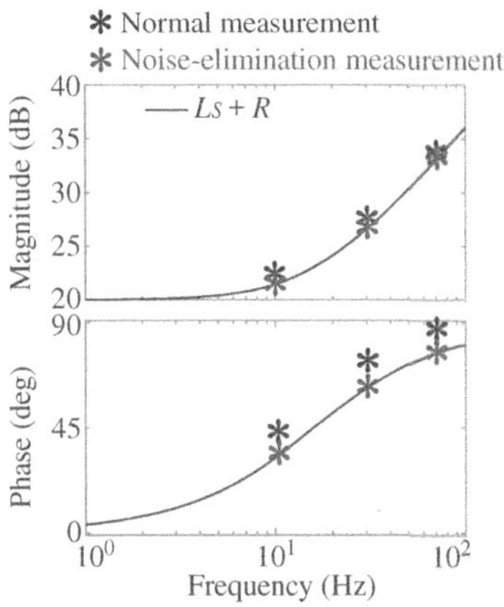

Figure 8.7 Comparison between noise-elimination measurement and normal measurement ($L = 0.1$, $R = 10$).

8.3.2 Impact of Noise on AC Impedance Measurements and Elimination Methods

A normal AC impedance measurement relies on the assumption that no other source operates at the frequency of injected voltage, which means that $\widetilde{\boldsymbol{v}}_{\text{con,s}} = 0$ and $\widetilde{\boldsymbol{v}}_{\text{g}} = 0$. Therefore, the impedance can be calculated as v/i mechanism at the injection point. If taking the grid side as an example, based on the normal impedance measurement described in Section 8.2.2 and Figure 8.4, the following relations are presented:

$$\begin{aligned} \widetilde{\boldsymbol{v}}_{n,1} &= \boldsymbol{Z}_{\text{g},n}\widetilde{\boldsymbol{i}}_{\text{r},n,1} \\ \widetilde{\boldsymbol{v}}_{n,2} &= \boldsymbol{Z}_{\text{g},n}\widetilde{\boldsymbol{i}}_{\text{r},n,2} \end{aligned} \tag{8.20}$$

If there is noise at the same frequency, which can be equivalently represented in the form shown in Figure 8.8, then Equation 8.20 is modified as follows:

$$\begin{aligned} \widetilde{\boldsymbol{v}}_{n,1} &= \boldsymbol{Z}_{\text{g},n}\widetilde{\boldsymbol{i}}_{\text{r},n,1} + \widetilde{\boldsymbol{v}}_{\text{noi},n} \\ \widetilde{\boldsymbol{v}}_{n,2} &= \boldsymbol{Z}_{\text{g},n}\widetilde{\boldsymbol{i}}_{\text{r},n,2} + \widetilde{\boldsymbol{v}}_{\text{noi},n} \end{aligned} \tag{8.21}$$

where, to align with the dq representation of Equation 8.20, $\widetilde{\boldsymbol{v}}_{\text{noi},n}$ is represented as the dq-values $[\widetilde{v}^{d}_{\text{noi},n}; \widetilde{v}^{q}_{\text{noi},n}]$.

If the impedance calculation still follows Equation 8.10, errors will arise due to $\widetilde{\boldsymbol{v}}_{\text{noi},n}$.

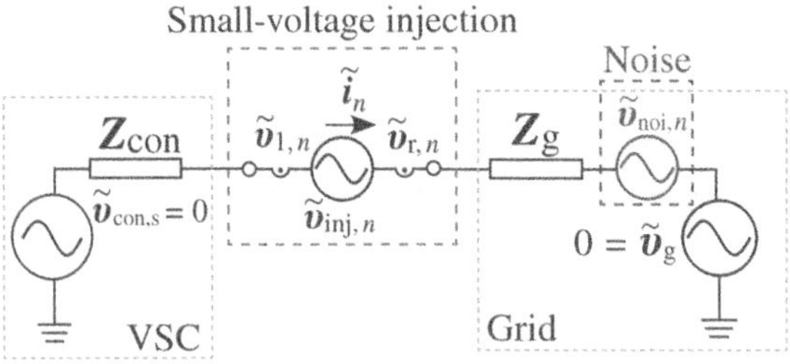

Figure 8.8 Equivalent circuit of noise impact on AC impedance measurement.

To avoid impedance errors, a noise-elimination method is suggested, which requires a third measurement at the measurement frequency, as represented below:

$$\tilde{\boldsymbol{v}}_{n,3} = \boldsymbol{Z}_{\text{g},n}\tilde{\boldsymbol{i}}_{\text{r},n,3} + \tilde{\boldsymbol{v}}_{\text{noi},n} \tag{8.22}$$

To eliminate the noise impact, subtracting Equation 8.22 from Equation 8.21 yields:

$$\begin{aligned} \tilde{\boldsymbol{v}}_{n,1} - \tilde{\boldsymbol{v}}_{n,3} &= \boldsymbol{Z}_{\text{g},n}(\tilde{\boldsymbol{i}}_{\text{r},n,1} - \tilde{\boldsymbol{i}}_{\text{r},n,3}) \\ \tilde{\boldsymbol{v}}_{n,2} - \tilde{\boldsymbol{v}}_{n,3} &= \boldsymbol{Z}_{\text{g},n}(\tilde{\boldsymbol{i}}_{\text{r},n,2} - \tilde{\boldsymbol{i}}_{\text{r},n,3}) \end{aligned} \tag{8.23}$$

Therefore, the noise-elimination impedance can be calculated as:

$$\boldsymbol{Z}_{\text{g},n} = [\tilde{\boldsymbol{v}}_{n,1} - \tilde{\boldsymbol{v}}_{n,3},\ \tilde{\boldsymbol{v}}_{n,2} - \tilde{\boldsymbol{v}}_{n,3}][\tilde{\boldsymbol{i}}_{\text{r},n,1} - \tilde{\boldsymbol{i}}_{\text{r},n,3},\ \tilde{\boldsymbol{i}}_{\text{r},n,2} - \tilde{\boldsymbol{i}}_{\text{r},n,3}]^{-1} \tag{8.24}$$

In the third measurement, it is not essential to inject a small signal; measuring only the noise at the injection point is also effective. If further simplification is desired, noise can be identified via FFT across a wide range of frequencies in a single measurement, eliminating the need for the third measurement at each frequency. This approach assumes that noise remains constant during each measurement.

The same impedance measurement case from Section 8.3.1 is used here to illustrate how noise affects measurement accuracy and how noise-elimination techniques can mitigate this impact. The DC system is changed to a three-phase system with same impedance $R + Ls$.

A three-phase controllable voltage source with $0.01V_{\text{g}}[\cos(10 \times 2\pi t + \pi/3) + \cos(30 \times 2\pi t + \pi/3) + \cos(70 \times 2\pi t + \pi/3)]$ is added to the circuit to simulate same-frequency noise during measurement. The measurement results are summarized in Table 8.2 and illustrated in Figure 8.9. It is found that noise introduces errors in each normal measurement. However, the noise-elimination measurement mitigates the noise impact, providing accurate impedance measurements.

Table 8.2 Measured AC impedances Ω in *dq*-form at 70 Hz.

Z_g^{dd}	Z_g^{dq}	Z_g^{qd}	Z_g^{qq}
Grid impedance			
10+43.98j	−31.41	31.41	10+43.98j
Noise-elimination measurement			
10+43.98j	−31.41	31.41	10+43.98j
Normal measurement			
−6.7+54.39j	−35.9-12.96j	41.82+16.71j	−2.96+48.47j

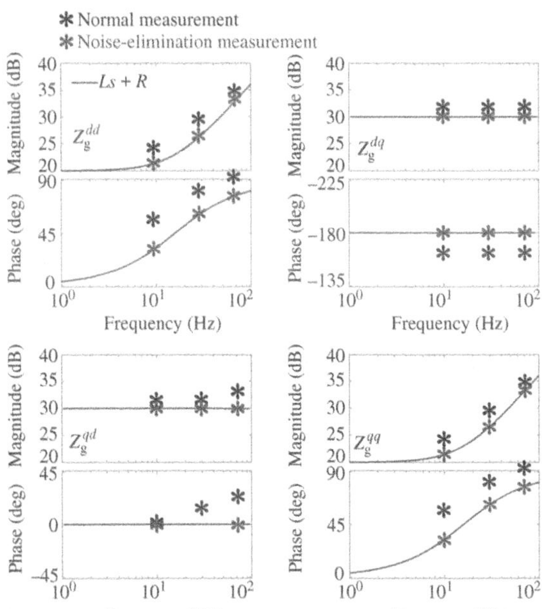

Figure 8.9 Comparison between noise-elimination AC impedance measurement and normal AC impedance measurement.

8.4 Impedance Measurement of *dq*-sequence Impedance or $\alpha\beta$ Impedance

Since a *dq*-signal can be transformed into *dq*-sequence signals, and subsequently into $\alpha\beta$-signals, the *dq*-sequence impedance and $\alpha\beta$ impedance are proposed, as introduced in Sections 5.2.4 and 5.2.5. It is shown in Section 5.2.5 that the *dq*-sequence impedance and $\alpha\beta$ impedance are nearly identical. This section primarily focuses on the measurement of *dq*-sequence impedance.

Based on the equivalent transformation of the above signals, the *dq*-sequence impedance can be measured from either *dq*-signals or $\alpha\beta$-signals. The transformation process is presented in Figure 8.10. Therefore, measurements in both the *dq*-frame and the $\alpha\beta$-frame are introduced, as shown in Figure 8.11. Basically, the injection requirements of both parallel current injection and series voltage injection can be maintained same. Same current or voltage injection method is applied as explained by Equations 8.8 and 8.13. Taking $\boldsymbol{Z}_{\mathrm{con}}^{\mathrm{pn}}$ measurement as an example as shown in Figure 8.12, the measurement process and transformation for both *dq*-frame and $\alpha\beta$-frame are explained below:

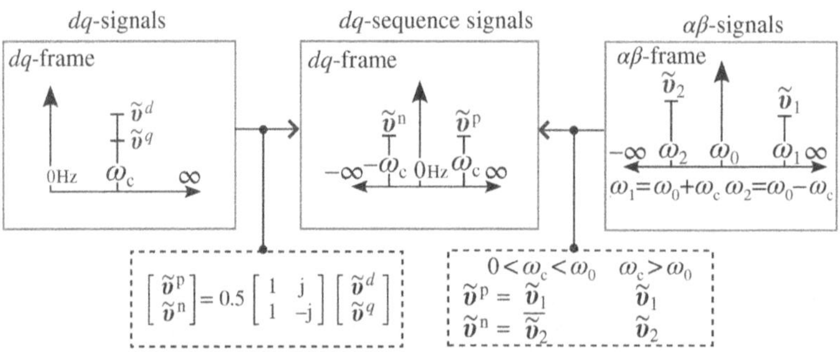

Figure 8.10 Transformation from *dq*-signals or $\alpha\beta$-signals to *dq*-sequence signals.

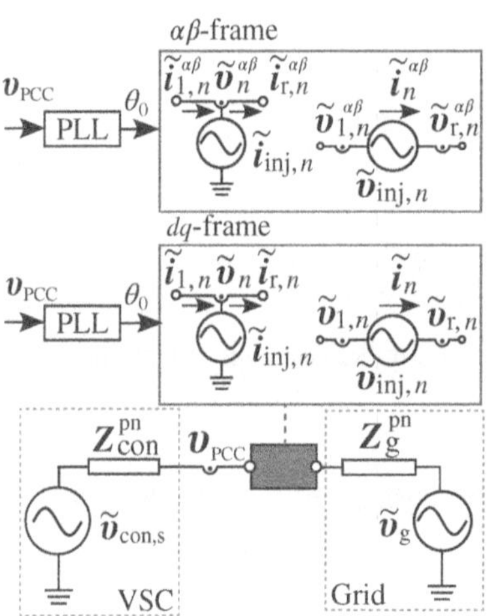

Figure 8.11 The *dq*-sequence impedance measurement via *dq*-signals or $\alpha\beta$-signals.

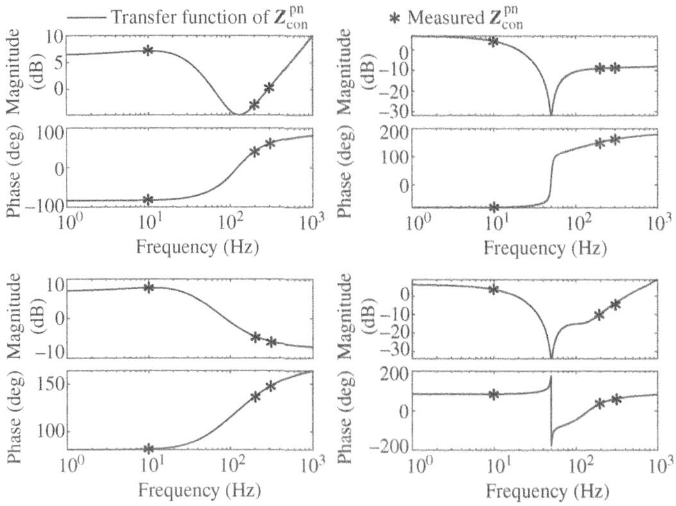

Figure 8.12 Measured *dq*-sequence impedances of a PLL-based converter.

- the *dq* measurement has been explained in detail in Section 8.2.2. For the measured *dq*-signals ($\tilde{\boldsymbol{i}}_{1,n,1}$, $\tilde{\boldsymbol{i}}_{1,n,2}$, $\tilde{\boldsymbol{v}}_{n,1}$, $\tilde{\boldsymbol{v}}_{n,2}$), *dq*-sequence signals ($\tilde{\boldsymbol{i}}^{pn}_{1,n,1}$, $\tilde{\boldsymbol{i}}^{pn}_{1,n,2}$, $\tilde{\boldsymbol{v}}^{pn}_{n,1}$, $\tilde{\boldsymbol{v}}^{pn}_{n,2}$) have been presented in Figure 8.10, or as below taking $\tilde{\boldsymbol{i}}_{1,n,1}$ as an example:

$$\begin{bmatrix} \tilde{i}^{p}_{1,n,1} \\ \tilde{i}^{n}_{1,n,1} \end{bmatrix} = 0.5 \begin{bmatrix} 1 & j \\ 1 & -j \end{bmatrix} \begin{bmatrix} \tilde{i}^{d}_{1,n,1} \\ \tilde{i}^{q}_{1,n,1} \end{bmatrix} \tag{8.25}$$

  where $\tilde{\boldsymbol{i}}_{1,n,1} = [\tilde{i}^{d}_{1,n,1}; \tilde{i}^{q}_{1,n,1}]$ and $\tilde{\boldsymbol{i}}^{pn}_{1,n,1} = [\tilde{i}^{p}_{1,n,1}; \tilde{i}^{n}_{1,n,1}]$.
- for the $\alpha\beta$ measurement, we need to measure the $\alpha\beta$-signals ($\tilde{\boldsymbol{i}}^{\alpha\beta}_{1,n,1}$, $\tilde{\boldsymbol{i}}^{\alpha\beta}_{1,n,2}$, $\tilde{\boldsymbol{v}}^{\alpha\beta}_{n,1}$, $\tilde{\boldsymbol{v}}^{\alpha\beta}_{n,2}$), and the *dq*-sequence signals ($\tilde{\boldsymbol{i}}^{pn}_{1,n,1}$, $\tilde{\boldsymbol{i}}^{pn}_{1,n,2}$, $\tilde{\boldsymbol{v}}^{pn}_{n,1}$, $\tilde{\boldsymbol{v}}^{pn}_{n,2}$) have been presented in Figure 8.10, or as below taking $\tilde{\boldsymbol{i}}^{pn}_{1,n,1}$ as an example:

$$\begin{array}{l|l} 0 < \omega_c < \omega_0 & \omega_c > \omega_0 \\ \tilde{i}^{p}_{1,n,1} = \tilde{i}_{1-1,n,1} & \tilde{i}^{p}_{1,n,1} = \tilde{i}_{1-1,n,1} \\ \tilde{i}^{n}_{1,n,1} = \bar{\tilde{i}}_{2-1,n,1} & \tilde{i}^{n}_{1,n,1} = \tilde{i}_{2-1,n,1} \end{array} \tag{8.26}$$

  with $\tilde{\boldsymbol{i}}^{\alpha\beta}_{1,n,1} = [\tilde{i}_{1-1,n,1}; \tilde{i}_{2-1,n,1}]$.

Therefore, the *dq*-sequence impedance can be calculated as:

$$\boldsymbol{Z}^{pn}_{con,n} = [-\tilde{\boldsymbol{v}}^{pn}_{n,1} \quad -\tilde{\boldsymbol{v}}^{pn}_{n,2}][\tilde{\boldsymbol{i}}^{pn}_{1,n,1} \quad \tilde{\boldsymbol{i}}^{pn}_{1,n,2}]^{-1} \tag{8.27}$$

9

Stability Analysis of Grid-following Converters

Jun Liang[1,*] *and Chuanyue Li*[2]

[1] *School of Engineering, Cardiff University, Cardiff, UK*
[2] *School of Electric Power Engineering, South China University of Technology, Guangzhou, China*

*Corresponding author: LiangJ1@cardiff.ac.uk

9.1 Instability Causes of PLL-based Control

9.1.1 Stability Performance Comparison Between Current Control and PLL

For a weak grid connection, the point of common coupling (PCC) voltage is not firm enough to be immunized against the control interaction. For a phase-locked loop (PLL)-based current control, as shown in Figure 9.1, where the outer loop is not applied, only PLL tuning is not enough to stabilize a weak grid connection well. An overall consideration of the control system is required. Both small-signal stability performance and time-domain performance are assessed and compared for current control and PLL. Within this performance comparison, their general tuning methods are applied, which are shown below.

The relation between current reference $\boldsymbol{i}_{\mathrm{con}}^{*}$ and converter output current $\boldsymbol{i}_{\mathrm{con}}$ is given below:

$$\boldsymbol{i}_{\mathrm{con}} = \frac{(k_{\mathrm{p}}^{\mathrm{c}} s + k_{\mathrm{i}}^{\mathrm{c}})/L_{\mathrm{f}}}{s^2 + \underbrace{[(R_{\mathrm{f}} + k_{\mathrm{p}}^{\mathrm{c}})/L_{\mathrm{f}}]}_{2\zeta^{\mathrm{c}}\omega_n^{\mathrm{c}}} s + \underbrace{k_{\mathrm{i}}^{\mathrm{c}}/L_{\mathrm{f}}}_{\omega_n^{\mathrm{c}\,2}}} \boldsymbol{i}_{\mathrm{con}}^{*} \tag{9.1}$$

where $\boldsymbol{i}$ in bold is a matrix that contains dq-components $[i_d; i_q]$ and $k_{\mathrm{p}}^{\mathrm{c}}$ and $k_{\mathrm{i}}^{\mathrm{c}}$ are the gains of the proportional–integral (PI) controller applied in the current control.

Stability Analysis of Converter-Rich Power Grids, First Edition. Jun Liang and Chuanyue Li.
© 2026 The Institute of Electrical and Electronics Engineers, Inc. Published 2026 by John Wiley & Sons, Inc. All rights reserved.

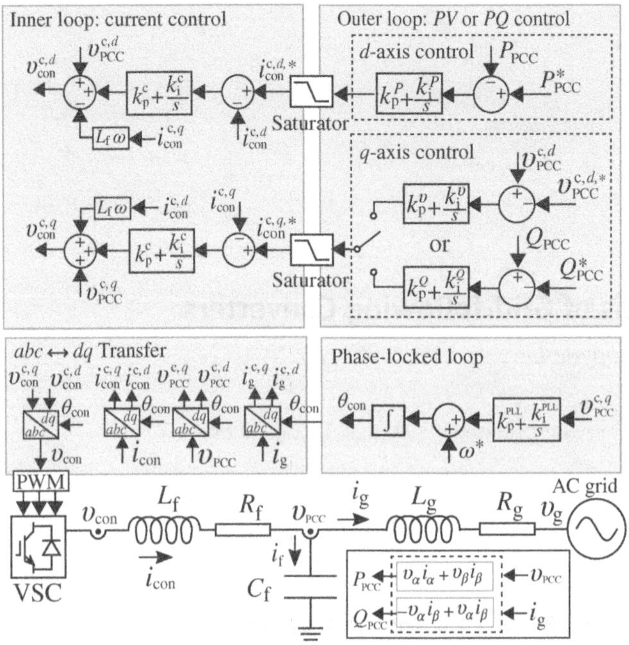

Figure 9.1 Grid-following (GFL) converter: PLL-based control.

The phase tracking process of PLL is formulated as below:

$$\theta_{\text{PLL}} = \frac{k_{\text{p}}^{\text{PLL}} s + k_{\text{i}}^{\text{PLL}}}{s^2 + \underbrace{k_{\text{p}}^{\text{PLL}}}_{2\zeta^{\text{PLL}}\omega_n^{\text{PLL}}} s + \underbrace{k_{\text{i}}^{\text{PLL}}}_{\omega_n^{\text{PLL}^2}}} \theta_{\text{PCC}} \tag{9.2}$$

where θ_{PLL} is the tracked phase and θ_{PCC} is the phase at PCC.

General tuning methods for current control and PLL are based on Equations 9.1 and 9.2, which tune their PI gains (k_{p}^{c}, k_{i}^{c}, $k_{\text{p}}^{\text{PLL}}$, and $k_{\text{i}}^{\text{PLL}}$) to achieve their desired natural frequencies ω_n^{c} and ω_n^{PLL} and maintain their damping ratios ζ^{c} and ζ^{PLL} at 0.707.

To enhance the stability of a very weak grid connection, a significant reduction for their natural frequencies ω_n^{c} and ω_n^{PLL} is required and its stabilization effectiveness is presented below using small-signal modeling method.

When a VSC with $i_{\text{con}}^{\text{c},d,*} = 1$ p.u. and $v_{\text{PCC}} = 1$ p.u. connects to a very weak grid such that the short-circuit ratio (SCR) = 1, its stability analysis results using pole map are presented in Figure 9.2. Both current control's and PLL's ω_n are 310 at an initial state where this VSC loses stability. It is found that ω_n reduction of both loops can pull the right half-plane (RHP) poles to the left half-plane (LHP), which concludes that both current control and PLL can contribute to stabilizing the very weak grid connection.

Although this VSC is stabilized with this very weak grid connection by ω_n reduction of both loops, the poles are still close to $x = 0$ as shown in Figure 9.2a, b, which indicates that this very weak grid connection is not well stabilized.

To further compare the stability performance of both loops, a time-domain simulation is conducted to find how the PCC voltage is influenced by VSC current regulation. When SCR = 1, $i_{\text{con}}^{\text{c},d,*} = 1$ p.u. regulation with constant $i_{\text{con}}^{\text{c},q,*}$ will cause a significant impact on PCC voltage. To avoid this significant impact and only present the control impact on PCC voltage, the state, in which $i_{\text{con}}^{\text{c},d,*} = 0.4$ p.u. only drops PCC voltage from 1 p.u. to 0.96 p.u., is selected.

At 1 s, a step change from 0 to 0.4 p.u. is applied to $i_{\text{con}}^{\text{c},d,*}$, the comparison study between current control and PLL is simulated, and the results are shown in Figure 9.3. It is found that even though both current control and PLL keep low ω_n, PCC voltage fluctuation still exists at a relatively high level. By applying the same ω_n reduction, compared to PLL, current control takes more impact (improvement) on PCC voltage fluctuation.

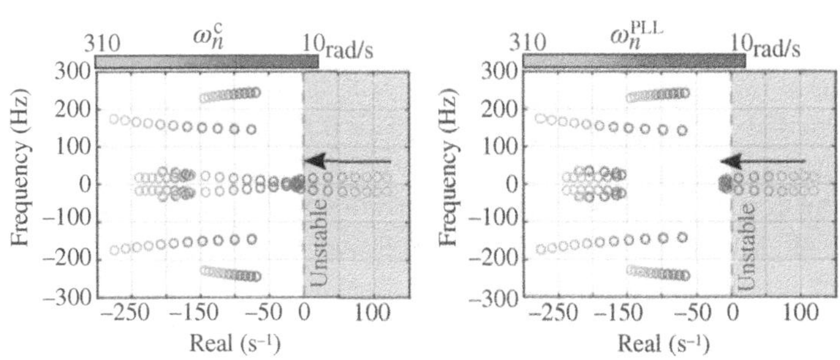

(a) Current control (b) PLL

Figure 9.2 Small-signal stability analysis when SCR = 1.

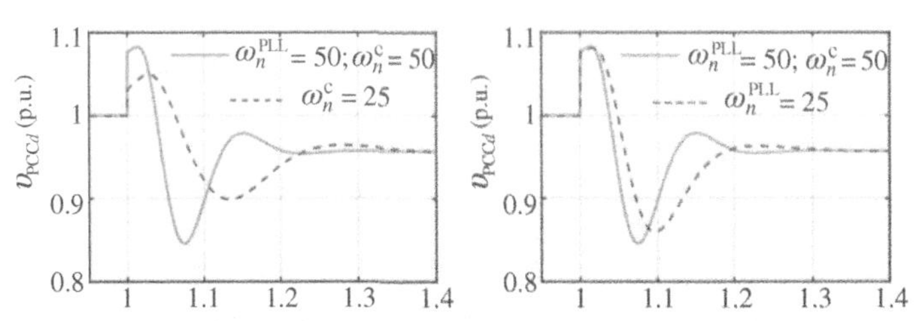

(a) Current control (b) PLL

Figure 9.3 Impact of $i_{\text{con}}^{\text{c},d,*} = 0.4$ p.u. regulation on PCC voltage when SCR = 1.

In sum, although both current control and PLL contribute to stabilizing the very weak grid connection based on small-signal stability analysis, PCC voltage fluctuation still exists at a relatively high level. Furthermore, it is indicated that current control dominates the impact of PCC voltage fluctuation based on time-domain simulation, which proves that the current control takes more impact on VSC instability performance.

9.1.2 Dominant Factor of Current Control Causing PCC Voltage Fluctuation

Based on the analysis above, PCC voltage fluctuation is significantly impacted by the current control. For a very weak grid connection, it is found below that current control tuning cannot improve this intractable instability issue well [1]. Furthermore, the dominant factor in the current control is identified.

To demonstrate how current control tuning impacts on PCC voltage fluctuation, at 1 s, $i_{\text{con}}^{\text{c,d,*}} = 0.4$ p.u. is applied and its simulation results are shown in Figure 9.4. It is found that reducing nature frequency ω_n^{c} helps to reduce the PCC voltage surge at 1 s, as shown in Figure 9.4a. However, the oscillation period increases because of this ω_n^{c} reduction. Increasing damping ratio ζ^{c} does help to damp this oscillation, as shown in Figure 9.4b. However, voltage surge increases because of the increasing damping ratio.

It concludes that tuning current control cannot well balance the voltage surge and oscillation period of PCC voltage fluctuation. This is the reason why VSC suffers from the intractable very weak grid instability issue.

To find out the dominant factor within this current control causing the above phenomena, relation between the current control and PCC voltage is derived below.

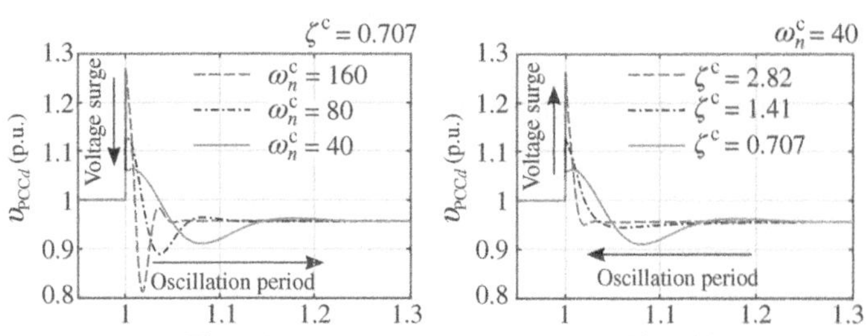

(a) Natural frequency tuning **(b)** Damping ratio tuning

Figure 9.4 PCC voltage fluctuation related to current control tuning when SCR = 1.

Within low-frequency range, C_f is ignored to assume $\boldsymbol{i}_{con} = \boldsymbol{i}_g$. Therefore, $\boldsymbol{v}_{PCC}$ is expressed as:

$$\boldsymbol{v}_{PCC} = \begin{bmatrix} L_g s + R_g & -\omega_0 L_g \\ \omega_0 L_g & L_g s + R_g \end{bmatrix} \boldsymbol{i}_{con} \tag{9.3}$$

where $\boldsymbol{v}_{PCC}$ in bold is the dq-matrix $[v^d_{PCC}; v^q_{PCC}]$, similar to the other bold parameters.

It is assumed that only $i^{c,d,*}_{con}$ or $i^{c,q,*}_{con}$ is regulated at the same time. Therefore, coupling parts $\omega_0 L_g$ of grid impedance are ignored. Substituting $\boldsymbol{i}_{con}$ in Equation 9.3 with Equation 9.1 yields:

$$\boldsymbol{v}_{PCC} = \frac{(k^c_p s + k^c_i)(L_g s + R_g)/L_f}{s^2 + [(R_f + k^c_p)/L_f]s + k^c_i/L_f} \boldsymbol{i}^*_{con} \tag{9.4}$$

By rearranging Equation 9.4, a clear relation between PCC voltage and current control is yielded:

$$\boldsymbol{v}_{PCC} = \left(\underbrace{\frac{k^c_p L_g}{L_f}}_{\text{coeff.a.1}} - \underbrace{\frac{k^c_p L_g}{L_f} \frac{[(k^c_p + R_f)/L_f - (R_g/L_g)]s + k^c_i/L_f}{s^2 + [(R_f + k^c_p)/L_f]s + k^c_i/L_f}}_{\text{coeff.a.2}} + \underbrace{\frac{k^c_i(L_g s + R_g)/L_f}{s^2 + [(R_f + k^c_p)/L_f]s + k^c_i/L_f}}_{\text{coeff.b}} \right) \boldsymbol{i}^*_{con} \tag{9.5}$$

At $t = 0$ s, $(-\text{coeff.a.2} + \text{coeff.b})i^*_{con}$ in Equation 9.5 is 0, which results in:

$$\boldsymbol{v}_{PCC}|_{t=0} = \frac{k^c_p L_g}{L_f} \boldsymbol{i}^*_c \tag{9.6}$$

Based on Equations 9.5 and 9.6, how a current control disturbs the PCC voltage is found below:

i. weak grid (high L_g) causes significant voltage surge $\Delta \boldsymbol{v}_{PCC}|_{t=0}$ based on Equation 9.6;
ii. reducing k^c_p helps to mitigate this voltage surge based on Equation 9.6;
iii. reducing k^c_p is equivalent to reducing the damping ratio based on Equation 9.5's denominators coeff.a.2 and coeff.b, which may induce a significant PCC voltage oscillation.

A conflict appears on $k_{\mathrm{p}}^{\mathrm{c}}$ tuning as described in (ii) and (iii). Although the PCC voltage surge can be mitigated by reducing $k_{\mathrm{p}}^{\mathrm{c}}$, a new PCC voltage oscillation may be induced by this $k_{\mathrm{p}}^{\mathrm{c}}$ reduction.

Time-domain validation of the above findings (i)–(iii) are implemented via a PLL-based current-controlled VSC.

To validate finding (i), a step change of $i_{\mathrm{con}}^{\mathrm{c},d,*}$ from 0 to 0.4 p.u. is applied when $t = 1$ s, as shown in Figure 9.5a. The voltage surge is increasing with the decrement of SCR, where $\mathrm{SCR} = 1/Z_{\mathrm{g}}$. The voltage surges of $\Delta v_{\mathrm{PCC}}^{d}$ are calculated as [0.26 0.13 0.065] based on Equation 9.6 with SCR = 1, 2, and 4, which match the simulation results of $\Delta v_{\mathrm{PCC}}^{d}$ at 1 s, as shown in Figure 9.5a.

Finding (ii) is used to suppress this voltage surge, $k_{\mathrm{p}}^{\mathrm{c}}$ is reduced based on Equation 9.6, and its simulation result is shown in Figure 9.5b. It is obvious that $k_{\mathrm{p}}^{\mathrm{c}}$ reduction helps to reduce the voltage surge at 1 s.

However, severe voltage oscillation occurs with $k_{\mathrm{p}}^{\mathrm{c}}$ reduction, as shown in Figure 9.5b. This low damping performance also fits the finding (iii).

In sum, current control over PLL dominates the PCC voltage fluctuation performance. A weaker grid causes a higher PCC voltage fluctuation. Furthermore, this PCC voltage impact cannot be eliminated by tuning current control when a very weak grid is connected such as SCR = 1. It is because a conflict appears on $k_{\mathrm{p}}^{\mathrm{c}}$ tuning based on findings (ii) and (iii), which is that reducing voltage surge will cause low damping capability and vice versa. This is the dominant instability mechanism of VSC connecting to a very weak grid.

9.1.3 Instability Cause of Outer Loop

When VSCs connect to a very weak grid, PCC voltage varies significantly with current regulation. To well preserve the PCC voltage during power regulation, power and voltage (*PV*) control is added as an outer loop, as shown in Figure 9.1. Power

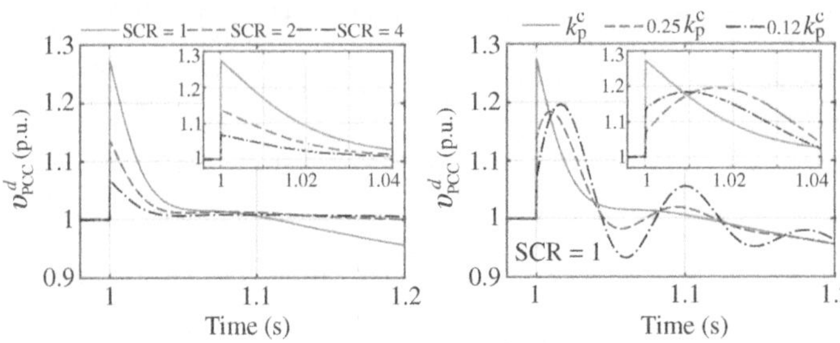

(a) Impacts of various SCRs **(b)** Impacts of various $k_{\mathrm{p}}^{\mathrm{c}}$

Figure 9.5 Voltage fluctuations caused by current control changing $i_{\mathrm{con}}^{c,d,*}$ from 0 to 0.4 p.u. when SCR = 1 (current control: $\zeta^{\mathrm{c}} = 0.707$, $\omega_n^{\mathrm{c}} = 160$; PLL: $\zeta^{\mathrm{PLL}} = 0.707$, $\omega_n^{\mathrm{PLL}} = 11$).

control and voltage control, respectively, generate d-axis and q-axis current references $i_{\text{con}}^{c,d,*}$ and $i_{\text{con}}^{c,q,*}$ for inner current control. To simply and clearly tune the outer loop, the following relation is defined as below:

$$k_{\text{p}}^{P} + \frac{k_{\text{i}}^{P}}{s} = r_{P\text{-c}}\left(k_{\text{p}}^{\text{c}} + \frac{k_{\text{i}}^{\text{c}}}{s}\right) \tag{9.7}$$

$$k_{\text{p}}^{v} + \frac{k_{\text{i}}^{v}}{s} \text{ or } k_{\text{p}}^{Q} + \frac{k_{\text{i}}^{Q}}{s} = r_{v\text{-c}}\left(k_{\text{p}}^{\text{c}} + \frac{k_{\text{i}}^{\text{c}}}{s}\right) \tag{9.8}$$

where $r_{P\text{-c}}$ and $r_{v\text{-c}}$ are, respectively, the response-speed ratios between power control and inner current control, and voltage control (or reactive power control) and inner current control.

The inner loop requires a fast-tracking ability for references from the outer loop. Generally, $r_{P\text{-c}} = 0.1$ and $r_{v\text{-c}} = 0.1$ are applied to let the inner loop be 10 times faster than the outer loop.

The effectiveness of an outer loop control actually relies on the inner current regulation. Therefore, how current injection impacts PCC voltage is explained via Figure 9.6, where the static dq-current operation area of a VSC connecting to an alternating current (AC) grid is presented. For a strong grid or even a weak grid, current injection from a VSC does not significantly impact PCC voltage. As shown in Figure 9.6a, when a weak grid SCR = 3 is connected, i_{con}^{d} increasing from 0 to 1 p.u. has a little impact on the PCC voltage. Therefore, only current control is enough to regulate the assigned power via i_{con}^{d}. However, if a very weak grid is connected, VSC's current injection has a significant impact on the PCC voltage. As shown in Figure 9.6b, at point C, only increasing i_{con}^{d} to 1 p.u. drops $\boldsymbol{v}_{\text{PCC}}$ to 0.24 p.u. Therefore, for a very weak grid connection, outer loop using voltage control is essential to preserve the PCC voltage at 1 p.u. during power regulation.

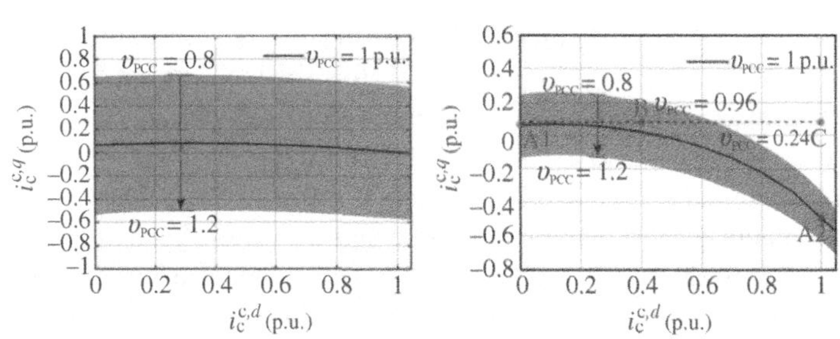

(a) Current operation area at SCR = 3 **(b)** Current operation area at SCR = 1

Figure 9.6 Relation between currents and PCC voltages under various SCRs.

An outer loop is added to form the *PV* control or *PQ* control, as shown in Figure 9.1. For a converter increasing its active power, the increased ΔP is shown below:

$$\Delta P \approx I_{\mathrm{g}}^{d} \Delta v_{\mathrm{PCC}}^{d} + V_{\mathrm{PCC}}^{d} \Delta i_{\mathrm{g}}^{d} \tag{9.9}$$

In an extremely weak grid, if the voltage support at q-axis is not fast enough, $\Delta i_{\mathrm{g}}^{d}$ results in a significant voltage drop $-\Delta v_{\mathrm{PCC}}^{d}$, which causes $\Delta P < 0$ based on Equation 9.9. This conflicts with the control target: increasing the active power.

In sum, for an extremely weak grid connection, both *PV* and *PQ* controls have the same root instability causes, as shown below:

- slow voltage support, which happens to outer loop: *PV* or *PQ* control. In an extremely weak grid, the fact that voltage drops with increasing current may conflict with the active power control. Slow voltage support via q-axis V control or Q control worsens this conflict.
- high natural frequencies of PLL and inner current control.

9.2 A Tuning Method for PLL-based Current Control

The instability causes of a VSC connecting to a very weak grid have been identified previously. Accordingly the following sections will provide corresponding tuning methods to effectively avoid these instability causes, which lead to a very weak grid connection eventually. Overall tuning strategies for VSC control are also provided to optimize the VSC stabilization. In this section, the overall tuning strategy for PLL-based current control is presented.

9.2.1 Solution for the Dominant Instability

As the dominant instability mechanism of the current control has been identified in Section 9.1.2, there could be many solutions to solve this instability conflict on $k_{\mathrm{p}}^{\mathrm{c}}$ as presented in findings (ii) and (iii). In this section, one simple and effective solution without any additional stabilization control blocks is proposed based on [2]. This solution is to restructure the PI controller in current control as an integral–proportional (IP) controller, as shown in Figure 9.7 (dashed line), which can simultaneously avoid the voltage surge and provide sufficient damping capability. Furthermore, this IP controller will not change the current control tuning method. The stabilization mechanism of this IP-based current control is introduced below.

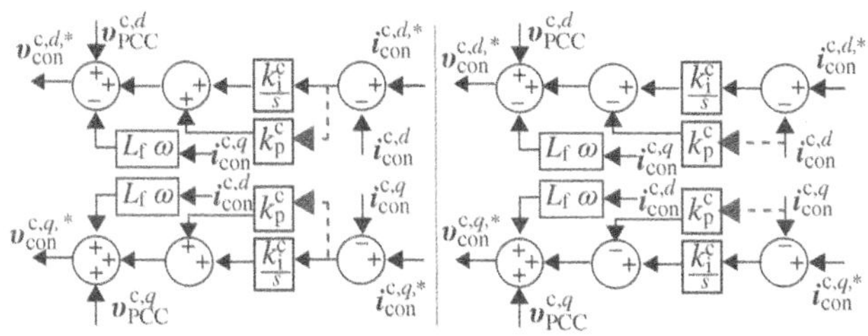

Figure 9.7 Current control: difference between a PI controller and an IP controller.

According to the same derivation process of Equation 9.1, the transfer function of current control based on this IP controller is derived below:

$$i_{\text{con}} = \frac{k_{\text{i}}^{\text{c}}/L_{\text{f}}}{s^2 + \underbrace{[(R_{\text{f}} + k_{\text{p}}^{\text{c}})/L_{\text{f}}]}_{2\zeta^{\text{c}}\omega_n^{\text{c}}}s + \underbrace{k_{\text{i}}^{\text{c}}/L_{\text{f}}}_{\omega_n^{\text{c}\,2}}} \boldsymbol{i}_{\text{con}}^{*} \tag{9.10}$$

Compared to the original transfer function (Equation 9.1) of the current control, Equation 9.10 has the same second-order denominator, which means this IP-based current control can still use the same tuning method by adjusting ζ^{c} and ω_n^{c}.

Following the same derivation process from Equation 9.3 to Equation 9.4 with Equation 9.10, the relation between PCC voltage and IP-based current control is yielded:

$$\boldsymbol{v}_{\text{f}} = \underbrace{\frac{k_{\text{i}}^{\text{c}}(L_{\text{g}}s + R_{\text{g}})/L_{\text{f}}}{s^2 + [(R_{\text{f}} + k_{\text{p}}^{\text{c}})/L_{\text{f}}]s + k_{\text{i}}^{\text{c}}/L_{\text{f}}}}_{\text{coeff.b}} \boldsymbol{i}_{\text{con}}^{*} \tag{9.11}$$

Thanks to this IP controller, coeff.a.1 and coeff.a.2 in Equation 9.5 are eliminated. Therefore, the voltage surge caused by PI controller via coeff.a.1 is avoided. k_{p}^{c} tuning only needs to consider the current damping capability.

A simulation result comparing the effectiveness of PI controller and IP controller is shown in Figure 9.8a. Same k_{p}^{c} and k_{i}^{c} are applied for both PI and IP controllers. After $i_{\text{con}}^{\text{c},d,*} = 0.4$ p.u. is applied, it is obvious that the voltage surge (solid line) at $t = 1$ s caused by the PI controller will not happen when an IP controller is applied (dashed line), which fits the analysis based on Equation 9.11.

A validation of Equation 9.11 is shown in the inset of Figure 9.8b, and it is found that the voltage overshoot that appears in simulation result matches Equation 9.11. Therefore, damping this voltage overshoot based on Equation 9.11 is effective.

The simulation result of enhancing the damping capability of the current controller is also shown in Figure 9.8b. $\zeta^{\text{c}} = 0.707$ of the current control is not enough

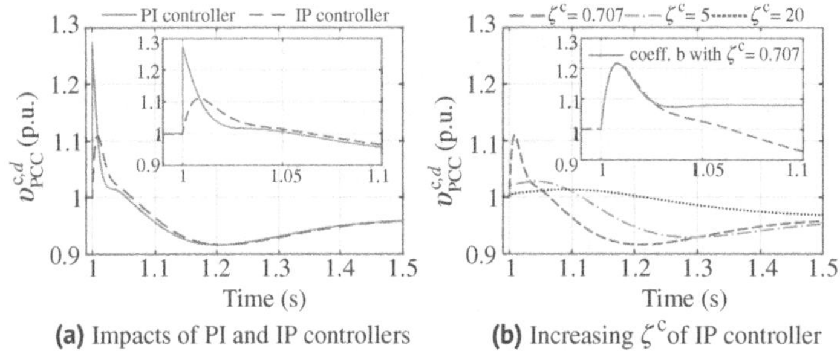

Figure 9.8 Comparison between PI controller and IP controller with $i_{\text{con}}^{c,d,*}$ step change from 0 to 0.4 p.u. when SCR = 1 (current control: $\zeta^c = 0.707$, $\omega_n^c = 160$; PLL: $\zeta^{\text{PLL}} = 0.707$, $\omega_n^{\text{PLL}} = 11$).

to damp the voltage overshoot (dashed line); this is because a weak grid (high inductance) will enlarge the voltage overshoot. Therefore, a higher damping ratio at $\zeta^c = 20$ is applied for current control, and this voltage overshoot is well damped.

Therefore, the conflict of k_p^c tuning is avoided by the IP controller for current control, and its advantages are emphasized below:

- IP controller does not change the tuning method of current control;
- voltage surge caused by the PI controller within current control is avoided;
- desired damping ratio can be applied without concern about significant voltage surge impact.

This proposed solution is validated in time-domain simulation. PLL-based current control is applied for the VSC. A comparison study between the conventional solution and the proposed solution is conducted when SCR = 1. The descriptions of both solutions are given below:

- conventional solution refers to current control using the PI controller. Based on its stability analysis, as shown in Figure 9.2, significantly reducing the natural frequency of current control and PLL may stabilize this VSC. The following parameters are used: $\zeta^c = 0.707$, $\omega_n^c = 11$ for current control; $\zeta^{\text{PLL}} = 0.707$, $\omega_n^{\text{PLL}} = 11$ for PLL.
- proposed solution refers to current control using the IP controller. Based on the previous analysis, the following parameters are used: $\zeta^c = 20$, $\omega_n^c = 160$ for current control; $\zeta^{\text{PLL}} = 0.707$, $\omega_n^{\text{PLL}} = 11$ for PLL.

It should be noted that with very weak grid connection, regulating i_{con}^d and i_{con}^q in a step-change way also contributes to this PCC voltage fluctuation. Step change over ramp change is selected for current regulation or power regulation in the following sections because it can present more dynamics of PCC voltage fluctuation.

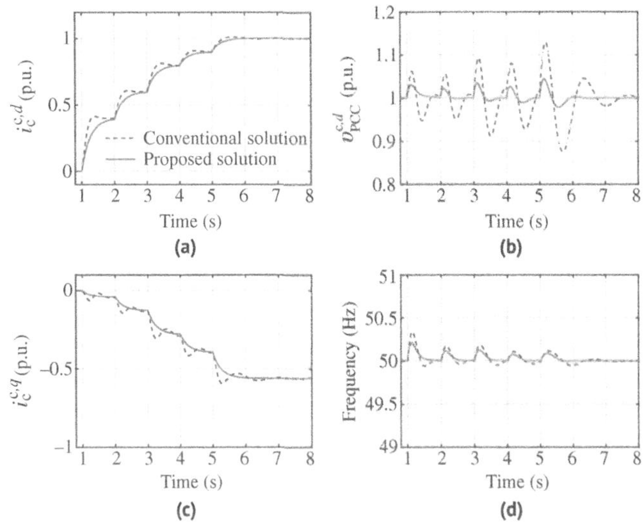

Figure 9.9 Comparison between conventional solution and proposed solution for rated current injection when SCR = 1.

The simulation results are presented in Figure 9.9. As shown in Figure 9.9a, c, both i_{con}^{d} and i_{con}^{q} are regulated simultaneously to support desired power and maintain the PCC voltage at 1 p.u. Both solutions have a similar performance on frequency variance, as shown in Figure 9.9d. However, the stabilization effectiveness of the conventional solution is not well so its PCC voltage oscillation is immoderate, as shown in Figure 9.9b. Current regulation causes a maximum of 12% voltage variance, and this voltage oscillation lasts for a relatively long period. In contrast, the proposed solution well damps this voltage oscillation, including lower voltage variance (maximum 5%) and shorter oscillation period, as shown in Figure 9.9b.

A very weak grid SCR = 1 tested above is the worst case for VSC integration, and the proposed solution is also validated to be suitable for connecting VSCs to any grids with SCR>1.

9.2.2 The Relationship Between Current Control and PLLs

For a VSC requiring current regulation, current control and PLL are applied as the control system, as shown in Figure 9.1 (no outer loop). To define their relation, the response-speed ratio is proposed and explained below:

$$r_{\text{PLL-c}} = \frac{t_{\text{s}}^{\text{c}}}{t_{\text{s}}^{\text{PLL}}} \tag{9.12}$$

where r_{PLL-c} is the response-speed ratio, t_s^{PLL} and t_s^c are, respectively, the response settling times of PLL and current control. For example, $r_{PLL-c} = 2$ means PLL is two times faster than current control.

As analyzed in Section 9.2.1, $\zeta^c = 20$ is a proper damping ratio for current control; therefore, its response settling time t_s^c is formulated based on an over-damped system:

$$t_s^c = \frac{4}{\left(-\zeta^c + \sqrt{{\zeta^c}^2 - 1}\right)\omega_n^c} \tag{9.13}$$

If r_{PLL-c} is defined, based on Equation 9.12, t_s^{PLL} is calculated as:

$$t_s^{PLL} = \frac{t_s^c}{r_{PLL-c}} \tag{9.14}$$

By setting $\zeta^{PLL} = 0.707$, its response settling time is formulated as:

$$t_s^{PLL} = \frac{4}{\zeta^{PLL}\omega_n^{PLL}} \tag{9.15}$$

After rearranging Equation 9.15, ω_n^{PLL} is calculated as:

$$\omega_n^{PLL} = \frac{4}{\zeta^{PLL} \times t_s^{PLL}} \tag{9.16}$$

Based on Equations 9.12–9.16, if r_{PLL-c} and ω_n^c are defined, all ζ^c, ω_n^c, ζ^{PLL}, and ω_n^{PLL}, for current control and PLL are known so that their controller's gains can be calculated based on Equations 9.2 and 9.10.

To achieve excellent stabilization effectiveness, the following objectives need to be solved:

- besides current control, how PLL impacts the PCC voltage?
- what response-speed ratio r_{PLL-c} between current control and PLL should be defined?
- what natural frequency range of current control ω_n^c should be defined?

(Control parameters for following stability analysis are presented below: current control: $\zeta^c = 20$, $\omega_n^c = 160$; PLL: $\zeta^{PLL} = 0.707$, $r_{PLL-c} = 2$.)

PCC voltage is proportional to I_{con}^d and its initial dynamic is dominated by current control, as shown in the dashed line in Figure 9.10a. However, PCC voltage eventually drops with increasing VSC current injection, which is caused by PLL. For example, as shown in Figure 9.10a, at initial state $v_{PCC}^d = 1$ p.u., after $i_{con}^{c,d,*} = 0.4$ p.u., v_{PCC}^d drops to 0.96 p.u. This mechanism is explained below. At steady state tracked by PLL, the following relations are presented:

$$\begin{aligned} V_{PCC}^d &= R_g I_g^d - \omega_0 L_g I_g^q + V_g^d \\ V_{PCC}^q &= R_g I_g^q + \omega_0 L_g I_g^d + V_g^q \end{aligned} \tag{9.17}$$

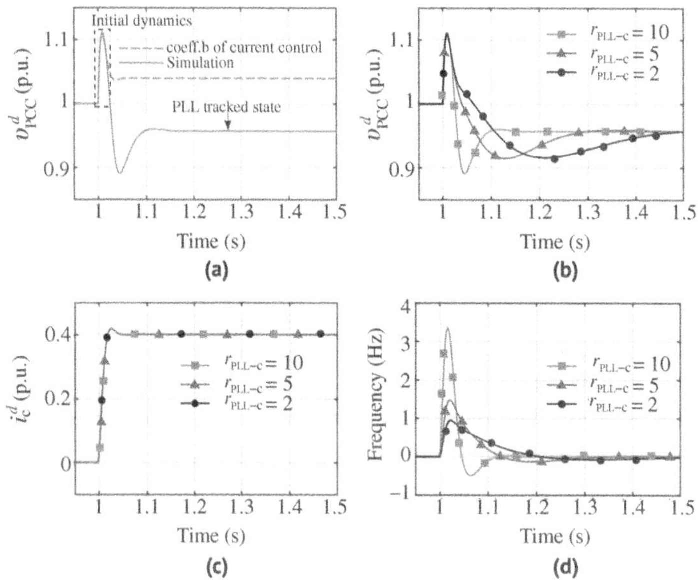

Figure 9.10 Impact of PLL on the system stability performance at SCR = 1. (In this study, $i_c^d = 0.4$ p.u. does not cause system instability when $r_{PLL-c} = 10$.)

where PCC voltage is at steady state, which is denoted as the upper-case parameter v_{PCC}, similar to I_g and V_g.

V_{PCC}^q is 0 because of PLL tracking and assume $V_g^{d^2} + V_g^{q^2} = 1$, $I_{con} = I_g$, and $R_g = 0$, then Equation 9.17 is simplified as:

$$v_{PCC}^d = -\omega_0 L_g I_{con}^q + \sqrt{1 - (\omega_0 L_g I_{con}^d)^2} \tag{9.18}$$

Equation 9.18 clearly explains why PCC voltage drops with I_{con}^d increasing, which is caused by PLL.

It is found that further increasing response speed of PLL still has little impact on the initial dynamics of v_{PCC}^d, as shown in Figure 9.10b. Although PLL is 10 times faster than the current control, the initial dynamics of v_{PCC}^d still dominates by the current control. Faster PLL helps to shorten the fluctuation period of v_{PCC}^d after $i_{con}^{c,d,*} = 0.4$ p.u. is applied. The current build-up process is not at all impacted by faster PLL, as shown in Figure 9.10c. PCC voltage fluctuation also causes the fluctuation of frequency, as shown in Figure 9.10d, due to which faster PLL exacerbates the frequency overshoot. It is concluded that even though the PLL does not have a significant impact on PCC voltage fluctuation, it should be slowed down to avoid excessive impact on frequency variance.

To assess the proper response-speed ratio r_{PLL-c} between PLL and current control, r_{PLL-c} ranging from 1/10 to 10 is tested based on stability analysis, and its results are shown in Figure 9.11. This stability analysis is conducted under the situation that the VSC works at $I_{con}^{d} = 1$ p.u. and $v_{PCC} = 1$ p.u. with SCR = 1. It is found that over-fast PLL causes the VSC instability such as $r_{PLL-c} = 10$, as shown in Figure 9.11a. If the PLL is much slower than the current control, this VSC still risks instability, as shown in Figure 9.11b. For example, when $r_{PLL-c} = 0.1$, poles are close to $x = 0$. It is found that $r_{PLL-c} = 2$ stays the VSC at a proper stability position, as shown in Figure 9.11b.

After defining properly $r_{PLL-c} = 2$ above and $\zeta^{c} = 20$, the natural frequency of current control ω_n^c is assessed by stability analysis to optimize the overall stability performance. As shown in Figure 9.12a, high ω_n^c leads the VSC to instability. Although ω_n^c reduction pulls the poles back to the LHP, the middle poles are pushed toward the RHP. To make a trade-off in the above situation, a range of $\omega_n^c \in [160\ 320]$ is selected to retain this VSC in a good stable condition, as shown in Figure 9.12b.

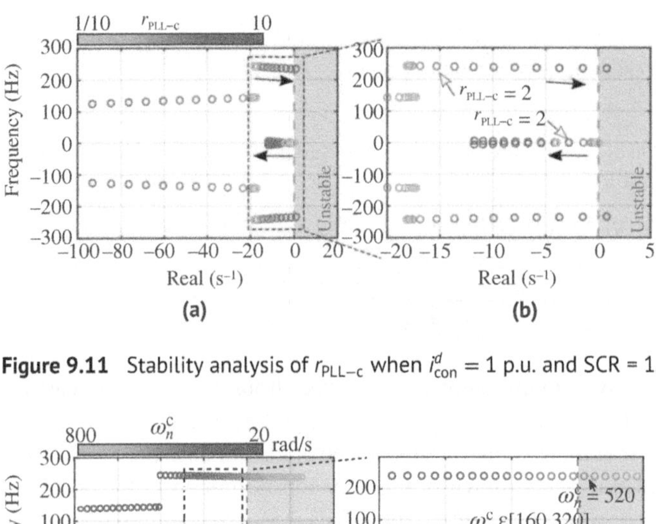

Figure 9.11 Stability analysis of r_{PLL-c} when $i_{con}^{d} = 1$ p.u. and SCR = 1.

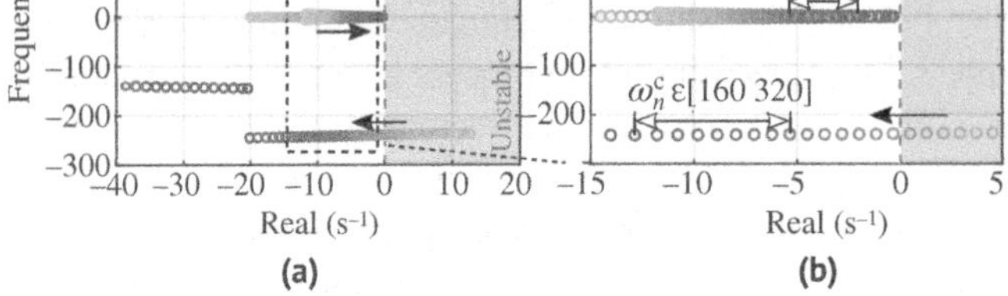

Figure 9.12 Stability analysis of ω_n^{PLL} when $i_{con}^{d} = 1$ p.u. and SCR = 1.

9.2.3 Tuning Strategies

Based on the above stability analysis, the tuning strategy for this PLL-based current-controlled VSC is proposed and shown in Figure 9.13. Current control as the dominant instability cause is tuned first and PLL tuning is followed based on its response-speed relation $r_{\text{PLL-c}}$, with current control. The predefined parameters as the input for this tuning strategy are quantified based on the above stability analysis and shown below:

1. for current control, $\zeta^{\text{c}} = 20$ and $\omega_n^{\text{c}} \in [160\ 320]$; its IP controllers' gains are calculated using ζ^{c} and ω_n^{c} based on Equation 9.10;
2. for PLL, $\zeta^{\text{PLL}} = 0.707$ and $r_{\text{PLL-c}} = 2$; its PI controllers' gains are calculated using ζ^{PLL} and $r_{\text{PLL-c}}$ based on Equations 9.2, 9.14, and 9.16.

It should be noted that improper design of control parameters still risks instability, as indicated in Figure 9.12 when $\omega_n^{\text{c}} = 520$, and the simulation result in Figure 9.14 (grey line) also proves this instability analysis.

A time-domain validation for the above tuning strategy is presented in Figure 9.14. The regulation of $i_{\text{con}}^{\text{c},d,*}$ and $i_{\text{con}}^{\text{c},q,*}$ follows the route in Figure 9.6 from the point A1 to A2, where $i_{\text{con}}^{\text{c},q,*}$ is regulated to maintain the PCC voltage at 1 p.u. To validate ω_n^{c} of current control in a full range [160 320], $\omega_n^{\text{c}} = 160$ and 320 are selected. The corresponding controllers' gains are calculated via the above tuning strategy. It is found that both $\omega_n^{\text{c}} = 160$ and 320 enable the VSC support 1 p.u. power at SCR = 1 stably. Using a higher natural frequency of current control can shorten the response time of current regulation. For example, if using 160 or 320, the response time of the current regulation is 1 or 0.5 s. It should be noted that a faster response time will lead to higher frequency variance. Within [160 320], 0.2 p.u. current step change causes the maximum 5% voltage variance and 0.5 Hz frequency variance, as shown in Figure 9.14a, c. It is considered that the proposed tuning strategy has a good performance in the stabilization of VSC's very weak grid connection.

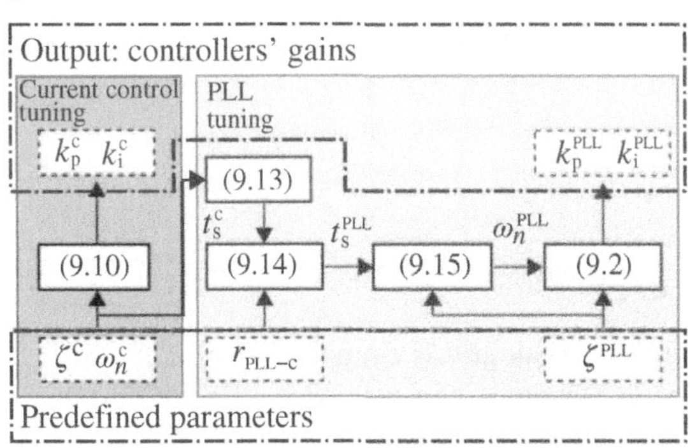

Figure 9.13 Tuning strategy for PLL-based current control.

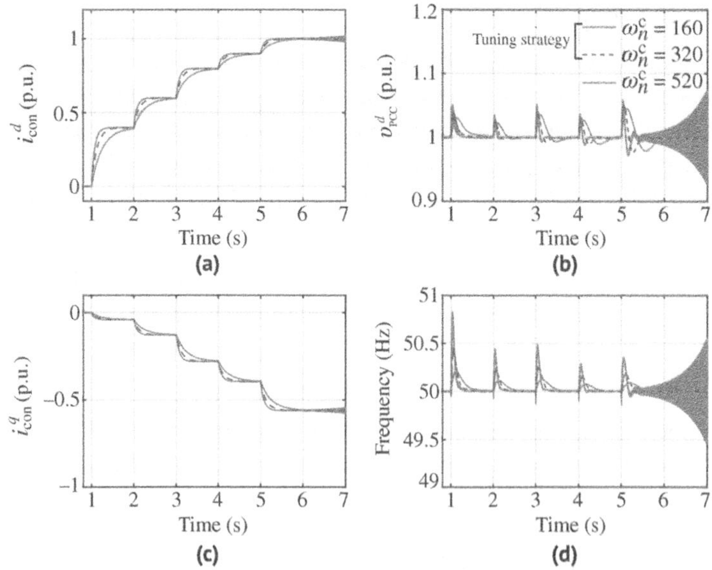

Figure 9.14 Time-domain validation of proposed tuning strategy: rated current injection of VSC at SCR = 1.

9.3 Overall Tuning Strategy for PLL-based *PV* Control Systems

In an outer loop, the relation between its *PV* controls is not well and clearly identified, which is critical for PCC voltage stabilization. This relation is defined below:

$$r_{v-P} = \frac{r_{v-c}}{r_{P-c}} \tag{9.19}$$

where r_{v-P} is the response-speed ratio between voltage control and power control.

Therefore, based on Equations 9.7–9.19, the controllers' gains of voltage control can be recalculated using r_{v-P} as:

$$k_{\mathrm{p}}^{v} + \frac{k_{\mathrm{i}}^{v}}{s} \text{ or } k_{\mathrm{p}}^{Q} + \frac{k_{\mathrm{i}}^{Q}}{s} = r_{v-P}\left(k_{\mathrm{p}}^{P} + \frac{k_{\mathrm{i}}^{P}}{s}\right) \tag{9.20}$$

9.3.1 Outer-loop Tuning Strategy

As explained in Figure 9.6b, if a very weak grid is connected, both I_{con}^{d} and I_{con}^{q} have a significant impact on the PCC voltage. For well-controlling PCC voltage at 1 p.u. via q-axis current, Δv_{PCC} supported by I_{con}^{q} must be more sufficient than Δv_{PCC} impacted by I_{con}^{d}, where the following relation should be satisfied:

$$\underbrace{\frac{\Delta i_{\text{con}}^{\text{c},q,*}}{\Delta i_{\text{con}}^{\text{c},d,*}}}_{r_{v-P}} > \frac{\Delta I_{\text{con}}^{q}}{\Delta I_{\text{con}}^{d}} = 1.8 \tag{9.21}$$

where it is assumed that the VSC operates at point C ($v_{\text{PCC}} = 1$ p.u. and $I_{\text{con}}^{d} = 1$ p.u.) in Figure 9.6b, which leads to $\Delta I_{\text{con}}^{d}/\Delta I_{\text{con}}^{q} = 1.8$; r_{v-P} is response-speed ratio between voltage control and power control.

Within Equation 9.21, $\Delta i_{\text{con}}^{\text{c},q,*}/\Delta i_{\text{con}}^{\text{c},d,*}$ is equivalent to the response-speed ratio r_{v-P}, which means voltage control should be at least 1.8 times faster than power control. To further identify the proper r_{v-P}, stability analysis for r_{v-P} ranging from 1 to 10 is conducted, where power control is 10 times slower than current control ($k_{P-c} = 0.1$) is applied. The stability analysis result is shown in Figure 9.15. When $r_{v-P} = 1$, some poles are in the RHP as shown in Figure 9.15b, which indicates that this VSC is unstable. Higher r_{v-P} does help to stabilize this VSC as shown in Figure 9.15a. It is found that when $r_{v-P} = 5$, the poles in the middle are the farthest from $x = 0$, which mostly stabilizes this VSC. Therefore, $r_{v-P} = 5$, which requires voltage control five times faster than power control, is selected for outer loop tuning. This tuning method is suitable for both *PV* and *PQ* controls at the outer loop.

Based on the analysis above, a sufficiently fast voltage support tuning is applied with $r_{v-P} = 5$ for both *PV* and *PQ* controls. As shown in Figure 9.16, the converter stably connects to the weak grid at SCR = 1, while slow voltage support $r_{v-P} = 1$ does cause the instability for both controls.

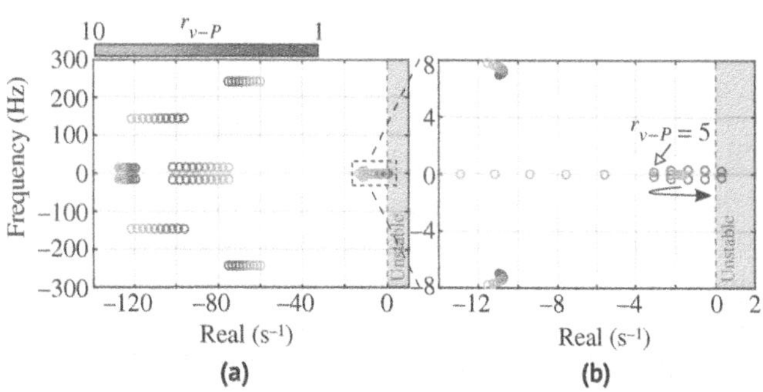

Figure 9.15 Stability analysis of response-speed ratio r_{v-P} between voltage control and power control at SCR = 1.

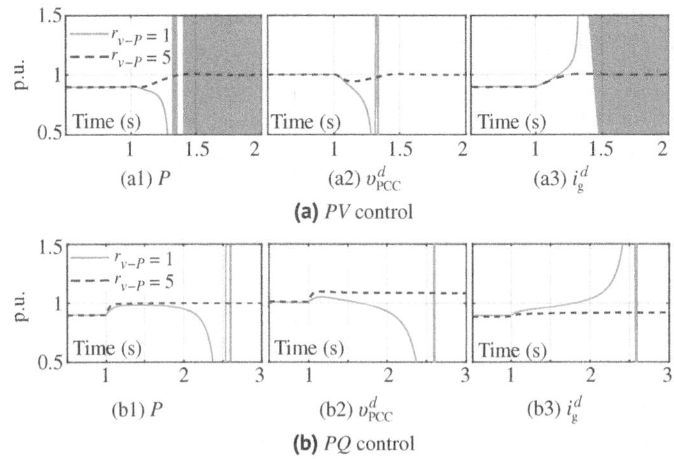

Figure 9.16 Effectiveness of fast voltage support on both *PV* and *PQ* controls when the converter connects to an extremely weak grid with SCR = 1.

9.3.2 Coordinated Tuning of PLL and Current Control

PLL and current control tuning relies on the stability analysis, as shown in Figure 9.17. Slowing both PLL and inner current control, which equivalently reduces ω_n of Equations 9.1–9.2, helps to stabilize the converter. PLL's ω_n^{PLL} reduction has the same stabilization effectiveness for both *PV* and *PQ* controls and must be limited below 110. Because of the outer loop, this inner current control does not require the high damping capability as described in Section 9.2.1, so $\zeta^{\mathrm{c}} = 0.707$ is applied. The inner current control's ω_n^{c} reduction has the slightly different stabilization effectiveness for *PV* and *PQ* controls and must be limited within 560 and 720, respectively. For a proper stability margin and response speed, $\omega_n^{\mathrm{c}} = 320$ is selected for both controls. More detailed tuning can be followed from Section 9.2.2.

9.3.3 Overall Tuning Strategy

Based on the above stability analysis, the tuning strategy for this PLL-based *PV*- and *PQ*-controlled VSC is proposed and shown in Figure 9.18. Current control as the dominant instability cause is tuned first, then PLL tuning and *PV* control tuning are followed based on their response-speed relations $r_{\mathrm{PLL-c}}$, $r_{P-\mathrm{c}}$, and r_{v-P} with the current control. The predefined parameters as the input for this tuning strategy are quantified based on the above stability analysis and shown below:

1. for current control, $\zeta^{\mathrm{c}} = 0.707$ and $\omega_n^{\mathrm{c}} \in [160\ 320]$; its IP controllers' gains are calculated using ζ^{c} and ω_n^{c} based on Equation 9.1;

2. for PLL, $\zeta^{\mathrm{PLL}} = 0.707$ and $r_{\mathrm{PLL-c}} = 2$; its PI controllers' gains are calculated using ζ^{PLL} and $r_{\mathrm{PLL-c}}$ based on Equations 9.2, 9.14, and 9.16;
3. for power and voltage control, $r_{P-\mathrm{c}} = 0.1$ and $r_{v-P} = 5$; its PI controllers' gains are calculated using $r_{P-\mathrm{c}}$ and r_{v-P} based on Equations 9.7 and 9.20.

By following the above tuning method of speeding up the voltage support ($r_{v-P} = 5$), and slowing current control to $\omega_n^{\mathrm{c}} = 320$, a two-level VSC can connect to an extremely weak grid (SCR = 1) with rated power injection using PLL-based *PQ* control or *PV* control, as shown in Figure 9.19.

As shown in Figure 9.19, it is validated at SCR = 10 that the proposed tuning method works well for a strong grid. It should be noted that the PLL-based current control is also validated with a strong grid, which is not presented here.

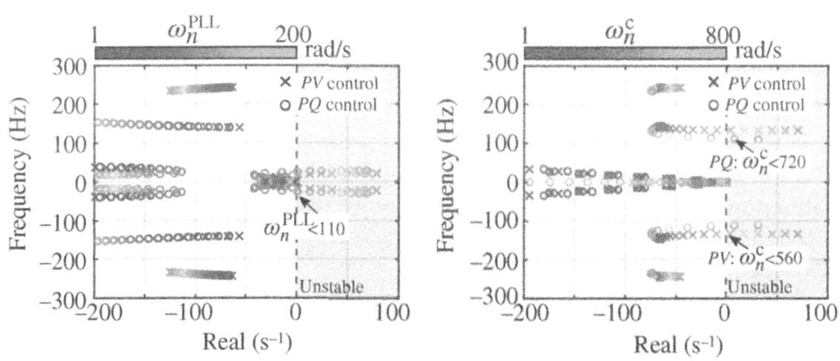

(a) Stability analysis of ω_n^{PLL} (b) Stability analysis of ω_n^{c}

Figure 9.17 Effectiveness of ω_n reduction for both PLL and current control on the extremely weak grid stabilization at SCR = 1 ($r_{v-P} = 5$).

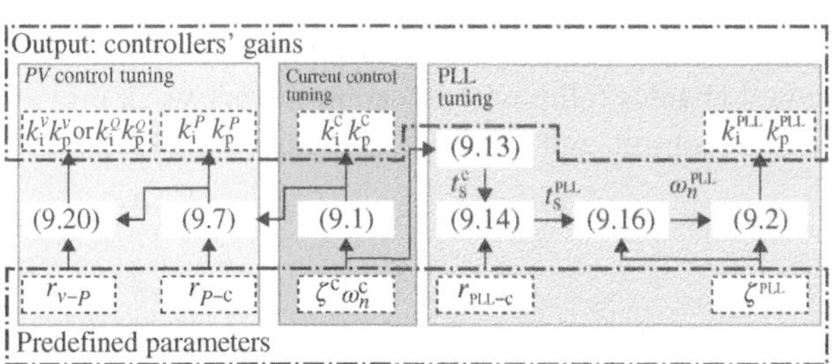

Figure 9.18 Tuning strategy for PLL-based *PV* (or reactive power) control.

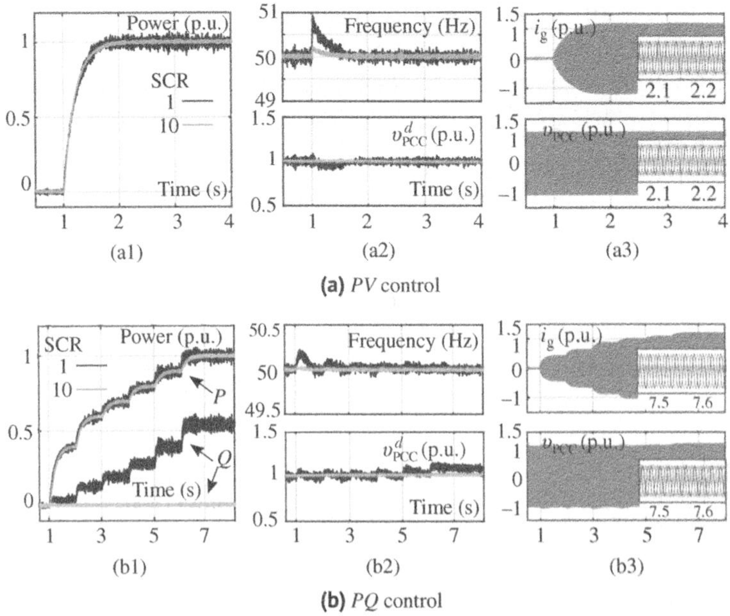

Figure 9.19 Rated power injection of a two-level VSC with the proposed tuning method when connecting to an extremely weak grid with SCR = 1 (current control: $\omega_n^c = 320$, $\zeta^c = 0.707$; outer loop: $r_{v-P} = 5$).

9.4 Summary

Fundamental causes of weak grid instability within each part of the PLL-based control are comprehensively revealed. Accordingly, an overall tuning method for extremely weak grid connection is developed for comprehensively tuning each part of the control, including PLL, inner current control loop, and outer control loop. This tuning method enables full power injection to very weak grid at SCR = 1 for current control, *PV* control, and *PQ* control relying on the PLL.

References

1 C. Li, S. Wang, F. Colas, and J. Liang. Dominant instability mechanism of VSI connecting to a very weak grid. *IEEE Transactions on Power Systems*, 37(1): 828–831, 2022. https://doi.org/10.1109/TPWRS.2021.3115968.

2 C. Li, S. Wang, and J. Liang. Tuning method of a grid-following converter for the extremely-weak-grid connection. *IEEE Transactions on Power Systems*, 37(4): 3169–3172, 2022. https://doi.org/10.1109/TPWRS.2022.3167899.

10

Stability Analysis of Grid-forming Converters

Jun Liang[1,*] *and Chuanyue Li*[2]

[1] *School of Engineering, Cardiff University, Cardiff, UK*
[2] *School of Electric Power Engineering, South China University of Technology, Guangzhou, China*

*Corresponding author: LiangJ1@cardiff.ac.uk

10.1 Development of Grid-forming Converters

Grid-forming (GFM) capability supported by voltage source converters (VSCs) has been discussed since the mid of 2010s due to the concern of declining system inertia and weak grid stability. In 2019 and 2020, inertia provision and black start capability have been demonstrated by Scottish Power Renewables in collaboration with Siemens Gamesa at the application of Dersalloch Wind Farm in Scotland, which presents promising results by adopting GFM converters in power systems. In 2022, "Grid Code Issue 6 Revision 13" stated that GFM capability can be implemented via GFM converters to satisfy the power system stability requirements. The GFM capability of VSCs is realized through the implementation of specific control systems. The VSC, being equivalently a controllable voltage source, provides extensive flexibility for control design, which brings many GFM control variants. This section will summarize the evolution of GFM control and categorize all variants to provide a comprehensive glance at GFM converters.

10.1.1 Connection with Synchronous Generators

Power systems have been formed by synchronous generators (SGs) for many decades. Characteristics of SGs heavily inspire the control design for GFM capability of converters. In this section, the characteristics of SGs imitated by converters are introduced, aiming to facilitate a comprehensive overview of GFM converters.

Stability Analysis of Converter-Rich Power Grids, First Edition. Jun Liang and Chuanyue Li.
© 2026 The Institute of Electrical and Electronics Engineers, Inc. Published 2026 by
John Wiley & Sons, Inc. All rights reserved.

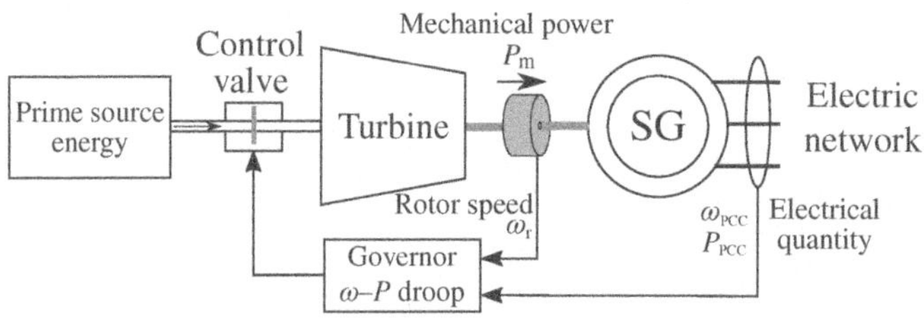

Figure 10.1 An SG system.

A typical configuration of SGs includes two parts: a governor and a turbine-driven SG, as shown in Figure 10.1. The governor must follow the order of ω–P droop control by adjusting rotor speed ω_r, and their relation is simply represented as below:

$$P_m = P^*_{PCC} + \frac{1}{d_P}(\omega^*_r - \omega_r) \tag{10.1}$$

where P^*_{PCC} is the preset power reference, ω^*_r is the reference of rotor speed, and d_P is the power droop coefficient.

The governor controls the rotor speed at ω^*_r by adjusting the control valve to output appropriate torque. The detailed dynamics of governor, which are referred to in [1], will not be presented here as they are simplified in the GFM converter. The turbine controlled by the governor is dedicated to the SG response to load changes. The power transfer through the motion between the turbine and the SG is presented as a swing equation:

$$2H\frac{d\omega}{dt} + D\omega = P_m - P_{PCC} \tag{10.2}$$

where H is the inertia constant and D is the damping ratio.

Assuming that ω_r is equal to the electrical frequency ω, replacing P_m in Equation 10.2 with Equation 10.1 yields:

$$2H\frac{d\omega}{dt} + D\omega = P^*_{PCC} + \frac{1}{d_P}(\omega_0 - \omega_{con}) - P_{PCC} \tag{10.3}$$

The swing equation added in the GFM control aims to support sufficient inertia to power systems via converters. Equation 10.3 represents the simplified mechanical dynamics of an SG and can be emulated by a GFM converter, as shown in Figure 10.2. Equation 10.3 can be further rearranged as follows:

$$\omega = (P^*_{PCC} - P_{PCC}) \times \underbrace{\frac{1}{2Hs + 1/d_P + D}}_{\text{Low-pass filter}} \tag{10.4}$$

All the forms of inertia illustrated in Figure 10.2 are mathematically equivalent, and Equation 10.4 presents a concise representation, which explains that the

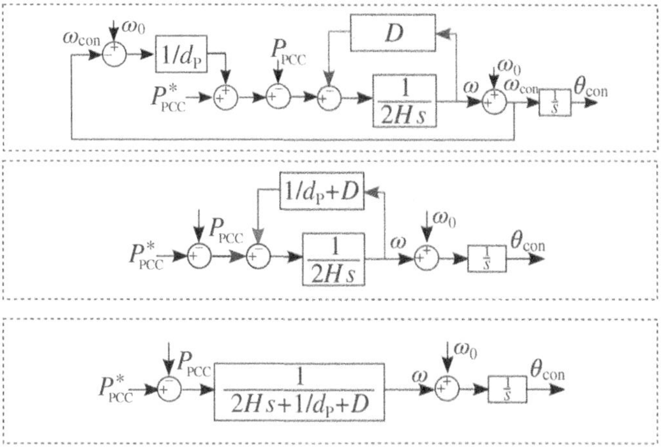

Figure 10.2 Different forms of inertia implemented in the droop control of a GFM converter.

inertia, damping, and droop gain can be represented by a low-pass filter in the droop control.

10.1.2 Development of GFM Control

In recent years, the concept of "grid-forming control" has emerged alongside the development of GFM converters. GFM control is a term used to encompass all types of controls that enable converters to be capable of GFM. This section introduces the evolution of GFM control by highlighting how it draws significant inspiration from the characteristics of SGs. While the term "GFM control" has only emerged in recent years, the control design of converters, specifically referring to droop control of SGs, can be traced back to the 1990s, as shown in Figure 10.3. At that time, to autonomously balance the load among multiple converters, the droop control [2] was adopted by converters and these converters could also form microgrids. In the late 2000s, the control "virtual synchronous generator/machine" (VSG/M) [3, 4] was proposed to enable converters to work like SGs to support weak grids and virtual inertia. At that time, the VSG mainly imitated the swing motion of SGs, as shown in Equation 10.2. In the late 2010s, increasing adoption of converters to integrate renewable energy sources resulted in emergence of a concept known as "GFM converters," which raised substantial interest due to their potential GFM capabilities. The VSG was considered as the earliest GFM control.

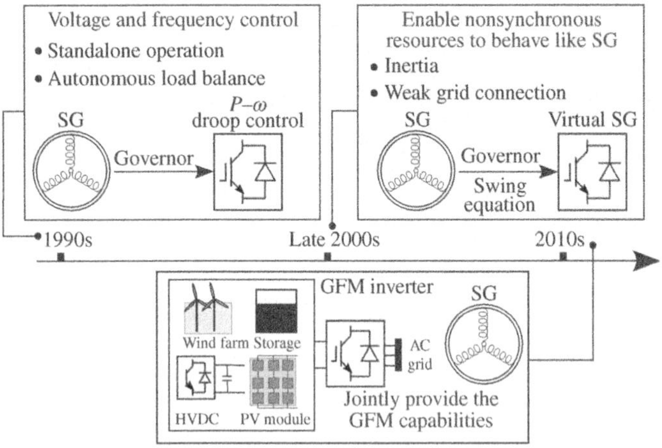

Figure 10.3 Development of GFM control.

10.1.3 GFM Capabilities

The term "grid forming" for a VSC was initially proposed to form a microgrid. Its functions completely differ from today's widely studied GFM converters in power systems. It is considered essential to clarify these distinctions and provide a more detailed specification of GFM functions and capabilities. For microgrids, a "grid-forming converter" is designed to generate specific voltage and frequency, behaving exactly like an alternating current (AC) voltage source. For power systems, the GFM converter is expected to share the GFM capabilities of SGs. Unlike the "grid-forming converter" in microgrids, only providing specific voltage and frequency functions does not fulfill the requirements of power systems to support its stability and security. To align with GFM capabilities of SGs, the functions of GFM converter are summarized but not limited to the following:

- droop functions for P–ω and Q–V to autonomously balance the power;
- sufficient inertia provision to limit the rate of change of frequency (ROCOF) of the system;
- black start and island operation;
- sufficient voltage support to maintain voltage strength;
- sufficient damping capability to avoid large oscillations and instabilities.

In recent years, as the concept of "grid-forming control" underwent extensive development, numerous variants have been proposed.

10.2 Definition and Variants of GFM Control

The definition of GFM control has not yet been widely agreed upon. Two definitions are provided below:

- some inclusive views consider islanding capability as sufficient for classifying a converter as GFM, which includes all GFM control variants, including phase-locked loop (PLL)-assisted converter control, as shown in Figures 10.4 and 10.5.
- the capability for open-circuit operation is considered an exclusive definition of GFM control, as it does not rely on any information from other systems. The variants under this definition are limited to those shown in Figure 10.4.

In island operation, although there is no other voltage source present besides the converter, this condition alone does not guarantee exclusive GFM behavior. In fact, a PLL-assisted converter control can also enable island operation, despite its inherent nature as a current-source control system, as shown in Figure 10.5a. It can fully possess the functions like island operation, inertia support, and autonomous power balance. The concept of a PLL-based grid-support control was first proposed in 2008 [5] and subsequently refined by incorporating the inertia function

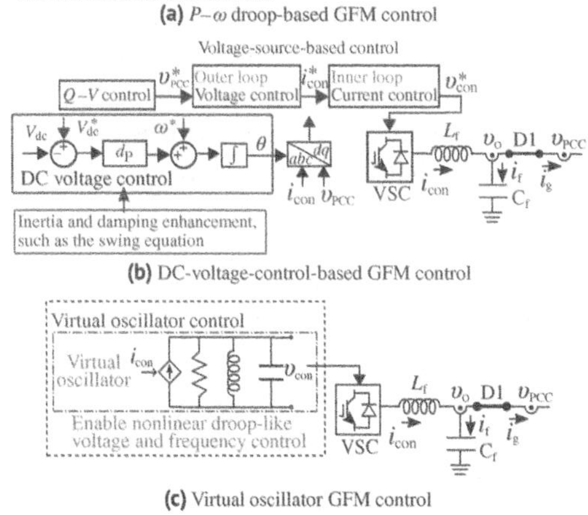

Figure 10.4 Variants of GFM control.

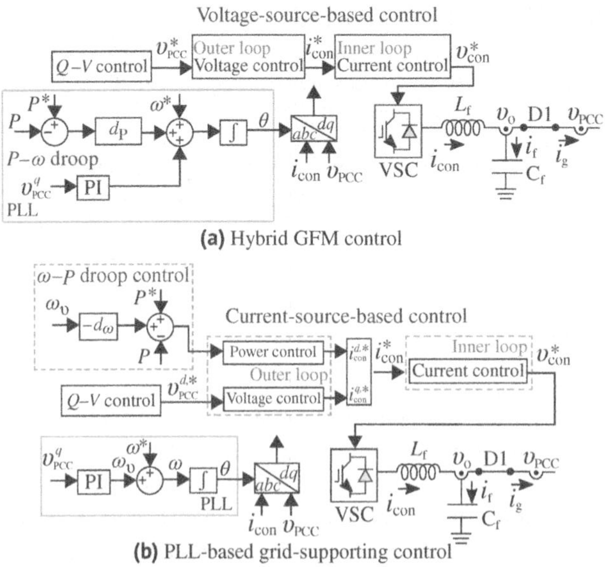

Figure 10.5 PLL-assisted converter control.

in 2009 [6]. The demonstration of a PLL-based grid-supporting control enabling island operation is provided in Section 10.4.

However, this control has received little attention due to concerns about instability arising from the weak grid performance of the PLL. To effectively combine the advantages of both droop and PLL, a hybrid GFM control is proposed [7], as shown in Figure 10.5b. The synchronization of the control system is achieved by a combination of both droop and PLL. Therefore, the hybrid GFM control has an advantage in overcurrent control capability as PLL can efficiently work with inner current control.

While the PLL cannot operate under open-circuit conditions due to the presence of a self-locking condition, as shown in Figure 10.6, it works well to lock onto the point of common coupling (PCC) voltage when a grid is present, as illustrated in Figure 10.6a. Even when the grid is disconnected, the current driven by a load can still shape the PCC voltage through the grid impedance and the load itself, enabling the PLL to maintain synchronization, as shown in Figure 10.6b. However, in the case of a true open circuit on the converter side, the PCC voltage is entirely formed by the converter, resulting in a self-locking condition where $v_{con} = v_{PLL}$. In other words, the PLL output becomes identical to its input, making it unable to function properly.

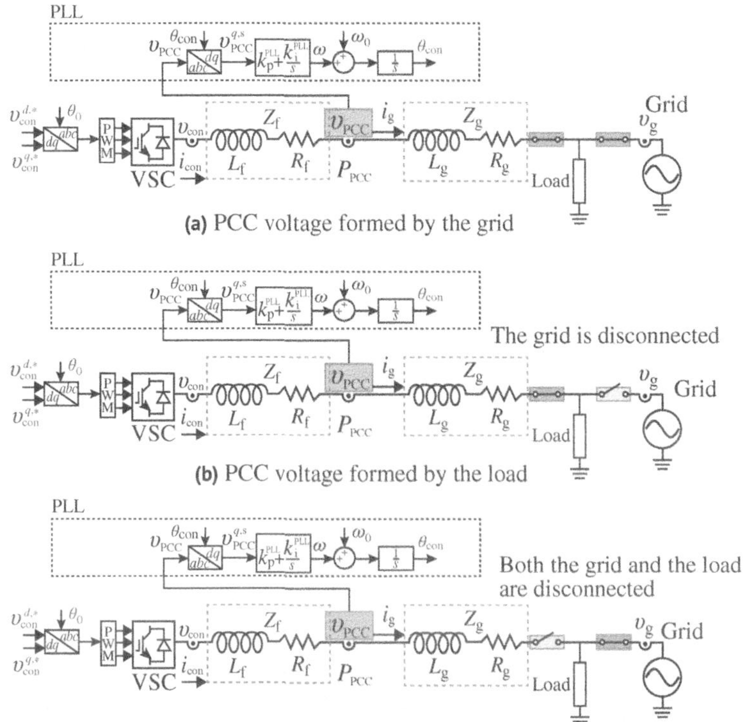

(a) PCC voltage formed by the grid

(b) PCC voltage formed by the load

(c) In open-circuit conditions, the PCC voltage is formed by the converter itself

Figure 10.6 PCC voltages under varying grid conditions.

For exclusive GFM control, which requires open-circuit operation, a PLL-based approach cannot support this functionality. Therefore, only the GFM control variants shown in Figure 10.4 are included in this category.

Up to this point, the exclusive GFM control variants can be classified into three categories, namely:

1. P–ω droop-based GFM control;
2. direct current (DC)-voltage-control-based GFM control;
3. virtual oscillator GFM control.

A basic P–ω droop-based GFM control is shown in Figure 10.4a, such as VSG [8], in both stationary frame [9, 10] and dq-frame [11]. The swing equation can be referred to in Equation 10.2. The dynamics of a governor can be simplified as a droop control, and the relation is shown below:

$$\omega = (P^* - P)d_P \tag{10.5}$$

The P–ω droop-based GFM control is not limited to the structure of a VSG; rather, it offers the flexibility to design inertia and damping independently. The swing equation is optional in this context, and alternative control strategies—such as lead–lag compensators—can achieve equivalent dynamic performance. In stationary-frame implementations, droop control serves to balance active power and establish a frequency reference for voltage generation. An inner voltage-source-based controller is then employed to track the resulting voltage reference.

For dq-frame control, droop serves to balance the load and maintain synchronization of the dq-frame frequency. The key distinction between GFM controls implemented in the stationary frame and the dq-frame lies in their responses to overcurrent conditions. When an overcurrent event occurs, the inner current control loop saturates and assumes control of the system to limit the current within a safe range. In stationary-frame control, the inner current loop is designed for a fixed rated frequency; therefore, any frequency deviation during an overcurrent condition may result in control inaccuracies. In contrast, for dq-frame control, the droop control is not disabled by the inner current control during overcurrent events, which can potentially lead to unstable interactions.

In summary, the P–ω droop-based GFM controls exhibit voltage source characteristics, enabling effective support of weak grids. However, when interfacing with a strong grid, it is essential to carefully evaluate the risk of instability. As previously noted, particular attention must also be given to the design and performance of current-limiting mechanisms.

To enable a GFM converter to control the DC voltage, a DC-voltage-control-based GFM scheme is developed, as shown in Figure 10.4b [12]. This approach utilizes the relationship between DC voltage and active power by modifying the conventional P–ω droop control into a DC voltage droop. This control strategy extends the application of GFM converters and offers the potential for high-voltage direct current (HVDC) links with GFM capabilities at both terminals—one implementing DC-based GFM control and the other enabling AC-side GFM control.

The virtual oscillator-based GFM control, as shown in Figure 10.4c, is founded on a different control mechanism compared to the above-mentioned GFM strategies. Briefly, it originates from the behavior of a linear, undamped LC resonant circuit to achieve nonlinear AC voltage control and has evolved to include other definable voltage amplitude methods [13]. This control approach has been applied to GFM systems in recent decades and is still under development, with ongoing efforts such as incorporating inertia [14].

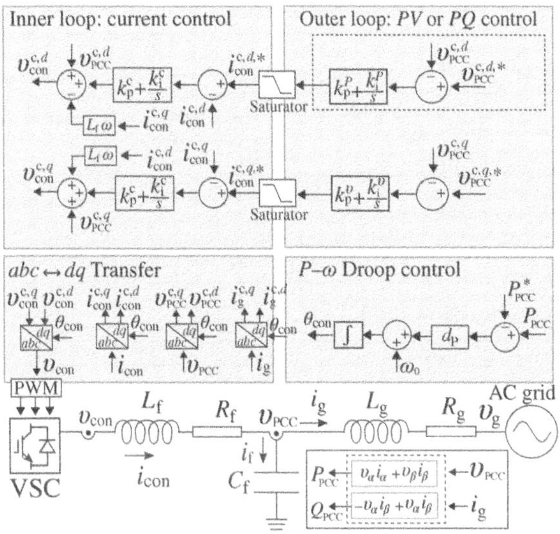

Figure 10.7 Typical P–ω droop-based GFM control.

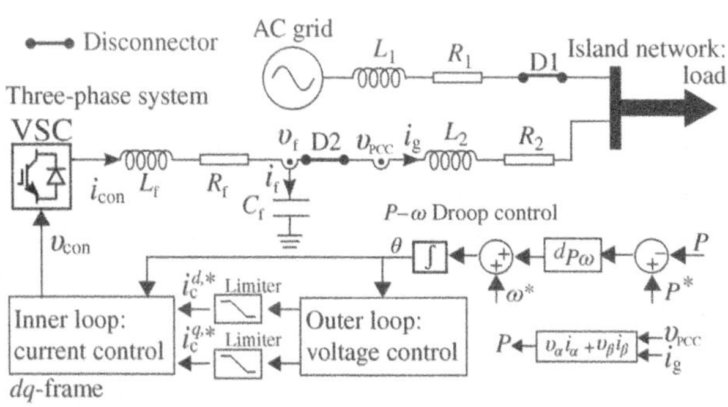

Figure 10.8 Island operation system for a grid-supporting converter.

10.3 Brief Comparative Analysis of Stability Characteristics: GFM Converters and SGs in Strong and Weak Grids

10.3.1 Causes of GFM Converter Instability in Strong Grids

Most GFM converters refer to the P–ω droop-based control, as it originates from the control principles of SGs. Some studies have observed that GFM converters may experience oscillations or instability when connected to a strong grid. This is attributed to the tight voltage control nature of GFM converters, which can interact strongly with a stiff grid, potentially leading to instabilities [15].

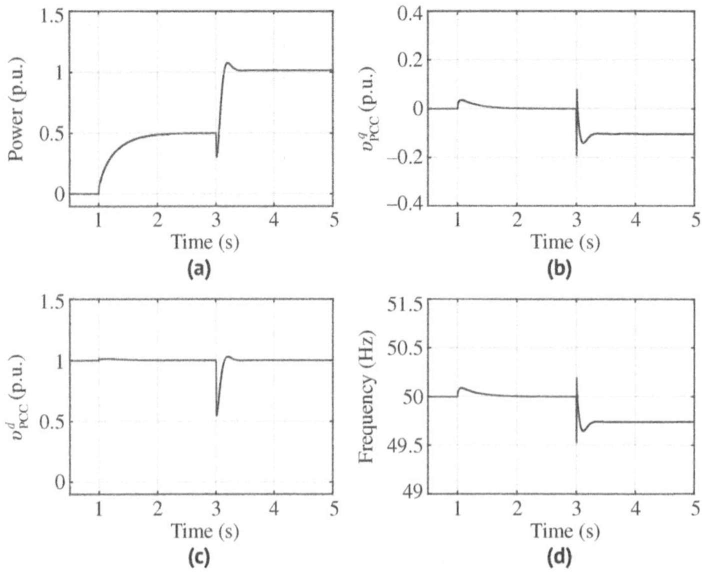

Figure 10.9 Island operation of a PLL-based grid-supporting converter.

However, it is uncommonly noted that an SG can also exhibit the same instability in a strong grid. Furthermore, the interaction between two voltage sources alone does not necessarily lead to oscillations. Instead, the primary cause lies in the control interactions of the GFM converter. A deeper understanding of these interactions is crucial for clearer insights into GFM control behavior. The following mechanism is identified:

Instability causes of GFM control in a strong grid: a small voltage deviation in a strong grid can cause a large current deviation, posing significant challenges for current regulation. Therefore, the inner-loop current control—commonly used as a control limitation strategy in converters—is the main cause of instability in strong grid conditions. In contrast, voltage control alone is not significantly affected by a strong grid.

The validation of the stability analysis is presented in Figure 10.10, including Control Option 1 with only voltage control and Control Option 2 with both current and voltage control applied to the GFM converter. With an increased short-circuit ratio (SCR) of 10, the grid becomes stronger. It is observed that the poles of Control Option 2 move to the right half-plane (RHP), while Control Option 1 is not significantly affected. This confirms that the inner current control is the main cause of instability in strong grid conditions.

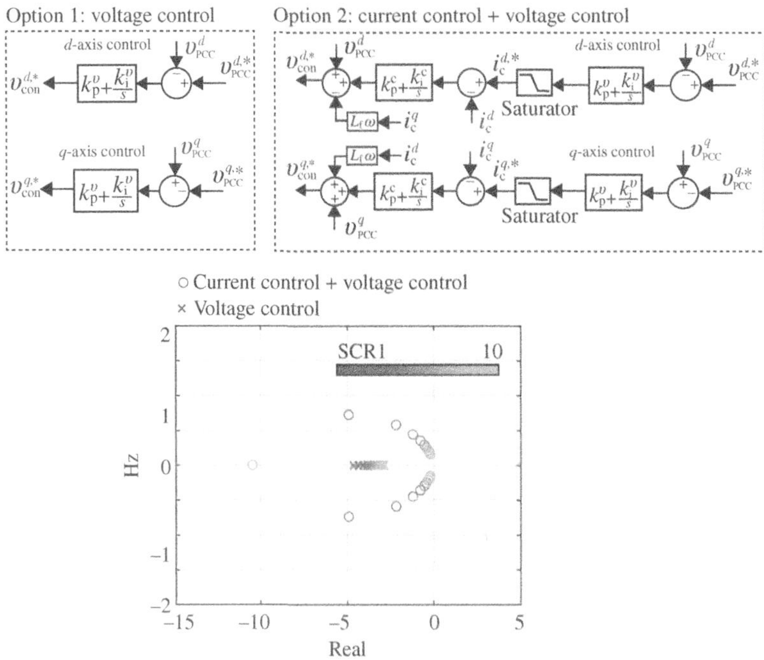

Figure 10.10 Impact of the inner current loop on strong grid instability.

10.3.2 A Brief Discussion on the Stability of SGs Under Weak Grid Conditions

While GFM control mainly imitates the mechanical dynamics of an SG, its voltage control differs from that of SGs, which is implemented through an excitation system. This leads to different stability performances. The stability dynamics of GFM converters—particularly those related to voltage control—cannot be considered equivalent to those of SGs. Even within GFM control itself, different voltage control structures result in different stability characteristics, as discussed in the previous sections.

An ideal SG model is shown in Figure 10.11, where the stator output voltage is regulated by controlling the excitation current i_f. This type of voltage control via the excitation system is fundamentally different from that of GFM converters. The nature of control interaction is also different, and thus their stability performance cannot be considered equivalent. Based on the stability study of SGs in [16], it has been shown that large grid reactance and high SG output can lead to negative damping due to the excitation system, which results in a similar instability situation as seen in GFL converters under weak grid conditions.

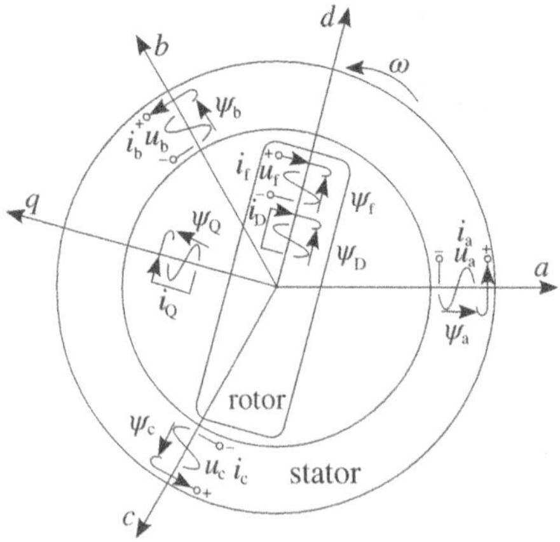

Figure 10.11 An ideal bipolar SG.

10.4 Comparative Analysis of PLL-based Grid-supporting Control and *P*–*ω* Droop-based GFM Controls

Compared to the VSG and the droop-based GFM control, PLL-based grid-supporting control has an advantage in overcurrent control capability as PLL can efficiently work with inner current control. At the moment, a GFM control commonly refers to a voltage-source-based control such as the three variants in Figure 10.4a, b. If the foundation of a GFM control is to form a grid, a PLL-based control in Figure 10.4c, which can achieve the same function, should belong to this GFM control family.

While this PLL-based grid-supporting control is normally characterized as a current source, a meager attention is paid to this GFM control, so this section is for comprehensively highlighting its distinction, merits, and challenges by comparing to voltage-source-based GFM controls.

10.4.1 Island Operation of a PLL-based Grid-supporting Converter

The control configuration, which features the characteristics of current source and only relies on frequency alignment of PLL, is categorized as a current-source-based control. Recently, the term "grid following" (GFL) is used to define this control configuration, which is widely agreed to be a distinguishable feature from "grid forming."

However, in fact, the mechanism of a current-source-based control is feasible enough for forming a grid, which can fulfill the requirements of GFM capability such as island operation. This mechanism is achieved by shifting the constant power control at outer loop to flexible ω–P droop control, as shown in Figure 10.4c, which forms the current-source-based GFM control. One main feature is that this control only relies on the PLL for frequency alignment during GFM, while the voltage-source-based GFM adopts P–ω droop control for frequency alignment.

To demonstrate the feasibility of this current-source-based GFM control, an island operation is simulated via a test circuit in Figure 10.8 and its results are shown in Figure 10.9. At 3 s, the AC grid is disconnected by disconnector D1 and the load is fully powered via the converter, as shown in Figure 10.9a. The PCC voltage is well maintained at 1 p.u., as shown in Figure 10.9b–d. The results also show that $\Delta P = 0.5$ p.u. results in $\Delta v_{\text{PCC}}^{q} = -0.1$ p.u. and $\Delta\omega = 0.25$ Hz, which match the preset parameters for ω–P droop control.

The potential of a current-source-based control to form a grid has been demonstrated above. The exploitation of inertia provision will not be discussed further.

10.4.2 Control Conflict Analysis: P–ω Droop and Inner Current Control in GFM Converters

The P–ω droop-based GFM control, widely used in GFM converters, is adopted as the voltage-source-based GFM control. Its overall control system is shown in Figure 10.7. The P–ω droop control is to balance the loads and maintain local frequency. The voltage at PCC is controlled via the outer loop. An inner current loop is essentially applied to protect the VSC from overcurrent. The outer loop and the inner loop work in the dq-frame (also known as a synchronous reference frame). The synchronization of this dq-frame relies on the P–ω droop control.

However, the P–ω droop control is a power-based synchronization, which brings two issues to GFM converters. First, the synchronization does not work under an open-line situation, such as opening the disconnector D2 in Figure 10.8. Second, the P–ω droop control may conflict with the inner current control after the current reference saturates, which could cause a wide voltage variance or further voltage collapse. This conflict brings a significant challenge for current limitation, and additional controller designs are required but it makes the overall control more complicated.

Under normal conditions, the PCC voltage is fixed at the desired value via the outer voltage control loop. Therefore, the power control via the P–ω droop is achieved by driving the current.

However, the aforementioned conflict could occur when current limiting starts via the inner current control and the outer voltage control saturates. The current reference ($i_{\text{con}}^{d,*}$ or $i_{\text{con}}^{q,*}$) is tightly held via the limiters. Therefore, the P–ω droop control has to drive the voltage to achieve the desired power output, as shown below:

$$\Delta P^* = 1.5(\Delta V_{\text{PCC}}^{d} I_{\text{g}}^{d} + \Delta V_{\text{PCC}}^{q} I_{\text{g}}^{q}) \tag{10.6}$$

where ΔP^* is the desired power change.

This could change the PCC voltage significantly, which is unacceptable from an AC network. A voltage collapse could be further induced by the P–ω droop control, which is explained below.

As shown in Figure 10.8, the P–ω droop control adjusts synchronizing phase θ to reach ΔP^*. Their relation is given below:

$$\Delta\theta = \int (d_{P\omega} \Delta P^*) \mathrm{d}t \tag{10.7}$$

The impact of $\Delta\theta$ on the PCC voltage $\Delta V_{\text{PCC}}^{d}$ is shown below based on the *abc*–*dq* transfer:

$$\Delta V_{\text{PCC}}^{d} = [\cos(\varphi - \theta - \Delta\theta) - \cos(\varphi - \theta)] V_{\text{PCC}} \tag{10.8}$$

where φ is the phase of PCC voltage, θ is the synchronized phase from the P–ω droop control, and $\Delta\theta$ can be referred from Equation 10.7.

If $(\varphi - \theta) \in (-\pi\ 0]$, ΔP^* results in $-\Delta V_{\text{PCC}}^{d}$, which explains why the P–ω droop control may cause the PCC voltage collapse.

10.4.3 Comparative Analysis of Grid-supporting and GFM Control Strategies

To explicitly present the difference between both converter controls, Table 10.1 summarizes both controls' characteristics.

Table 10.1 Comparison of PLL-based grid-supporting control and P–ω droop-based GFM control.

Characteristic	PLL-based grid-supporting control	P–ω droop-based GFM control
Island operation	Yes	Yes
Frequency-based power balance	Yes	Yes
Inertia provision	Yes	Yes
Open-circuit synchronization	No	Yes
Inner loop's current limitation	No	Yes

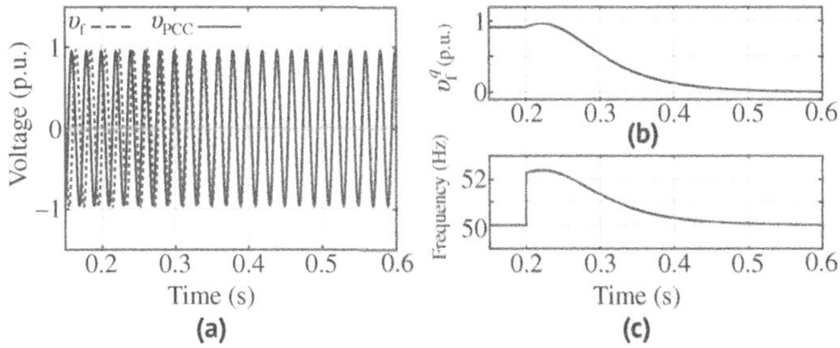

Figure 10.12 Open-line synchronization from 0.2 s (D2 in Figure 10.8 is open).

Both grid-supporting and GFM controls possess the first three characteristics, which are the basic GFM requirements as explained in the above section. One difference is that the synchronization of PLL relies on the voltage, which enables current-source-based GFM control to synchronize with an open-circuit network, such as when D2 is open in Figure 10.8, while the voltage-source-based GFM control cannot achieve this due to its power-based synchronization.

A test result of the open-line synchronization is shown in Figure 10.12. The synchronization starts from 0.2 s with D2 open. It is found that the v_f of the VSC effectively synchronizes with the grid voltage v_{PCC} due to the PLL. v_f^q and frequency of the VSC return to their set references at 0 p.u. and 50 Hz, respectively.

The second difference is about the way of current limitation, although both grid-supporting and GFM controls adopt inner current controls. For voltage-source-based GFM control, its current limitation via the inner loop may force the P–ω droop to destabilize voltage as mentioned in the above section, while the adoption of PLL in current-source-based GFM control effectively and simply enables this current limitation via the inner loop.

A test case based on Figure 10.8 for both controls is given, which is increasing the active power to 1.2 p.u. and the inner current control saturates at 1.1 p.u. The results are shown in Figure 10.13. It is found that when the current reaches 1.1 p.u., P–ω droop-based GFM control causes the VSC voltage collapse, as shown in Figure 10.13c, while the PLL-based grid-supporting control well limits the current at 1.1 p.u. and the VSC still works well with the grid.

10.4.4 Challenges of Current-source-based GFM Control

The challenges are split into two categories: (i) specification of GFM capability; (ii) concern of adopting current-source-based control. Actually, challenge (i) is for all types of GFM controls and is being discussed by the German Technical

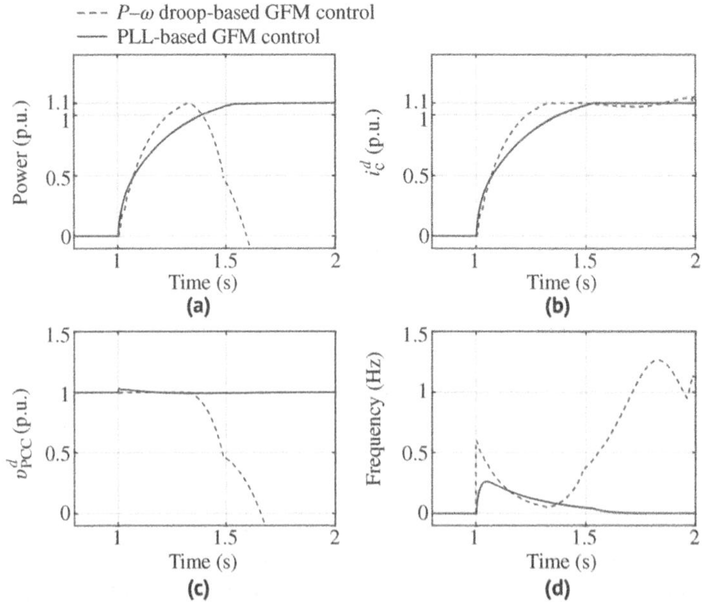

Figure 10.13 Current limitation via the inner current control.

Regulator VDE FNN [17] and the UK National Grid [18]. The recent discussion brings following considerations:

- what functions the GFM capability can serve, which can include frequency-power regulation, inertia, island operation, damping, and voltage support?
- how to qualify those GFM functions for grid code compliance?
- how to maintain GFM capability of low-current-withstanding converters during transients?

If the foundation of GFM relies on a current-source-based control, one enormous challenge for this type of GFM control is:

- the widely raised concern about unstable weak grid connection of current-source-based control.

After extensive review, only a few researches [19, 20] attempted to dismiss this concern, but their results did not give a fully satisfied solution. Stable rated power injection to an extremely weak grid is not validated.

Although a current-source-based GFM control takes advantages of open-circuit synchronization, simple current limitation design, and stable interaction with a strong grid, the concern of weak grid instability significantly distracts the relevant attention and hinders its further development.

10.5 Comparative Analysis of Stability in GFL and GFM Converters

For a grid-connected converter requiring control over the AC voltage magnitude and active power, both GFL control and GFM control, as shown in Figures 9.1 and 10.7, can be applied. However, the function of island operation, which is not applicable to GFL control, is not mandatory in this context. The emerging issue is how both controls interact with the power system and what differences they exhibit in terms of stability.

Therefore, a comprehensive stability comparison study between the GFL control and GFM control for regulating the AC voltage magnitude and active power is essential. This section presents the following aspects of the stability comparison:

- grid strength: the stability influence of the connected grid is studied, ranging from a strong grid to a weak grid.
- control parameter impact: both controls require an inner current control loop. Stability comparisons are conducted based on various cutoff frequencies of this inner current loop.
- measures for stability enhancement: different types of virtual impedance control methods for enhancing the stability of both controls are proposed and compared.

The small-signal modeling and stability analysis methods for both the GFL converter and the GFM converter were introduced in Chapters 4 and 5 and will not be presented again. The parameters of both controls are shown in Table 10.2.

Table 10.2 Parameters of both GFL control and GFM control.

	GFL control	
ω_c^c	Cutoff frequency of closed-loop current control	800 rad/s
k_p^c, k_i^c	Proportional–integral (PI) for current control	0.382, 4
k_p^P, k_i^P	PI for power control	$0.1k_p^c, 0.1k_i^c$
k_p^v, k_i^v	PI for voltage control	$0.1k_p^c, 0.1k_i^c$
ω_c^{PLL}	Cutoff frequency of phase-locked loop	210 rad/s
k_p^{PLL}, k_i^{PLL}	PI for PLL	420, 44,100
	GFM control	
ω_c^c	Cutoff frequency of closed-loop current control	4,000 rad/s
k_p^c, k_i^c	PI for current control	1.91, 20
k_p^v, k_i^v	PI for voltage control	$0.1k_p^c, 0.1k_i^c$
d_P	Droop coefficient	$0.005\omega_0$

10.5.1 Stability Analysis Under Various Grid Strength Conditions

For a transmission network, the grid impedance is mainly inductive. The stability influence of the grid strength is via grid inductance X_{L_g} ranging from 0.05 to 0.95 p.u. on the grid-connected converter being studied. The grid resistance R_g maintains $0.1X_{L_g}$. The determinants of the impedance ratio matrixes of the grid-connected converters with the GFL control and the GFM control are shown in Figure 10.14. The pole loci close to $x = 0$ are shown. It is found that:

1. for a converter with the GFL control, increasing grid impedance makes the poles to move toward the RHP. The stability is therefore reduced. When $X_{L_g} >$ 0.75 p.u., the system becomes unstable as shown in Figure 10.14.
2. for a converter with the GFM control, it is more stable with the large grid impedance (weak grid), as shown in Figure 10.14.
3. for a converter with the GFM control, decreasing the grid impedance makes the poles to move toward the RHP. The stability is therefore reduced. When $X_{L_g} < 0.15$ p.u., the system becomes unstable as shown in Figure 10.14.

The impedance analysis is applied to a converter with GFM control. If the grid impedance is small, the magnitude of $\boldsymbol{Z}_{con} - \boldsymbol{Z}_g$ is small in the low-frequency range due to their inductive characteristics, which results in a large magnitude for $\frac{1}{\det(\boldsymbol{Z}_{con} - \boldsymbol{Z}_g)}$. As shown in Figure 10.15, the peak magnitude increases as the grid impedance decreases. In summary, the converter with GFM control may experience instability with a strong grid.

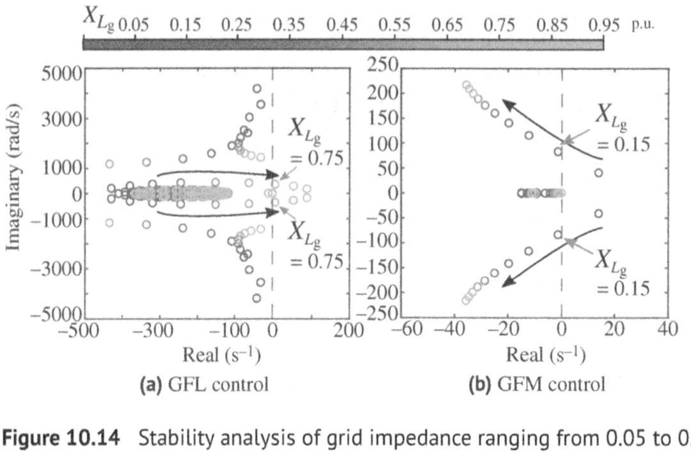

(a) GFL control (b) GFM control

Figure 10.14 Stability analysis of grid impedance ranging from 0.05 to 0.95 p.u. (X_{L_g}).

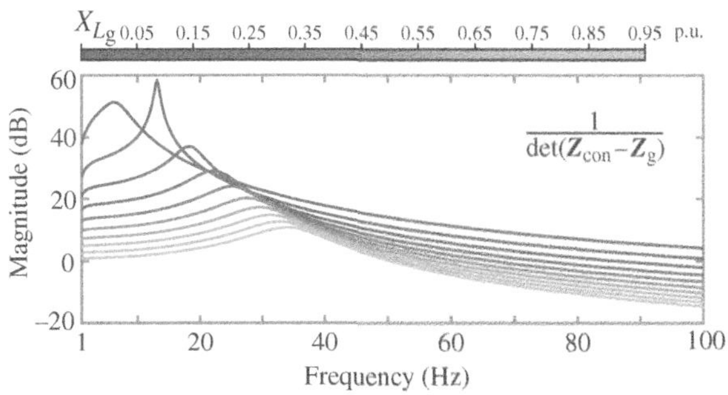

Figure 10.15 Bode plot of $\frac{1}{\det(\mathbf{Z}_{con} - \mathbf{Z}_g)}$ with the grid impedance ranging from 0.05 to 0.95 p.u.

10.5.2 Stability Analysis with Various Cutoff Frequencies of the Inner Current Control Loop

The closed-loop cutoff frequency (ω_c^c) is normally used to design the PI for the current loop. Both controls contain the inner current loop to avoid the overload. Therefore, the stability influences of ω_c^c ranging from 800 to 8,000 rad/s on both controls are analyzed. The outer-loop PI is 10 times slower. The determinants of the impedance ratio matrixes based on the grid-connected converters with the GFL control and the GFM control are shown in Figure 10.16. The pole loci close to $x = 0$ are shown. It is found that:

1. for a converter with the GFL control, increasing ω_c^c makes the poles to move toward the RHP. The instability happens with a high ω_c^c such as $\omega_c^c > 5{,}714$ rad/s, as shown in Figure 10.16a.
2. for a converter with the GFM control, increasing ω_c^c makes the poles to move away from the RHP. The instability happens with a low ω_c^c such as $\omega_c^c < 2{,}667$ rad/s, as shown in Figure 10.16b.

It is clear from Figure 10.16a, b that the design of the inner current loop for the GFL control is different from that of GFM control. Reducing ω_c^c helps to stabilize the converter with the GFL control. Conversely, increasing ω_c^c helps to stabilize the converter with the GFM control. The determinants of the impedance ratio matrixes for both controls are also shown as Bode plots to find out their resonant frequencies when $\omega_c^c = 6{,}154$ and $\omega_c^c = 2{,}667$, respectively, as shown in Figure 10.16c, d. It is found that ultrahigh magnitudes appear at 390 and 8.9 Hz, respectively, under the GFL control and the GFM control.

The stability analysis and resonant frequency analysis based on pole maps and Bode plots are validated via the simulations. Two poles that close to $x = 0$ are selected and pointed out in Figure 10.16a, b. Values of ω_c^c of the two poles are also

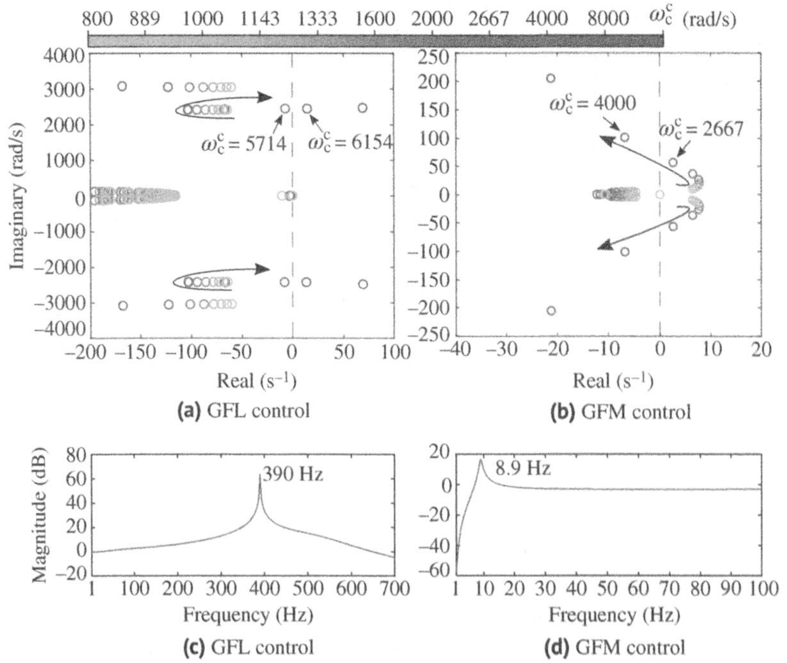

Figure 10.16 Stability analysis of various closed-loop cutoff frequencies (ω_c^c) of the inner current loop.

shown. The simulation results for the selected poles are shown in Figure 10.17. For the converter with the GFL control, the system is stable at $\omega_c^c = 5{,}714$ rad/s and unstable at $\omega_c^c = 6{,}154$ rad/s as shown in Figure 10.17a. Similar stability result is found for the converter with the GFM control, as shown in Figure 10.17b. The simulation results match the stability analysis based on the pole map. The Fourier analysis results are shown in Figure 10.17c, d. The main resonant frequencies after 1 s of i_g^d for both controls are 388 and 9 Hz, respectively, which also match the analysis results from Bode plots.

10.5.3 Stability Enhancement Using Virtual Impedance

The virtual impedances are added in the current control for modifying the converter impedance $\boldsymbol{Z}_{con}$ of the converter based on the Bode plots of $\boldsymbol{Z}_{con}$ and $\boldsymbol{Z}_g$. The ultra-small magnitude appearing in det($\boldsymbol{Z}_{con} - \boldsymbol{Z}_g$) should be avoided in order to enhance the stability. The Bode plots of $\boldsymbol{Z}_{con}$ for both controls are shown in Figures 10.18 and 10.19 based on the inductive grid impedance $\boldsymbol{Z}_g$. It is found that:

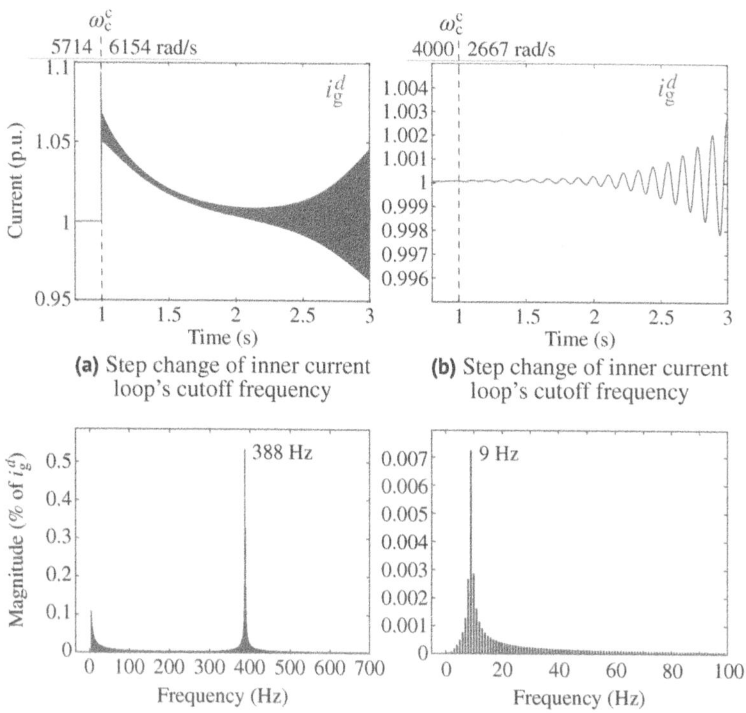

(a) Step change of inner current loop's cutoff frequency

(b) Step change of inner current loop's cutoff frequency

(c) Frequency analysis of i_g^d vector control

(d) Frequency analysis of i_g^d power synchronization control

Figure 10.17 Simulation validation of the inner current loop based on its cutoff frequency.

1. for the GFL control, as shown in Figure 10.18, $\boldsymbol{Z}_{\text{con}}$ behaves like a capacitive impedance. There are crossing points of $\boldsymbol{Z}_{\text{con}}$ and $\boldsymbol{Z}_{\text{g}}$ in the magnitudes of both *dd*- and *qq*-elements. If their phases are close at the crossing points, which results in a large magnitude of $\frac{1}{\det(\boldsymbol{Z}_{\text{con}} - \boldsymbol{Z}_{\text{g}})}$, the risk of instability increases. Especially for the *qq*-element, a larger grid impedance will make the magnitude crossing point more close to the phase crossing point. It can be avoided by adding the virtual inductor.
2. for the GFM control, as shown in Figure 10.19, $\boldsymbol{Z}_{\text{con}}$ behaves like an inductive impedance. There are no crossing points at the mid-frequency range. However, $\boldsymbol{Z}_{\text{con}}$ has a small magnitude at the low frequency, which results in a large magnitude of $\frac{1}{\det(\boldsymbol{Z}_{\text{con}} - \boldsymbol{Z}_{\text{g}})}$. If the grid impedance is reduced, the risk of the instability increases. The virtual resistance can be inserted to change the magnitude of $\boldsymbol{Z}_{\text{con}}$ in the low-frequency range to avoid the risk of the instability.

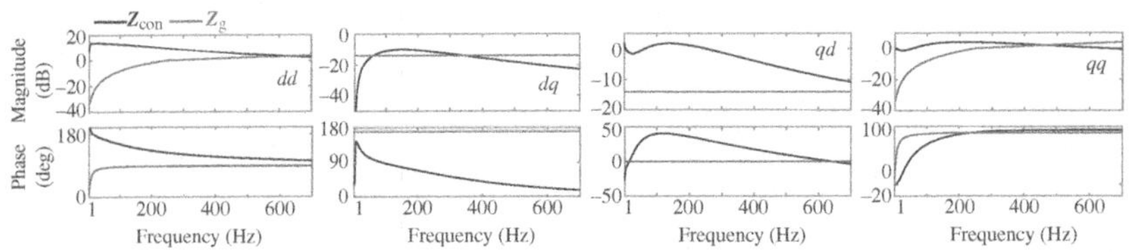

Figure 10.18 Impedances of GFL converter.

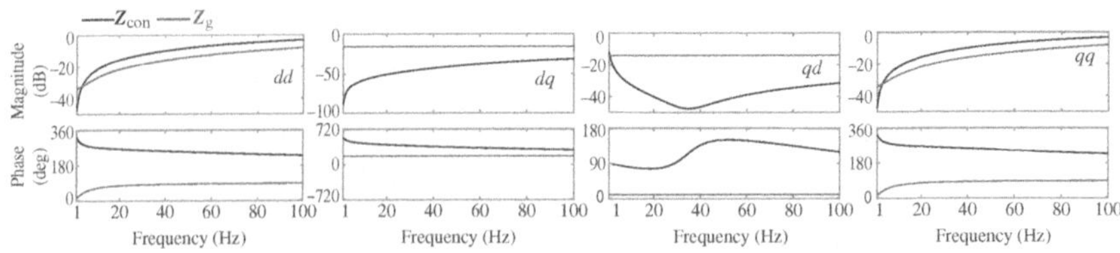

Figure 10.19 Impedances of GFM converter.

The unstable cases of both controls at $\omega_c^c = 6{,}154$ and 2,667 rad/s, respectively (as shown in Figure 10.17), are used to validate the effectiveness of virtual impedance. The Bode plots of $\frac{1}{\det(\boldsymbol{Z}_{con} - \boldsymbol{Z}_g)}$ for both controls are drawn in Figure 10.20a, b. It is found that both peak magnitudes are reduced significantly with the increasing virtual inductance and resistance. The determinants of the impedance ratio matrixes for both controls are also drawn. The pole loci close to $x = 0$ are shown in Figure 10.20c, d. Increasing virtual inductance and resistance makes the poles to move to the left half-plane (LHP) for the GFL control and the GFM control, respectively, which validates the effectiveness of the virtual impedance for enhancing the stability of a grid-connected converter.

The effectiveness of the virtual impedance for both controls is also validated via the simulations. The maximum values of the virtual impedance from Figure 10.20a, b are selected, where $L_v = 3.18 \times 10^{-4}$ p.u. and $R_v = 0.02$ p.u. When a step change of ω_c^c happens at 1 s, the converter is still stable with the inserted virtual impedance for both controls, as shown in Figure 10.21. On the contrary, the converters with both controls lose stability without the virtual impedances as shown in Figure 10.17a, b.

10.5.4 Summary

For a grid-connected converter required for the AC voltage magnitude and active power control, the stability comparison between the GFM and the GFL controls is achieved based on the *dq*-impedance stability analysis. Stability influences of

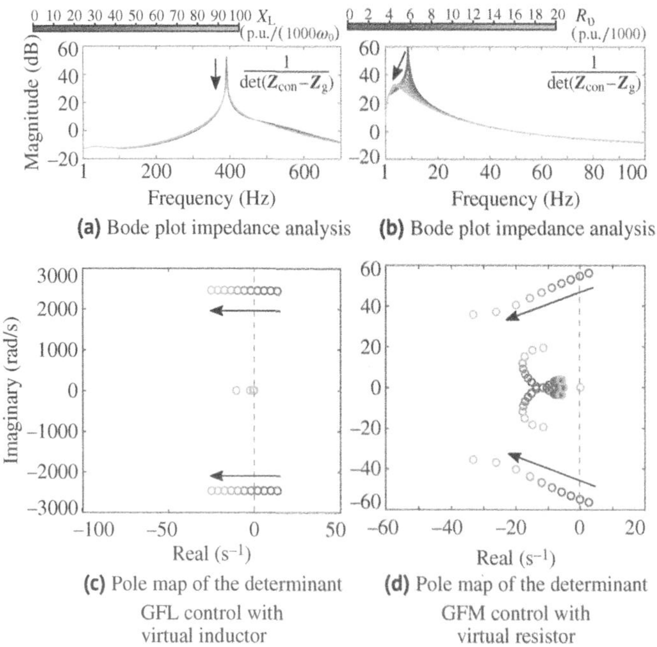

Figure 10.20 Simulation validation using virtual impedance.

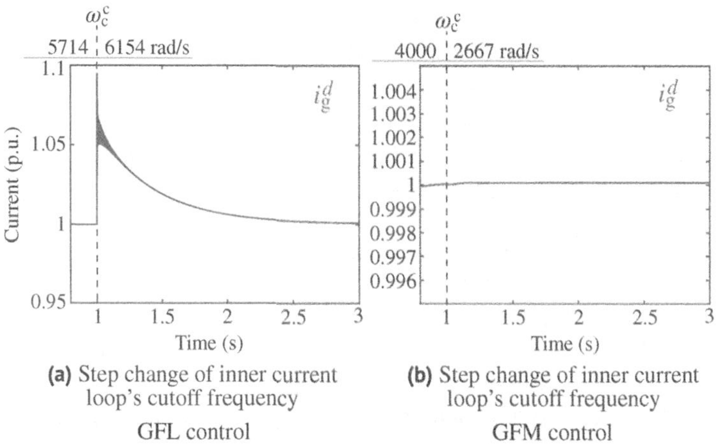

Figure 10.21 Simulation validation of the inner current loop based on its cutoff frequency.

three factors including the grid impedance, the inner current loop, and the virtual impedance on both controls are analyzed and compared. It is concluded that:

1. the GFM could suffer instabilities when a strong grid is connected. Because its impedance model behaves capacitively, which is small at low-frequency range. On the contrary, the stability of the converter with the GFL control is weakened when the grid impedance increases.
2. increasing the cutoff frequency of the inner current loop helps to stabilize the converter with the GFM control, which is converse to the converter with GFL control.
3. it is found that the impedances of the GFL control and the GFM are capacitive and inductive, respectively. Therefore, the virtual resistance helps to stabilize the converter with the GFM control. The virtual inductance helps to stabilize the converter with the GFL control. Value of the virtual impedance can be determined via the derived impedance models.

References

1 I. Boldea. *Synchronous Generators*. CRC Press, Boca Raton, FL, 2016.

2 M.C. Chandorkar, D.M. Divan, and R. Adapa. Control of parallel connected inverters in standalone AC supply systems. *IEEE Transactions on Industry Applications*, 29(1): 136–143, 1993. https://doi.org/10.1109/28.195899.

3 H.-P. Beck and R. Hesse. Virtual synchronous machine. In *2007 9th International Conference on Electrical Power Quality and Utilisation*, pages 1–6, 2007. https://doi.org/10.1109/EPQU.2007.4424220.

4 J. Driesen and K. Visscher. Virtual synchronous generators. In *2008 IEEE Power and Energy Society General Meeting – Conversion and Delivery of Electrical Energy in the 21st Century*, pages 1–3, 2008. https://doi.org/10.1109/PES.2008.4596800.

5 I.A. Hiskens and E.M. Fleming. Control of inverter-connected sources in autonomous microgrids. In *2008 American Control Conference*, pages 586–590, 2008. https://doi.org/10.1109/ACC.2008.4586555.

6 M.P.N. van Wesenbeeck, S.W.H. de Haan, P. Varela, and K. Visscher. Grid tied converter with virtual kinetic storage. In *2009 IEEE Bucharest PowerTech*, pages 1–7, 2009. https://doi.org/10.1109/PTC.2009.5282048.

7 L. Harnefors, J. Kukkola, M. Routimo, M. Hinkkanen, and X. Wang. A universal controller for grid-connected voltage-source converters. *IEEE Journal of Emerging and Selected Topics in Power Electronics*, 9(5): 5761–5770, 2021. https://doi.org/10.1109/JESTPE.2020.3039407.

8 J. Liu, Y. Miura, and T. Ise. Comparison of dynamic characteristics between virtual synchronous generator and droop control in inverter-based distributed

generators. *IEEE Transactions on Power Electronics*, 31(5): 3600–3611, 2016. https://doi.org/10.1109/TPEL.2015.2465852.

9 W. Zhang, A.M. Cantarellas, J. Rocabert, A. Luna, and P. Rodriguez. Synchronous power controller with flexible droop characteristics for renewable power generation systems. *IEEE Transactions on Sustainable Energy*, 7(4): 1572–1582, 2016. https://doi.org/10.1109/TSTE.2016.2565059.

10 M.G. Taul, X. Wang, P. Davari, and F. Blaabjerg. Current limiting control with enhanced dynamics of grid-forming converters during fault conditions. *IEEE Journal of Emerging and Selected Topics in Power Electronics*, 8(2): 1062–1073, 2020. https://doi.org/10.1109/JESTPE.2019.2931477.

11 M. Guan, W. Pan, J. Zhang, Q. Hao, J. Cheng, and X. Zheng. Synchronous generator emulation control strategy for voltage source converter (VSC) stations. *IEEE Transactions on Power Systems*, 30(6): 3093–3101, 2015. https://doi.org/10.1109/TPWRS.2014.2384498.

12 L. Huang, H. Xin, Z. Wang, K. Wu, H. Wang, J. Hu, and C. Lu. A virtual synchronous control for voltage-source converters utilizing dynamics of DC-link capacitor to realize self-synchronization. *IEEE Journal of Emerging and Selected Topics in Power Electronics*, 5(4): 1565–1577, 2017. https://doi.org/10.1109/JESTPE.2017.2740424.

13 M. Lu. Virtual oscillator grid-forming inverters: state of the art, modeling, and stability. *IEEE Transactions on Power Electronics*, 37(10): 11579–11591, 2022. https://doi.org/10.1109/TPEL.2022.3163377.

14 J. Li, J.E. Fletcher, D.G. Holmes, and B.P. McGrath. Developing a machine equivalent inertial response for a virtual oscillator controlled inverter in a machine-inverter based microgrid. In *2020 IEEE Energy Conversion Congress and Exposition (ECCE)*, pages 4314–4321, 2020. https://doi.org/10.1109/ECCE44975.2020.9236073.

15 C. Li, J. Liang, L.M. Cipcigan, W. Ming, F. Colas, and X. Guillaud. Dq impedance stability analysis for the power-controlled grid-connected inverter. *IEEE Transactions on Energy Conversion*, 35(4): 1762–1771, 2020. https://doi.org/10.1109/TEC.2020.2989855.

16 P. Kundur. *Power System Stability and Control*. McGraw-Hill Education, New York, 1994.

17 VDE FNN. FNN Guideline: Grid Forming Behaviour of HVDC Systems and DC-connected PPMs. VDE FNN Guideline, August 2020.

18 National Grid ESO. GC0137: Minimum Specification Required for Provision of GB Grid Forming (GBGF) Capability (formerly Virtual Synchronous Machine/VSM Capability). Final Modification Report GC0137, November 2021.

19 M.N. Ambia, K. Meng, W. Xiao, A. Al-Durra, and Z.Y. Dong. Interactive grid synchronization-based virtual synchronous generator control scheme on weak grid integration. *IEEE Transactions on Smart Grid*, 13(5): 4057–4071, 2022. https://doi.org/10.1109/TSG.2021.3138999.

20 M. Zhang and L. Sun. PLL and additional frequency control constituting an adaptive synchronization mechanism for VSCs. *IEEE Transactions on Power Delivery*, 35(6): 4920–4923, 2020. https://doi.org/10.1109/TPWRS.2020.3020725.

11

Transient Stability Analysis of Grid-following and Grid-forming Converters

Jun Liang and Jinlei Chen*

School of Engineering, Cardiff University, CF24 3AA, Wales, Cardiff, Queen's Buildings, The Parade, UK

*Corresponding Author: Prof Jun Liang; LiangJ1@cardiff.ac.uk

11.1 Transient Stability of Conventional Synchronous Generators Based Power Systems

11.1.1 Introduction of Transient Stability

In conventional power systems, SGs are expected to maintain synchronism under severe disturbances such as short circuits, generator outages, or sudden load variations. Their transient response is primarily governed by the balance between mechanical input power and electrical output, as depicted by the SG's power-angle characteristic (see Figure 11.1). Achieving resynchronization and sustained stability requires adequate synchronizing torque along with sufficient damping to counteract the accelerating effects caused by the disturbance.

To evaluate this behavior, a variety of analytical approaches have been developed. Classical methods such as the equal area criterion (EAC), phase portrait analysis, and Lyapunov-based techniques—including their extended variants—offer valuable insights for both single-machine and multi-machine systems. Moreover, transient stability can be improved through several enhancement strategies, including regulating generator input power, accelerating fault detection and protection tripping, and incorporating supplementary devices to provide voltage support. The detailed analysis methods and enhancement measures are discussed in Sections 11.1.2 and 11.1.2.1.

Stability Analysis of Converter-Rich Power Grids, First Edition. Jun Liang and Chuanyue Li.
© 2026 The Institute of Electrical and Electronics Engineers, Inc. Published 2026 by John Wiley & Sons, Inc. All rights reserved.

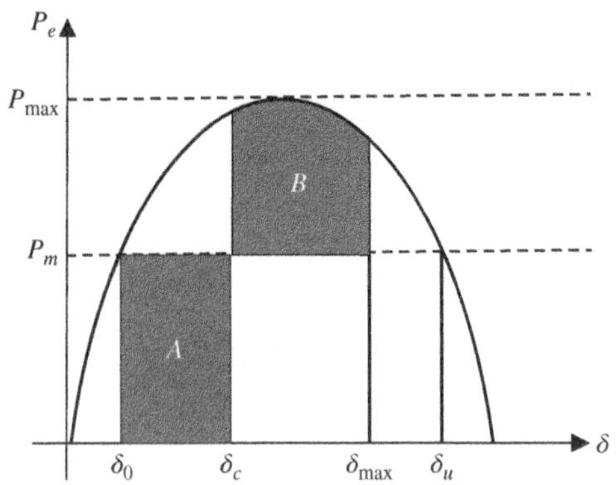

Figure 11.1 Power-angle curve of the SG.

11.1.2 Transient Stability Analysis Approaches

Accurate assessment of transient stability not only is crucial for the safe operation of the system but also directly affects fault-clearing strategies and the effectiveness of control system design.

Among the various analysis methods, time-domain simulation is a fundamental and direct approach. It involves numerically integrating the system's differential equations to obtain the time-varying trajectories of generator rotor angles, speeds, and power outputs. This method offers high accuracy and strong adaptability, making it suitable for analyzing all types of faults and complex system structures. As such, it is widely used in industrial practice. However, due to its high computational cost and limited interpretability, it is less suited for rapid stability boundary determination or sensitivity analysis.

To improve analytical efficiency and deepen understanding, researchers have developed a range of model-based and theoretically grounded transient stability assessment methods. The most representative among these include the EAC, Lyapunov direct method, and phase portrait method.

11.1.2.1 Equal Area Criterion

As aforementioned, the EAC is a classic and intuitive technique for assessing transient stability in the single-machine infinite-bus (SMIB) context. By mapping mechanical and electrical power against rotor angle, one compares the accelerating and decelerating energy areas to judge stability: if the deceleration area outweighs the acceleration area, the system will return to synchronism. It is visually appealing for simple systems but not directly applicable to multi-machine networks. Also, the influence of damping effects is overlooked in the EAC.

To bridge the gap between SMIB methods and multi-machine dynamics, the extended equal area criterion (EEAC) can be adopted. It aggregates machines into an OMIB (one-machine infinite-bus) equivalent, distinguishing a "critical cluster" that may lose synchronism and a "non-critical" group acting like an infinite bus. This approximation allows application of EAC principles to multi-machine behavior efficiently. EEAC is particularly valuable in scenarios requiring quick screening of numerous contingencies, before dedicating resources to detailed time-domain simulations.

11.1.2.2 Lyapunov Direct Methods

Lyapunov direct methods determine stability without solving time-domain dynamics by constructing an energy-like function that decreases over time within a region around a stable equilibrium. When such a function exists, the system is locally stable. In conventional power systems, the Lyapunov function is constructed as shown in Equation 11.1. The total energy consists of potential energy and kinetic energy:

$$E_{\text{tot}} \triangleq E_k + E_p = \frac{1}{2}J\omega_c^2 + \int_{\delta_0}^{\delta_c} (P_e - P_m)d\delta \tag{11.1}$$

This method can be used for transient stability analysis of multi-machine systems. However, the challenges lie in how to accurately construct the Lyapunov function and how to determine the boundary of its stable region. In practice, two prominent Lyapunov boundary methods are used:

- **PEBS (Potential Energy Boundary Surface):** approximates the stability boundary using potential energy surfaces. It is computationally lightweight but can be conservative in accuracy;
- **UEP (Unstable Equilibrium Point) Method:** identifies the nearest unstable equilibrium and computes the corresponding system energy as the critical threshold. It offers higher precision but involves more complex computations.

These methods are especially useful for multi-machine systems and have been integrated into control design and real-time stability monitoring frameworks.

11.1.2.3 Phase Portrait Method

The phase portrait method visualizes system dynamics by plotting state trajectories—such as rotor angle versus angular speed—over time. It offers a powerful, intuitive picture of how systems evolve under disturbance. The system is described by the following ordinary differential equations:

$$\begin{cases} \dfrac{dx}{dt} = \dot{x} \\ \dfrac{d\dot{x}}{dt} = f(x, \dot{x}) \end{cases} \tag{11.2}$$

The relationship between x and $\dot{x}$ can be used to describe the system behavior. Taking x and $\dot{x}$ as the coordinates of the plane, this forms the phase plane. While ideal for low-order or single-machine models, applying this method to high-dimensional, multi-machine systems typically requires dimension reduction techniques such as Poincaré sections or focusing on representative machine clusters.

11.1.3 Transient Stability Enhancement Methods

In order to enhance the capability of the system to withstand and recover from such disturbances, a variety of traditional techniques have been developed [1, 2]. These measures can be broadly categorized into three main objectives: reducing the input power of SGs, enhancing their maximum power capability, and shortening the fault clearing time. Collectively, they serve to enlarge the decelerating area and/or to reduce accelerating area in the EAC, thereby improving transient stability.

To reduce the input power of SGs, one approach is generator tripping, which alleviates excessive power imbalance during fault conditions. Another technique is fast valving control, where turbine valves are rapidly closed and reopened to lower the accelerating power of generators after a disturbance. As illustrated in Figure 11.2, these methods effectively decrease the input power from P_{m0} to P_{m1}.

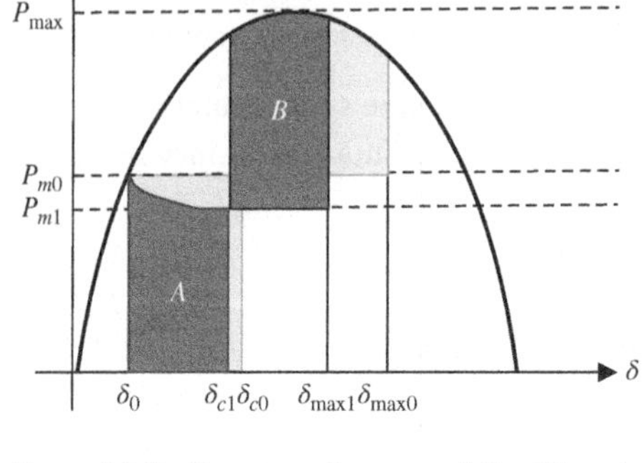

Figure 11.2 Power-angle curve of the SG under a reduced input power.

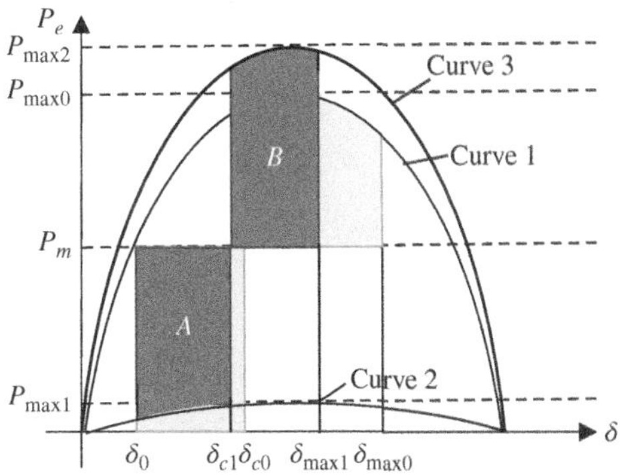

Figure 11.3 Power-angle curve of the SG under an increased maximum power.

To enhance the maximum power capability of SGs, one approach is to reinforce the system infrastructure—for example, by adding parallel AC circuits, installing series capacitor compensation on transmission lines, or using transformers with lower leakage reactance to reduce grid impedance. In addition, the integration of supplementary devices with voltage support capability—such as synchronous condensers, FACTS devices (e.g., thyristor-controlled series capacitors and static synchronous compensators), as well as HVDC converters—can further enhance the power capability. Excitation control also plays an important role to raise the generator terminal voltage after a fault, thus increasing the maximum power. The principle of these methods is illustrated in Figure 11.3. If the fault does not occur directly at the converter's output terminal, the fault voltage will remain above zero, as shown by curve 2. Once the fault is cleared, the operating point shifts from curve 2 to curve 3. Owing to the voltage support effect, the maximum power of curve 3 is slightly higher than that of curve 1.

To shorten the fault clearing time, rapid fault detection, diagnosis, and fast AC circuit breaker tripping schemes such as AC traveling-wave protection can achieve this. Since the kinetic energy gained by generators during a fault is proportional to the fault duration, quicker fault clearing directly reduces the risk of instability. This method is illustrated in Figure 11.4.

11.2 Transient Stability of Grid-following Converters

11.2.1 Overview of Transient Stability of Grid-following Converters

Transient stability of grid-following (GFL) converters is governed by their ability to maintain synchronism with the grid through the phase-locked loop (PLL) during and after large disturbances. The PLL can enable GFL converters to

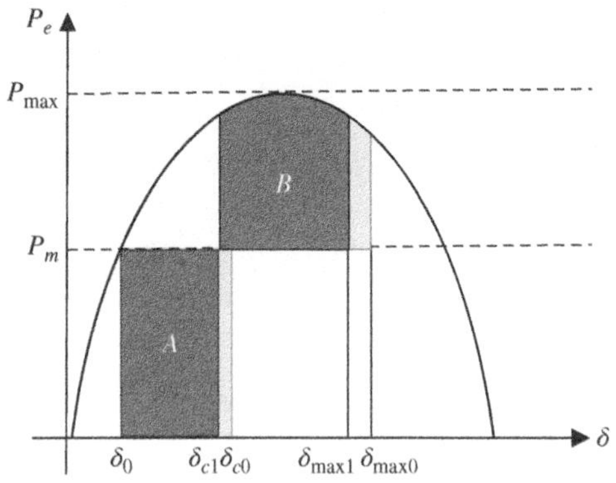

Figure 11.4 Power-angle curve of the SG under a short fault clearing time.

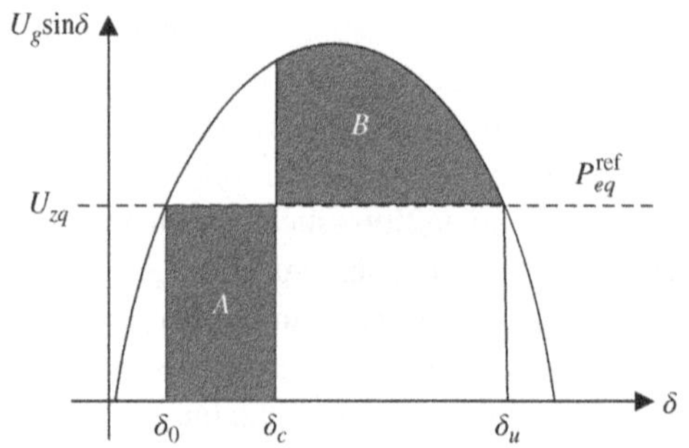

Figure 11.5 Voltage-angle curve of the GFL converter.

accurately track the grid voltage phase for resynchronization. GFL converters can be modeled using a second-order nonlinear differential equation. This model can be further reformulated into a structure similar to the swing equation used for SGs [3], allowing the application of conventional stability analysis methods such as the EAC, phase portraits, and Lyapunov-based approaches. Instead of P–δ curve in the conventional swing equation, the voltage-angle curve ($U_g \sin \delta$–δ) is used to describe the behaviors of GFL converters, which is shown in Figure 11.5. The equivalent power reference P_{eq}^{ref} is related to the d-axis current and is represented as $P_{eq}^{ref} = \omega L_g i_d$.

The transient stability of GFL converters is influenced by the tuning of the PLL parameters [3]. Specifically, increasing the proportional gain of the PLL enhances system damping, improving stability, whereas increasing the integral gain reduces both the system's equivalent inertia and damping, which can weaken the system's transient stability. In [4], an analysis of GFL converters using the Lyapunov energy method shows that GFL converters experience negative damping during the operation, especially under large phase differences, which can drive the system toward instability. A more detailed analysis in [5] highlights that GFL converters are vulnerable in weak grid environments. The study also shows that increasing the virtual inertia in either converter type can reduce the impact of faults on the synchronization angle, helping to maintain stability by preventing large angular deviations from the pre-fault equilibrium. The paper also compares the transient responses between GFL and grid-forming (GFM) converters. In GFL converters, damping is a sinusoidal function of the phase angle, which can become negative and destabilize the system when no equilibrium point exists. In contrast, GFM converters maintain consistently positive and higher virtual damping, contributing to more robust transient stability.

For transient enhancement measures of GFL converters, a common approach to enhance transient stability is by adjusting the current references. For example, [6] proposes reducing the active current proportionally to the voltage drop during a fault, while [7] suggests increasing the active power reference based on the PLL frequency error. In [8], the current vector is aligned with the line impedance angle, and [9] sets the reference power equal to the actual power, eliminating the accelerating and decelerating areas in the EAC. All these strategies rely on measurements of AC voltage, frequency, or both for calculating the current references during the fault. A general control structure is illustrated in Figure 11.6.

A second group of stabilization methods improves transient stability by tuning PLL parameters. One approach freezes the PLL during severe faults, allowing the converter to operate with pre-fault frequency and phase, ensuring stability even under zero-voltage conditions, though it cannot track phase or frequency changes [10]. In [11] and [12], methods to reschedule the PLL's proportional gain and the integral gain are proposed to enhance damping.

11.2.2 Control Structure

The standard control structure with decoupled active and reactive power regulation is illustrated in Figure 11.7. The inner loop uses current vector control in the

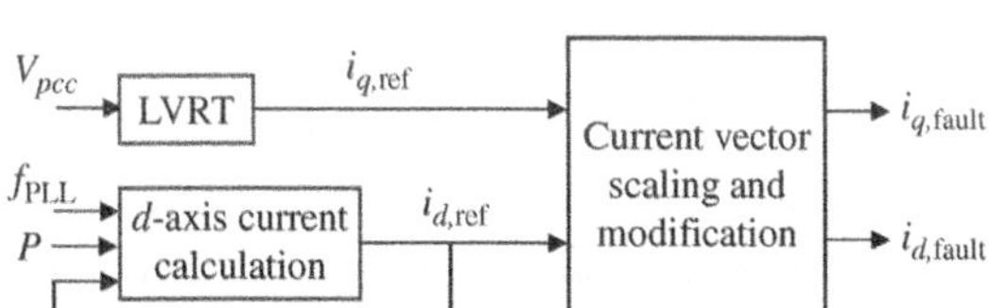

Figure 11.6 Current reference adjustment for GFL converters to enhance transient stability.

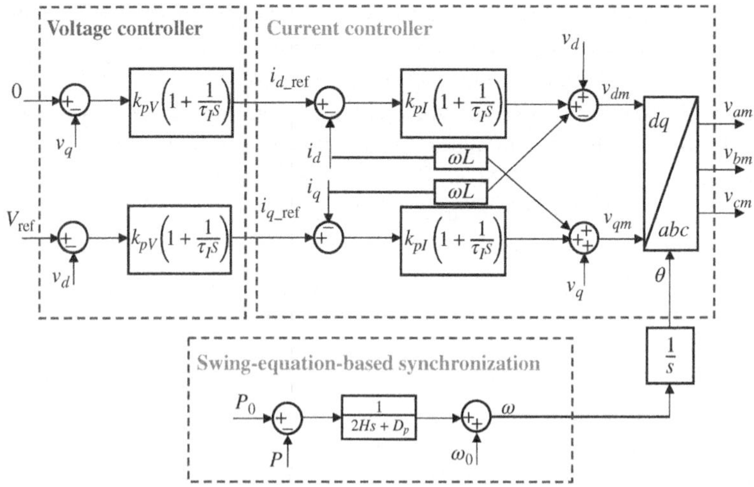

Figure 11.7 Control structure of the GFL converter.

d–q reference frame, where aligning the terminal voltage vector with the d-axis allows the d-axis current to control active power and the q-axis current to control reactive power. The outer loop generates current references based on either DC voltage or active power control for the d-axis, and reactive power or AC voltage control for the q-axis. To achieve proper alignment, a PLL is used to track the terminal voltage angle. The PLL must be fast and accurate to ensure stable operation. Additionally, to enhance frequency response, primary frequency control and virtual inertia can be implemented via an additional outer loop, whose output is combined with the main power references.

11.2.3 Transient Stability Analysis

Figure 11.8 presents the $U_g \sin \delta$–δ characteristic of GFM converters under different conditions. As shown in Figure 11.8a, a larger voltage drop shrinks the decelerating area, increasing the likelihood of transient instability. Likewise, Figure 11.8b shows that raising the power reference during normal operation, or increasing the d-axis current under current-limited conditions, also reduces the decelerating area, further compromising transient stability.

Figure 11.9 presents a comparison of phase portraits between a reduced-order model (the dynamics of the power and current controllers is omitted) and full-order simulations conducted in Power Systems Computer-aided Design (PSCAD) under a disturbance resulting from the voltage drop. The reduced-order model is

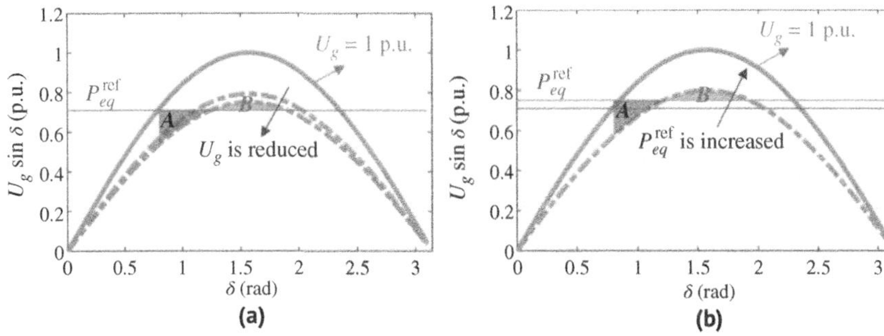

Figure 11.8 P–δ curve for GFL converters. (a) Curve under AC voltage change. (b) Curve power reference change.

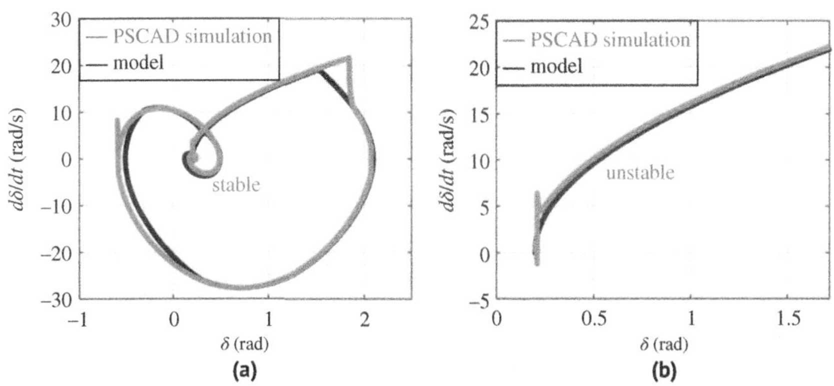

Figure 11.9 Phase portraits for analyzing transient stability of GFL converters. (a) Stable condition. (b) Unstable condition.

derived solely from the dynamics of the PLL, simplifying the system while capturing its core transient behavior. Despite this simplification, the trajectories under a large disturbance show good agreement between the two models. This close match confirms that the reduced-order model is accurate for analyzing transient stability as it can effectively represent the system's dynamic response in GFL converters.

Figures 11.10 and 11.11 demonstrate the influence of the inertia constant and damping coefficient in the inertia/damping controller (as shown in Figure 11.7) on transient stability. Under a small voltage drop, current saturation does not occur, so increasing either the damping or inertia enhances the stability margin by reducing the maximum rotor angle deviation (δ_m) in the figures. However, during a severe voltage drop, further increasing these parameters has negligible

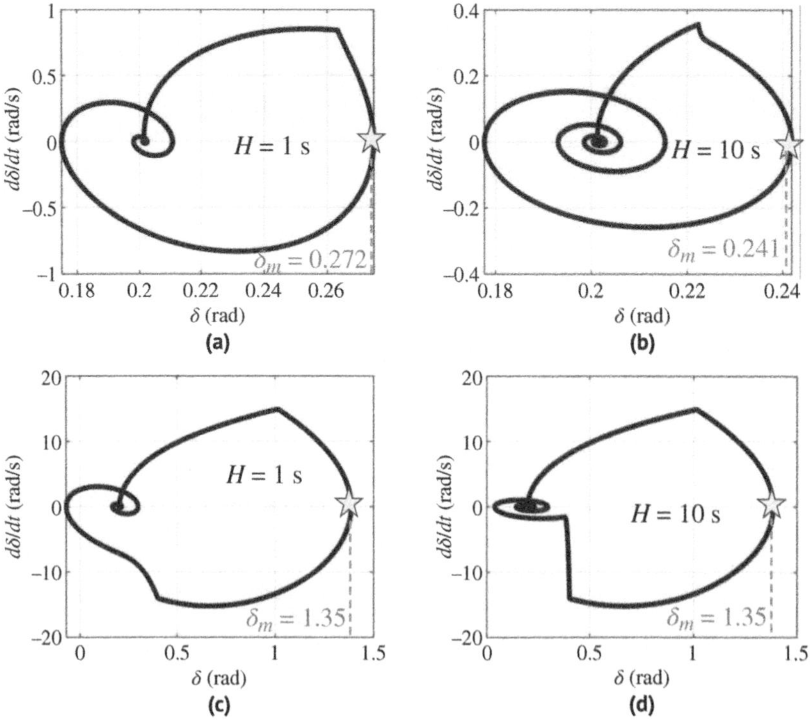

Figure 11.10 Effects of the inertia constant on transient stability. (a) $H = 1$ s under a small voltage drop. (b) $H = 10$ s under a small voltage drop. (c) $H = 1$ s under a severe voltage drop. (d) $H = 10$ s under a severe voltage drop.

effect on improving transient stability. This is because the output of the outer inertia/damping controller becomes saturated, limiting its ability to influence the system response.

Figures 11.12 and 11.13 analyze the impact of PLL parameters on transient stability. The results show that increasing the PLL's proportional gain enhances transient stability by improving system damping. In contrast, increasing the integral gain worsens stability, as it reduces the system's equivalent inertia and damping. Unlike the inertia/damping controller in Figures 11.10 and 11.11, whose effect diminishes under severe voltage drops due to output saturation, the influence of PLL parameters remains consistent across the full range of voltage sags, regardless of current saturation.

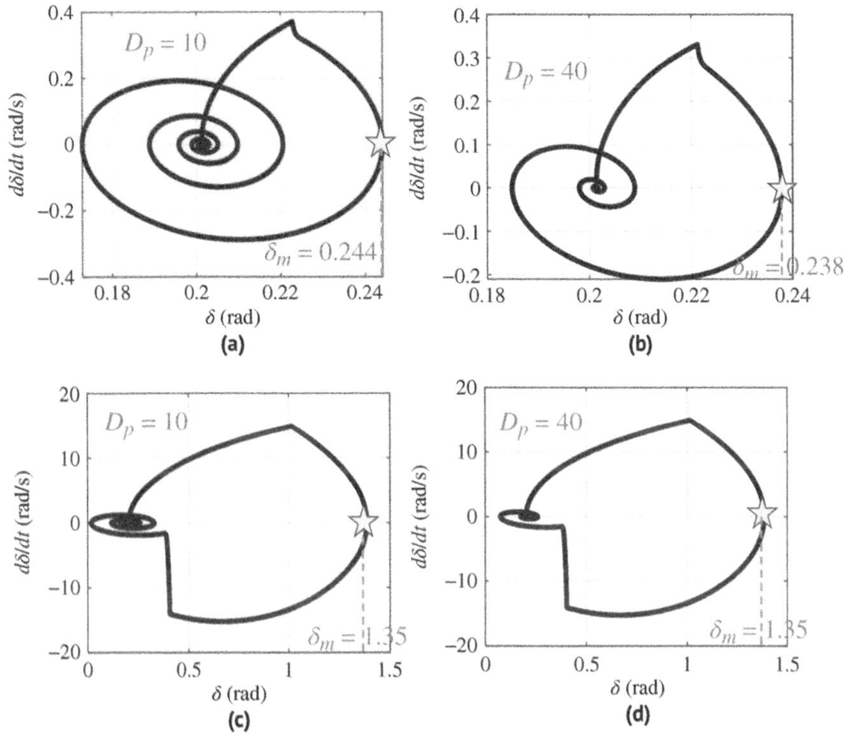

Figure 11.11 Effects of the damping coefficient on transient stability. (a) $D_p = 10$ under a small voltage drop. (b) $D_p = 40$ under a small voltage drop. (c) $D_p = 10$ under a severe voltage drop. (d) $D_p = 40$ under a severe voltage drop.

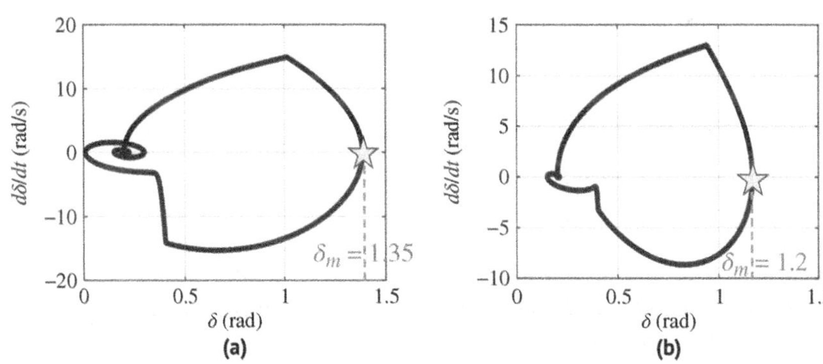

Figure 11.12 Effects of the proportional gain of PLL (k_{p_PLL}) on transient stability. (a) A smaller k_{p_PLL} under a severe voltage drop. (b) A larger k_{p_PLL} under a severe voltage drop.

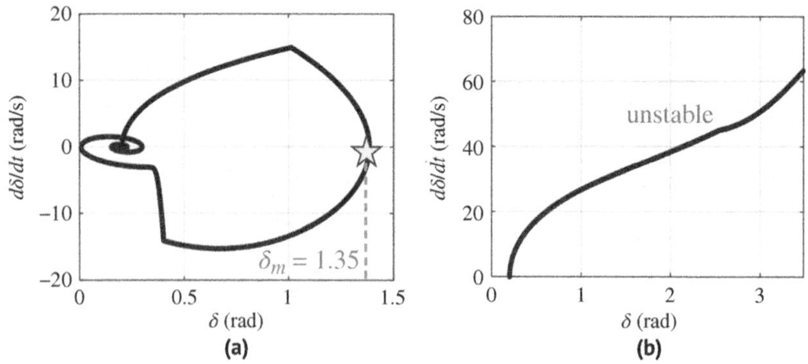

Figure 11.13 Effects of the integral gain of PLL (k_{i_PLL}) on transient stability. (a) A smaller k_{i_PLL} under a severe voltage drop. (b) A larger k_{i_PLL} under a severe voltage drop.

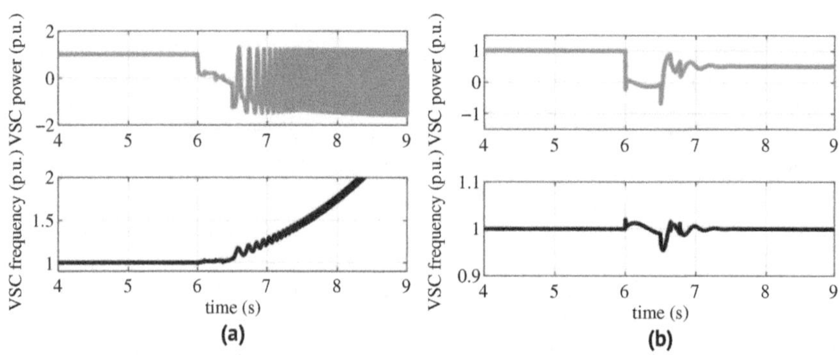

Figure 11.14 Results for an 80% voltage drop with a duration of 500 ms. (a) d-axis current reference is 1 p.u. under fault. (b) d-axis current reference is 0.5 p.u. under fault.

11.2.4 Simulation Verifications

The simulations are performed using PSCAD/EMT. Figure 11.14 presents the impact of different d-axis current reference settings on transient stability. At time 6 s, an 80% voltage drop is introduced, lasting for 500 ms. When the d-axis current reference is set to 1 p.u., the system fails to regain stability after the fault is cleared. In contrast, reducing the d-axis current reference to 0.5 p.u. during the fault allows the system to remain stable.

Figure 11.15 presents simulation results illustrating the impact of the inertia constant and damping coefficient on transient stability. Two scenarios are compared: one with inertia and damping enabled, and another without. Under a

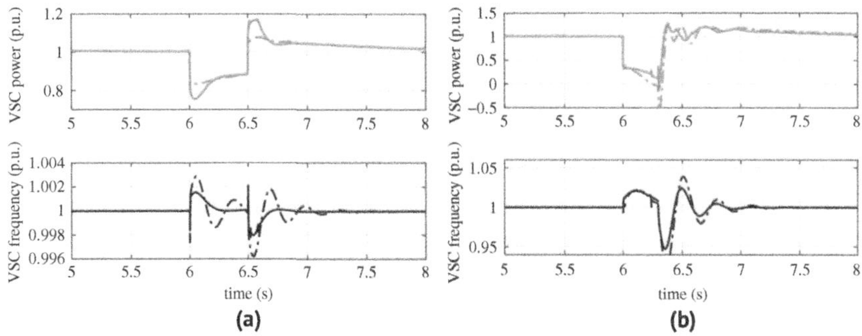

Figure 11.15 Results for effects of inertia and damping on transient stability (solid line: H and D_p are nonzero; dashed line: H and D_p are zero). (a) Under a 30% voltage drop. (b) Under an 80% voltage drop.

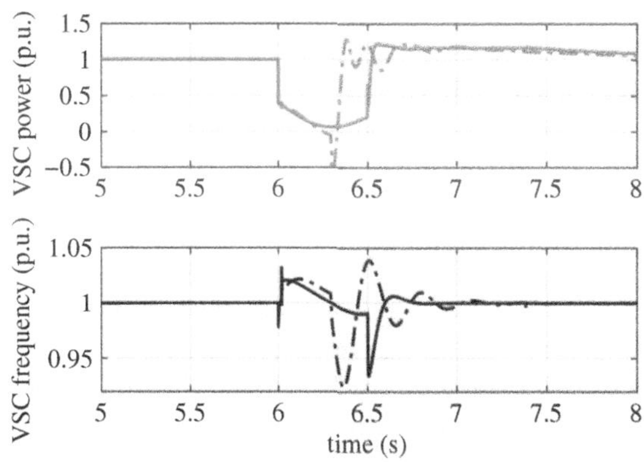

Figure 11.16 Results for effect of k_{p_PLL} on transient stability (solid line: $k_{p\_PLL} = 10$; dashed line: $k_{p\_PLL} = 30$).

moderate 30% voltage drop—where current saturation does not occur—the addition of inertia and damping clearly enhances stability. However, when the voltage drop increases to 80%, the results of both cases become nearly identical, as current saturation limits the effectiveness of the inertia and damping terms.

Figures 11.16 and 11.17 illustrate the effects of $k_{p,PLL}$ and $k_{i,PLL}$ on system stability. As shown, increasing $k_{p,PLL}$ effectively suppresses frequency deviations during a disturbance and enables faster recovery after the fault is cleared. In contrast, increasing $k_{i,PLL}$ leads to instability following the fault, confirming the findings made in the theoretical analysis.

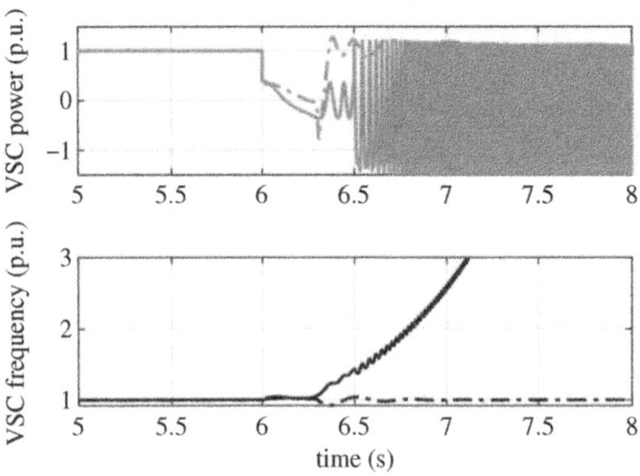

Figure 11.17 Results for effect of k_{i_PLL} on transient stability (solid line: $k_{i\_PLL} = 500$; dashed line: $k_{i\_PLL} = 1500$).

11.3 Transient Stability of Grid-forming Converters

11.3.1 Overview of Transient Stability of Grid-forming Converters

The transient stability of GFM converters is determined by their self-synchronizing capability, which is closely related to the swing equation. Owing to their mathematical similarity with SGs, classical methods such as the EAC, phase portrait analysis, and Lyapunov energy functions remain applicable. In this framework, the virtual inertia and damping coefficients play a critical role in influencing the transient stability [13]. In addition, a key challenge lies in the fact that the transient stability performance of GFM converters is heavily influenced by their maximum current limit. Therefore, the current limitation must be carefully considered when evaluating transient stability. Figure 11.18 illustrates the power-angle characteristic of the swing-equation-based GFM converter under a d-axis priority-based current-limiting strategy, which merges features of the traditional SG P–δ curve with the behavior of converters operating under current limitations [14]. If the reference of the current controller remains below the maximum value, the operating point will follow the conventional power-angle curve. Once the current reference reaches saturation, the operating point trajectory will transition to the saturated curve.

Apart from the d-axis priority strategy, q-axis priority and phase angle priority current-limiting strategies have been investigated in [15]. Based on the comparative study, it is found that the transient stability under fault conditions can be enhanced with increased value of φ. Thus, the q-axis priority with $\varphi = \pi/2$ is preferable when considering the transient stability margin only. However, the q-axis priority strategy may influence voltage control performance. Hence, the φ is

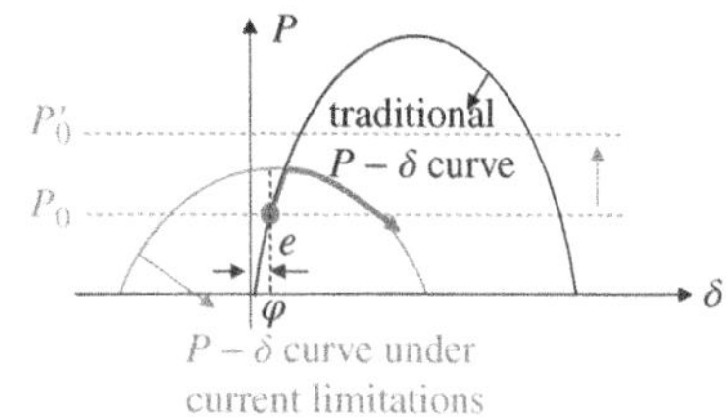

Figure 11.18 P–δ curve of the GFM converter.

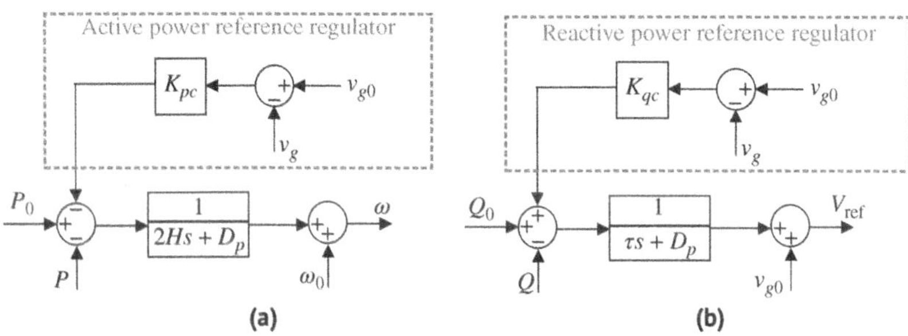

Figure 11.19 Change of power references for GFM converters. (a) Reduction in active power reference. (b) Increase in reactive power reference.

further optimally selected in [16] by concurrently considering the stability margin and the voltage control performance after a fault.

The transient stability enhancement strategies have been proposed in previous works. Refs. [17] and [18] demonstrate that transient stability can be enhanced by reducing the active power reference and increasing the reactive power reference during voltage sags (see Figure 11.19). These adjustments aim to decrease accelerating power and support voltage recovery, respectively. The key challenge lies in determining the appropriate reference changes, which typically require prior knowledge of the grid impedance.

In addition to tuning power references, controller parameters and structures can be adapted to enhance transient stability. For example, [19] proposes a gain-scheduled virtual inertia and damping control for GFM converters, replacing fixed values with adaptive ones that better respond to transient conditions. Meanwhile, [20] introduces a frequency-domain inertia design approach, which applies high inertia at low frequencies to dampen frequency deviations and lower inertia at high frequencies to enable faster system recovery. In [21], a mode-switching control strategy is proposed (see Figure 11.20), where a control gain K is introduced in the forward path of the control loop. This gain switches between +1 and −1 depending on the operating condition, effectively preventing positive-feedback behavior once the system crosses the unstable equilibrium point.

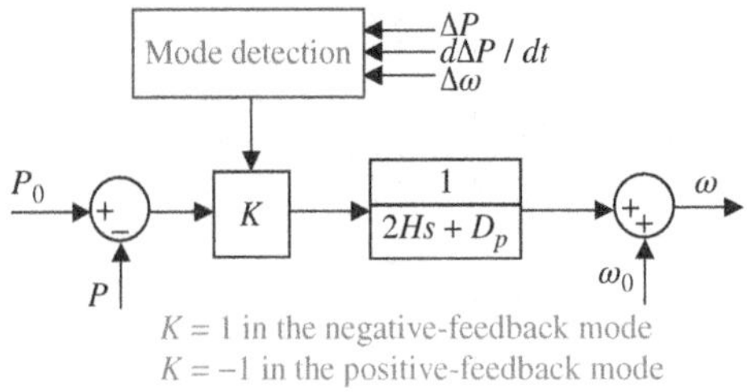

Figure 11.20 Mode adaptive control of the GFM converter.

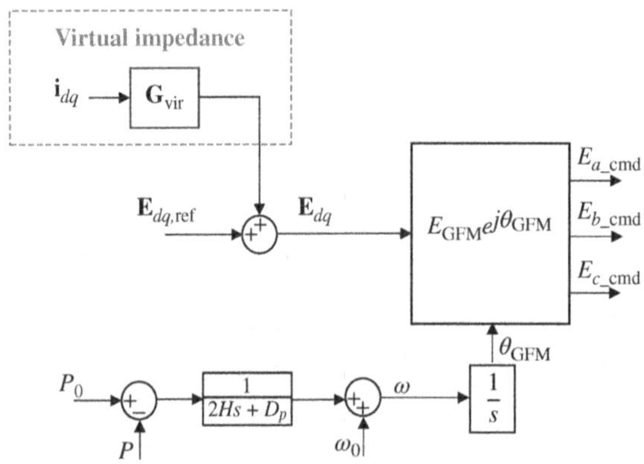

Figure 11.21 Current-limiting control through virtual impedance.

To mitigate the impact of current limitations, [22] proposes switching the control mode from grid-forming to grid-following during faults. To avoid this mode switching, [23] introduces a method where the outer droop controller's power references are replaced with attainable values based on the measured voltage drop. This allows the power-synchronization control to remain active while keeping the injected current within limits, avoiding controller saturation and integrator windup. Additionally, [24] suggests implementing current limitation using a virtual resistor, eliminating the need for inner current control loops (see Figure 11.21).

11.3.2 Control Structure

The GFM converter's control structure is illustrated in Figure 11.22. It consists of a swing-equation-based power synchronization loop (i.e., P–ω droop-based synchronization loop), an outer voltage controller and an inner current controller. The converter uses the swing equation—a second-order differential equation

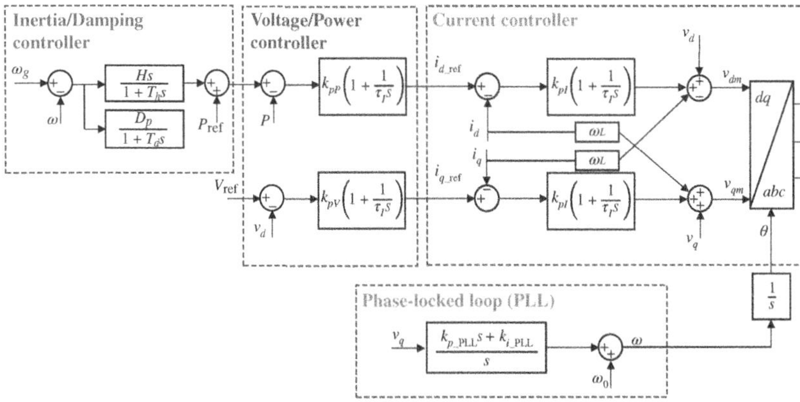

Figure 11.22 Control structure of the GFM converter.

describing rotor dynamics—to govern the evolution of its internal angle and frequency (H and D_p are the inertia constant and the damping coefficient). This enables the converter to inherently synchronize with the grid, respond to frequency deviations, and contribute to inertial and primary frequency control, thereby supporting stable and resilient grid operation in low-inertia, inverter-dominated systems.

11.3.3 Transient Stability Analysis

As seen in Figure 11.18, compared to conventional systems, the presence of current limits reduces the stable operating region. The angle φ, defined as the phase difference between the current-limiting vector and the d-axis, plays a significant role in determining transient stability. Basically, the current-limiting approach includes the d-axis priority, q-axis priority, and phase angle priority-based current-limiting algorithms (see Figure 11.23).

In Figure 11.24, it is seen that the decelerating area is increased with a larger φ. This implies that more energy can be utilized to facilitate the return of the operating point to its initial state, thereby augmenting transient stability. However, φ should not be excessively large, as it could impair voltage control capability in the post-fault stage, which is illustrated in Figure 11.24b. Therefore, φ should be selected considering both the transient stability margin and voltage control performance.

Figure 11.25 shows the influence of different current limitation values on the transient stability margin. It is found that the decelerating area is increased when the maximum current is set from 1.2 to 1.5 p.u., which demonstrates that the transient stability can be enhanced with a higher current limitation value.

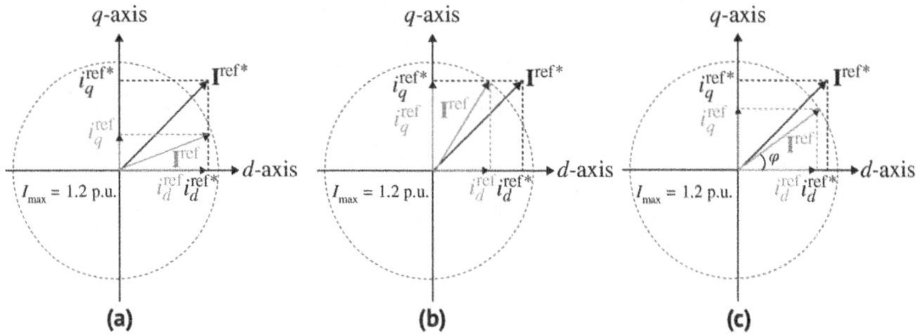

Figure 11.23 Priority-based current-limiting approach. (a) d-axis priority. (b) *q*-axis priority. (c) Phase angle priority.

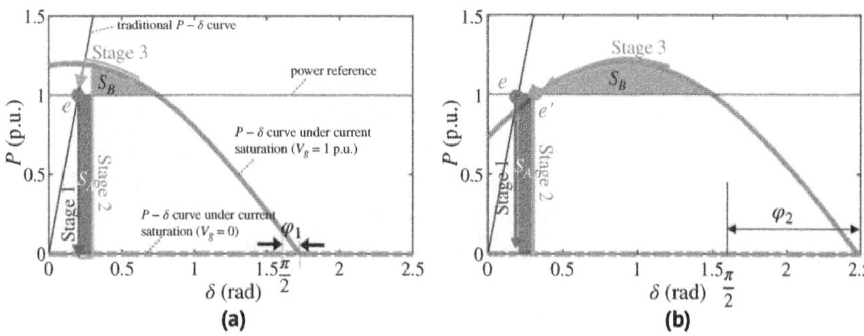

Figure 11.24 *P*–*δ* under different phase angle of current-limiting vector. (a) Small *ϕ*. (b) Large *ϕ*.

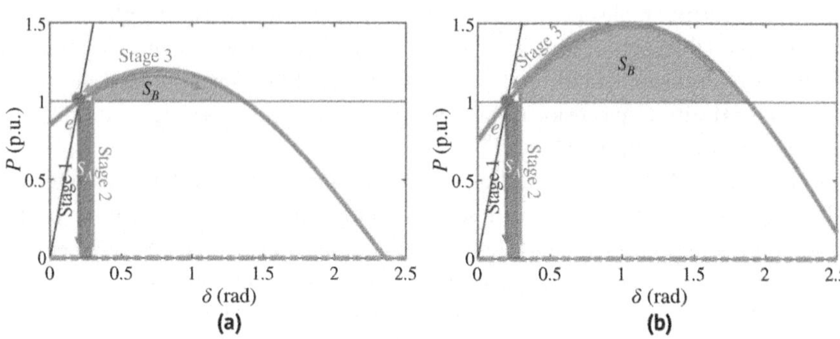

Figure 11.25 *P*–*δ* under different current limitation values. (a) Maximum current is 1.2 p.u. (b) Maximum current is 1.5 p.u.

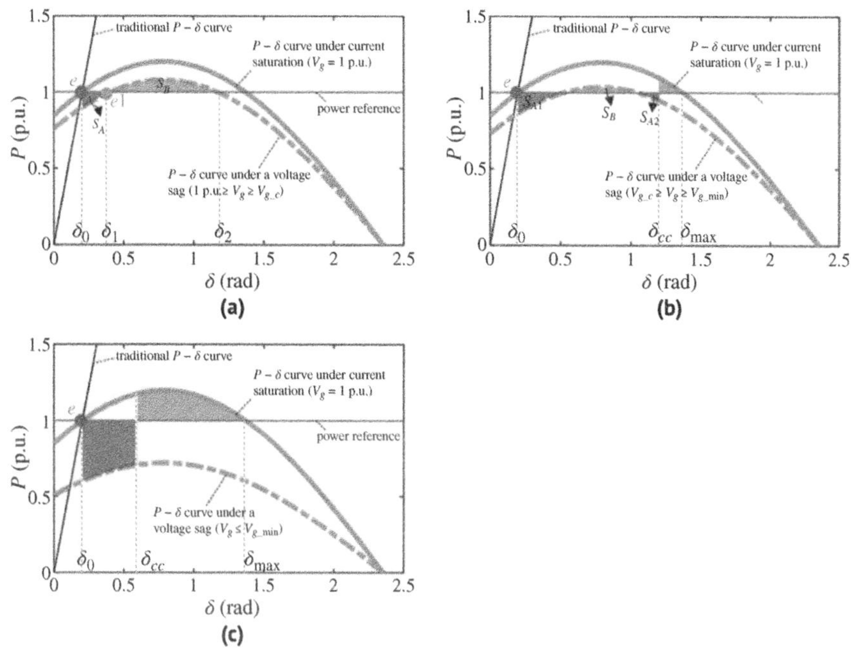

Figure 11.26 *P–δ* under different grid drops. (a) Slight voltage drop. (b) Medium voltage drop. (c) Severe voltage drop.

Figure 11.26 illustrates how varying levels of voltage drop affect system stability. Three distinct scenarios can be observed. In Figure 11.26a, a minor voltage drop still allows for the existence of an equilibrium point. During the fault, the decelerating area is larger than the accelerating area, enabling the operating point to return to equilibrium during system recovery. In Figure 11.26b, under a moderate voltage drop, the accelerating area exceeds the decelerating area. Although an equilibrium point still exists, the system may experience transient instability unless the fault is cleared quickly. The outcome depends on the system's damping capability. In Figure 11.26c, a severe voltage drop eliminates the equilibrium point during the fault. In this case, fault clearance is essential to avoid system instability.

To quantitatively identify the inertia and damping effects, the phase portraits are adopted for analysis. The phase portraits are derived from the dynamics of the swing equation, representing a reduced-order model of the converter. Since the synchronization loop operates at a slower response compared to the voltage and current control loops, the faster dynamics of the voltage and current controllers are neglected in the analysis.

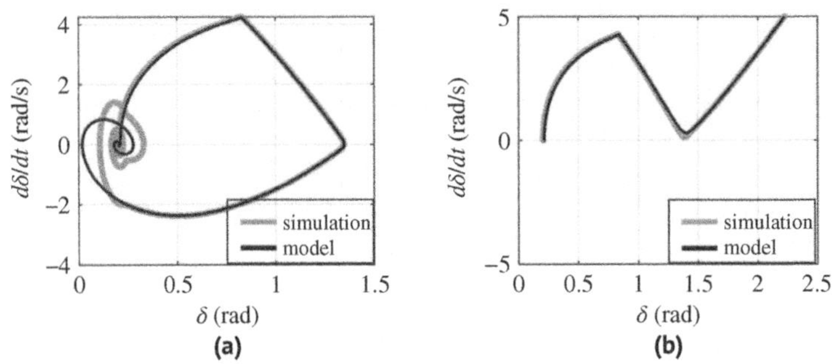

Figure 11.27 Phase portraits for analyzing transient stability of GFM converters. (a) Stable condition. (b) Unstable condition.

Figure 11.27 depicts comparisons of phase portraits under various conditions of voltage sag, inertia, and damping coefficients. The simulation results align closely with the model-based results during voltage sag events, with a minor discrepancy observed in regions where operating points approach equilibrium. This discrepancy is attributed to the dynamics of the voltage control loop. However, the accuracy of the critical clearing angle (CCA) derived from the model-based phase portrait remains unaffected by these dynamics.

Based on the reduced-order model, effects of inertia and damping on CCA and critical clearing time (CCT) are analyzed through phase portraits considering different voltage sags, with results depicted in Figures 11.28 and 11.29.

In Figure 11.28, the CCA (δ_{cc}^{act}) obtained from phase portrait analysis is compared with δ_{cc} obtained from the EAC method. The blue dots represent measured data obtained from iterative model runs, which are fitted using a polynomial to generate a three-dimensional surface. The values of D_p and H are chosen within a reasonable range. It is seen that δ_{cc}^{act} is greater than δ_{cc} under all scenarios of voltage sags, which implies that the results obtained for the phase portraits are more accurate than those obtained from the EAC method. In Figure 11.28a during a severe voltage sag (e.g., the AC short-circuit fault), δ_{cc}^{act} increases with higher damping, while δ_{cc}^{act} decreases inversely with inertia. This inverse relationship is more pronounced at higher damping values. Similar trends are observed in Figure 11.28b for a moderate voltage drop. However, the sensitivity of δ_{cc}^{act} to damping and inertia coefficients diminishes with smaller voltage drops in Figure 11.28c.

Figure 11.29 illustrates the corresponding t_{cc}^{act} to δ_{cc}^{act} in Figure 11.28, showing positive relationships of t_{cc}^{act} with both inertia and damping coefficients. However, these impacts are less pronounced under slight voltage drop conditions. Moreover, it is seen that the sensitivity of H to t_{cc}^{act} is higher than that of D_p. This suggests that the inertia coefficient has a greater impact on the CCT compared to the damping coefficient when subjected to the same incremental changes in values.

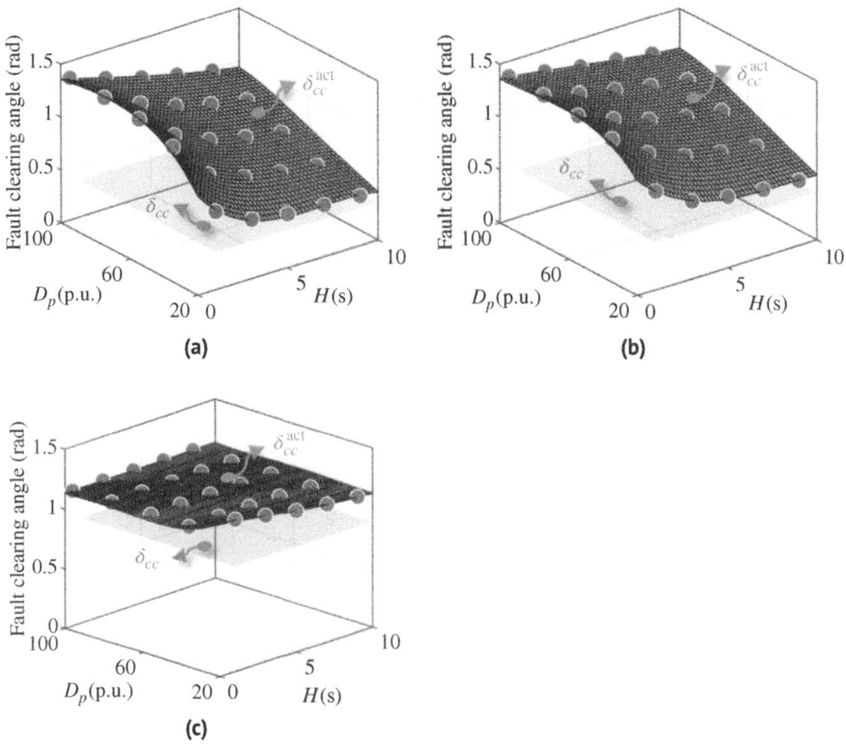

Figure 11.28 Fault clearing angles under different voltage drops (the dots are the data obtained from phase portraits, and the surface is the polynomial fitting based on the obtained data). (a) Severe voltage drop. (b) Medium voltage drop. (c) Slight voltage drop.

The trends of δ_{cc}^{act} and t_{cc}^{act} exhibit nonlinear characteristics among voltage sags, inertia, and damping coefficients, which make it difficult to design a unified polynomial fitting function for estimation. To address this complexity, an Artificial Neural Network (ANN)–based estimation method is utilized to predict the CCA and CCT in Figure 11.30.

The ANN consists of an input layer, a hidden layer, and an output layer. Three signals—grid voltage, damping, and inertia coefficients—are fed into the input layer. The hidden layer contains N neurons situated between the input and output layers. The output layer consists of two neurons, representing the actual CCA and CCT. For determining the optimal number of layers and neurons in an ANN, it is advisable to use fewer layers and neurons for smaller datasets, whereas larger datasets benefit from more layers and neurons. In many cases, one or two hidden layers are sufficient, as adding more layers can increase the complexity analytical

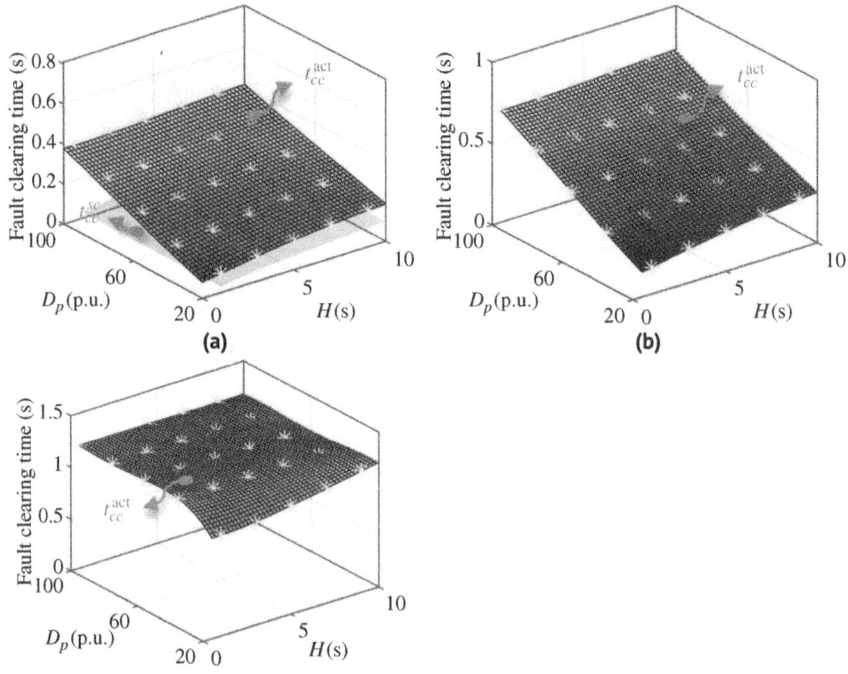

Figure 11.29 Fault clearing time corresponding to the fault clearing angle in Figure 11.28 under different voltage drops. (a) Severe voltage drop. (b) Medium voltage drop. (c) Slight voltage drop.

formulas. The number of neurons in each layer influences the ANN's complexity and its ability to represent patterns effectively. A suitable number of neurons can promote rapid pattern learning. However, using too many neurons can lead to overfitting during training, affecting the network's ability to generalize.

11.3.4 Simulation Verifications

Figures 11.31–11.33 verify the influence of different voltage drops on the transient stability. In Figure 11.31, a 10% voltage sag occurs at time 10 s with a fault duration of 3 s. Since grid voltage drops by less than the threshold value, the equilibrium point persists during the sag, ensuring system stability independent of the fault clearing time.

However, a 20% voltage drop must be restored within a specified time to prevent instability. As depicted in Figure 11.32, the actual critical clearing time t_{cc}^{act} is approximately 1 s. In the case of a 100% voltage sag in Figure 11.33, t_{cc}^{act} decreases to

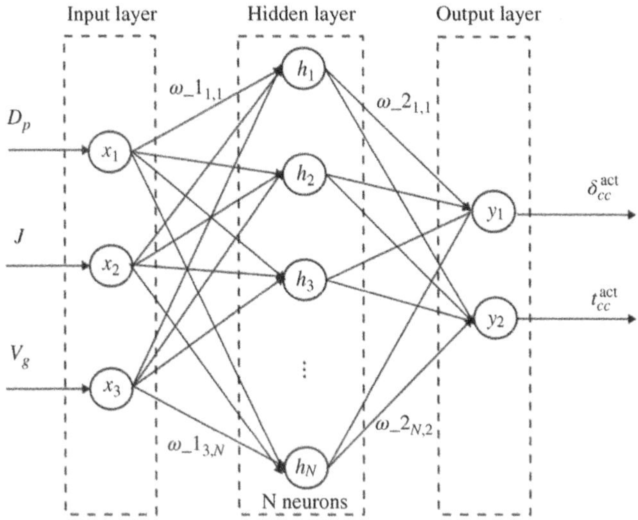

Figure 11.30 ANN method for estimating δ_{cc}^{act} and t^{act}.

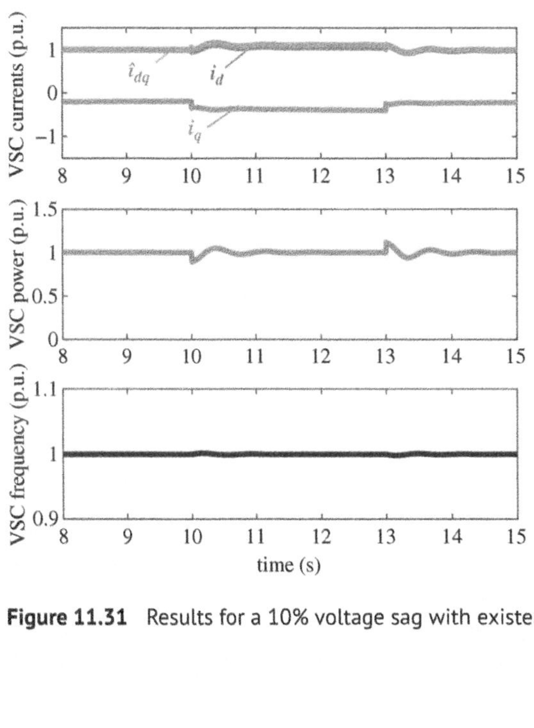

Figure 11.31 Results for a 10% voltage sag with existence of equilibrium points.

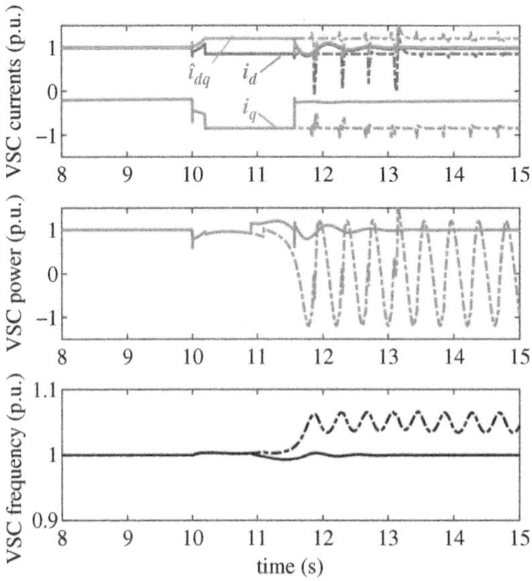

Figure 11.32 Results for a 20% voltage sag without existence of equilibrium points (solid line: fault duration is 0.9 s; dashed line: fault duration is 1.1 s).

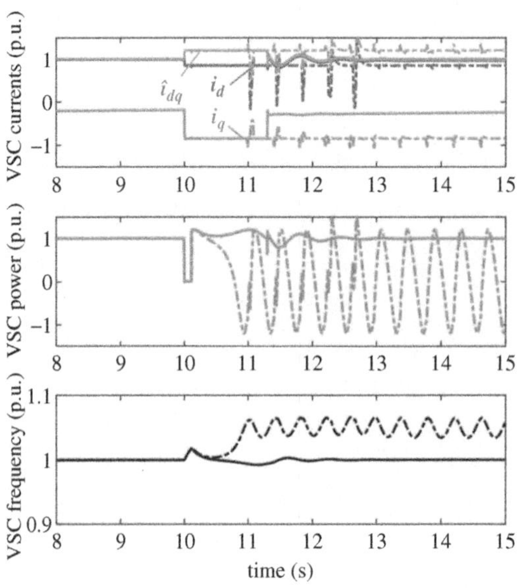

Figure 11.33 Results for a 100% voltage sag without existence of equilibrium points (solid line: fault duration is 0.11 s; dashed line: fault duration is 0.12 s).

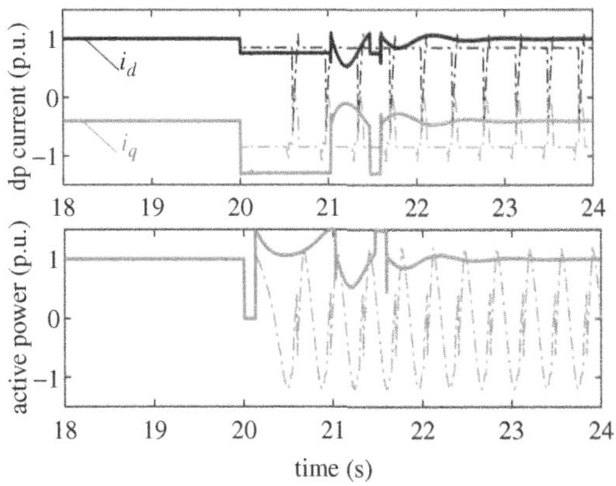

Figure 11.34 Transient stability under different current limitations (solid line: $I_{sat} = 1.5$ p.u.; dashed line: $I_{sat} = 1.2$ p.u.).

110 ms, indicating the fault clearance should be placed considerably earlier with a larger voltage sag.

Transient stability is verified under different current limitations, as shown in Figure 11.34. The current limitations are set at $I_{\text{sat}} = 1.2$ and 1.5 p.u., with a fault clearing time of 130 ms. As can be seen, under a 100% voltage sag, the system fails to restore stable operation when I_{sat} is 1.2 p.u. However, when I_{sat} is increased to 1.5 p.u., stable operation is recovered after fault clearance. This indicates that transient stability can be improved by increasing current limitations.

Figure 11.35 shows the results under a 50% voltage sag occurring at time 10 s with $H = 0.5$ s and $D_p = 20$ p.u. The low-voltage duration is selected as 250 ms, and the system becomes unstable after the fault is cleared.

Figure 11.36 illustrates the impact of increased inertia and damping on the low-voltage ride through capability. In this scenario, the inertia and damping values are doubled compared to those in Figure 11.35, while maintaining a low-voltage duration of 250 ms. The findings reveal that increasing both inertia and damping coefficients enhances system transient stability. Nonetheless, the higher inertia depicted in Figure 11.36b results in larger oscillations during the recovery stage, thereby extending the settling time.

For the ANN-based estimation approach, the MATLAB Neural Net Fitting toolbox is employed to build and train the ANN model and the weighting coefficients of each layer can be automatically extracted using this MATLAB toolbox. A single hidden layer consisting of 10 neurons is chosen for the network architecture. A dataset consisting of 269 data points is generated using model-based phase

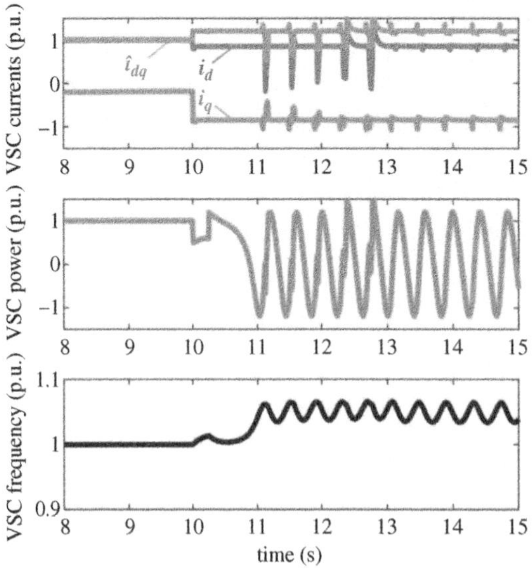

Figure 11.35 Results for a fault duration of 250 ms under a 50% voltage sag with $H = 2.5$ s and $D_p = 20$ p.u.

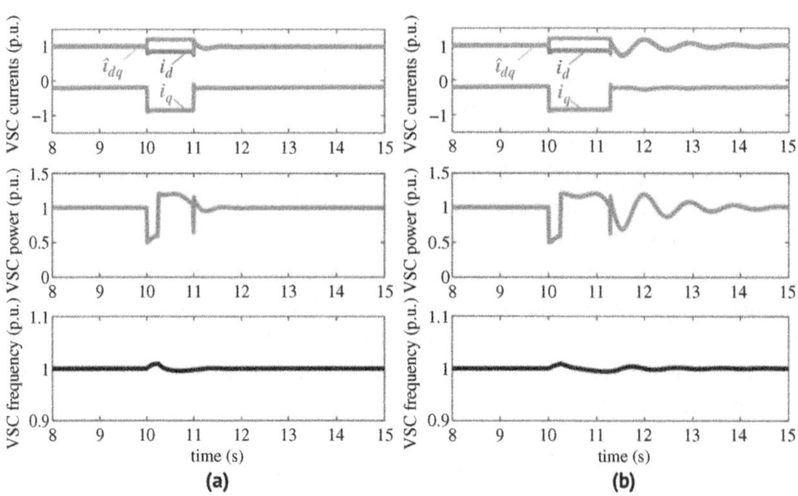

Figure 11.36 Results for a fault duration of 250 ms under a 50% voltage sag with increased inertia and damping. (a) $H = 2.5$ s and $D_p = 40$ p.u. (b) $H = 5$ s and $D_p = 20$ p.u.

Table 11.1 ANN training configurations and performance

Number of input data	Training data	Validation data	Test data	$E_{rr_{rms}}$ of δ_{cc}^{act}	$E_{rr_{rms}}$ of t_{cc}^{act}
269	80%	10%	10%	3.09%	3.54%

portraits and serves as the training data for the ANN model. As seen in Table 11.1, the relative root-mean-square errors of the estimated δ_{cc}^{act} and t_{cc}^{act} are 3.09% and 3.54%, respectively, which proves the effectiveness of the estimation method.

11.4 Summary

The transient stability of conventional synchronous generators is constrained by their inherent power-angle characteristics and mechanical input power, which can only be adjusted slowly. In contrast, power electronic converters provide much higher controllability, as both their input and phase angle can be flexibly and rapidly regulated. While this advantage improves their capability to support system stability, it is also restricted by factors such as the limited overload capacity of power electronic devices.

For GFL converters, classical modeling and transient stability analysis methods developed for SGs—such as EAC—can be extended with appropriate modifications. Similar influencing factors apply; for example, a grid voltage drop reduces the decelerating area and weakens stability—while the converter's specific dynamics must also be considered. The dynamics of GFL converters can be reformulated into an equivalent swing equation, where the equivalent input power is tied to the *d*-axis current. Reducing the *d*-axis current reference during faults enlarges the decelerating area, thereby improving transient stability. Moreover, GFL converters can contribute virtual inertia and damping, either by tuning PLL parameters or by introducing additional control loops, thus further enhancing stability.

For GFM converters, transient stability is significantly influenced by current-limiting behavior. The power-angle curve is determined not only by the magnitude of the current limit but also by the phase angle between the limiting current vector and the *d*-axis. Increasing both the current limit and the phase angle expands the decelerating area, improving the stability margin. In addition, GFM converters explicitly define the virtual inertia and damping coefficients in the swing equation, offering a direct means of regulating inertia and damping capability. In both GFL and GFM converters, enhancing equivalent damping and inertia helps to extend the critical fault clearing time.

Therefore, although converter-based systems inherit some fundamental stability principles from SGs, their unique dynamic characteristics, control flexibility, and current-limiting constraints necessitate dedicated analysis and design strategies to ensure robust transient stability.

References

1 Prabha Kundur. *Power System Stability and Control*. McGraw-Hill Education, New York, 1994.

2 A.A. Fouad and P.M. Anderson. *Power System Control and Stability*. IEEE Press, 2003.

3 Q. Hu, L. Fu, F. Ma, and F. Ji. Large signal synchronizing instability of PLL-based VSC connected to weak AC grid. *IEEE Transactions on Power Systems*, 34(4): 3220–3229, July 2019.

4 X. Fu, P. Wang, M. Cheng, Z. Liu, and B. Zhang. Large-signal stability of grid-forming and grid-following controls in voltage source converter: A comparative study. *IEEE Transactions on Power Electronics*, 36(7): 7832–7840, July 2021.

5 A. Hadjileonidas, Y. Li, and T.C. Green. Comparative analysis of transient stability of grid-forming and grid-following inverters. In *2022 IEEE International Power Electronics and Application Conference and Exposition (PEAC)*, Guangzhou, Guangdong, China, pages 296–301, 2022.

6 I. Erlich, F. Shewarega, S. Engelhardt, J. Kretschmann, J. Fortmann, and F. Koch. Effect of wind turbine output current during faults on grid voltage and the transient stability of wind parks. In *Proc. IEEE Power Energy Soc. Generation Meeting*, pages 1–8, 2009.

7 H. Geng, L. Liu, and R. Li. Synchronization and reactive current support of PMSG-based wind farm during severe grid fault. *IEEE Transactions on Sustainable Energy*, 9(4): 1596–1604, October 2018.

8 S. Ma, H. Geng, L. Liu, G. Yang, and B.C. Pal. Grid-synchronization stability improvement of large scale wind farm during severe grid fault. *IEEE Transactions on Power Systems*, 33(1): 216–226, January 2018.

9 X. He, H. Geng, R. Li, and B.C. Pal. Transient stability analysis and enhancement of renewable energy conversion system during LVRT. *IEEE Transactions on Sustainable Energy*, 11(3): 1612–1623, July 2019.

10 B. Weise. Impact of k-factor and active current reduction during fault-ride-through of generating units connected via voltage-sourced converters on power system stability. *IET Renewable Power Generation*, 9(1): 25–36, January 2015.

11 H. Wu and X. Wang. Transient stability impact of the phase-locked loop on grid-connected voltage source converters. In *Proc. Int. Power Electron. Conf.*, pages 2673–2680, 2018.

12 H. Wu and X. Wang. An adaptive phase-locked loop for the transient stability enhancement of grid-connected voltage source converters. In *Proc. IEEE Energy Convers. Congr. Expo.*, pages 5892–5898, 2018.

13 D. Pan, X. Wang, F. Liu, and R. Shi. Transient stability of voltage-source converters with grid-forming control: A design-oriented study. *IEEE Journal of Emerging and Selected Topics in Power Electronics*, 8(2): 1019–1033, June 2020.

14 H. Xin, L. Huang, L. Zhang, Z. Wang, and J. Hu. Synchronous instability mechanism of p-f droop-controlled voltage source converter caused by current saturation. *IEEE Transactions on Power Systems*, 31(6): 5206–5207, November 2016.

15 B. Fan, T. Liu, F. Zhao, H. Wu, and X. Wang. A review of current-limiting control of grid-forming inverters under symmetrical disturbances. *IEEE Open Journal of Power Electronics*, 3: 955–969, 2022.

16 E. Rokrok, T. Qoria, A. Bruyere, B. Francois, and X. Guillaud. Transient stability assessment and enhancement of grid-forming converters embedding current reference saturation as current limiting strategy. *IEEE Transactions on Power Systems*, 37(2): 1519–1531, March 2022.

17 Z. Shuai, C. Shen, X. Liu, Z. Li, and Z.J. Shen. Transient angle stability of virtual synchronous generators using Lyapunov's direct method. *IEEE Transactions on Smart Grid*, 10(4): 4648–4661, July 2019.

18 D. Pan, X. Wang, F. Liu, and R. Shi. Transient stability impact of reactive power control on grid-connected converters. In *Proc. IEEE Energy Convers. Congr. Expo.*, pages 4311–4316, 2019.

19 J. Alipoor, Y. Miura, and T. Ise. Power system stabilization using virtual synchronous generator with alternating moment of inertia. *IEEE Journal of Emerging and Selected Topics in Power Electronics*, 3(2): 451–458, June 2015.

20 S. Jiang, Y. Zhu, T. Xu, X. Wang, and G. Konstantinou. Frequency domain inertia design of grid-forming converters. *IEEE Transactions on Power Electronics*, 40(7): 8886–8898, July 2025.

21 H. Wu and X. Wang. A mode-adaptive power-angle control method for transient stability enhancement of virtual synchronous generators. *IEEE Journal of Emerging and Selected Topics in Power Electronics*, 8(2): 1034–1049, June 2020.

22 K. Shi, W. Song, P. Xu, R. Liu, Z. Fang, and Y. Ji. Low-voltage ride through control strategy for a virtual synchronous generator based on smooth switching. *IEEE Access*, 6: 2703–2711, December 2018.

23 M.G. Taul, X. Wang, P. Davari, and F. Blaabjerg. Current limiting control with enhanced dynamics of grid-forming converters during fault conditions. *IEEE Journal of Emerging and Selected Topics in Power Electronics*, 8(2): 1062–1073, June 2019.

24 C. Glockler, D. Duckwitz, and F. Welck. Virtual synchronous machine control with virtual resistor for enhanced short circuit capability. In *Proc. IEEE PES Innovative Smart Grid Technol. Conf. Eur.*, pages 1–6, 2017.

Appendix A

Small-signal Model of Grid-following Converters

An example of a small-signal model for grid-following (GFL) converters is provided in this appendix to help understand the modeling process from scratch. The state-space matrix and impedance of the GFL converter, as shown in Figure 9.1, are presented.

Under steady-state conditions, the state variables of the small-signal model for the GFL converter system are calculated as follows:

$$\begin{cases} V_{\text{con}}^{d} = R_{\text{f}} I_{\text{con}}^{d} - \omega_0 L_{\text{f}} I_{\text{con}}^{q} + V_{\text{PCC}}^{d} \\ V_{\text{con}}^{q} = R_{\text{f}} I_{\text{con}}^{q} + \omega_0 L_{\text{f}} I_{\text{con}}^{d} + V_{\text{PCC}}^{q} \\ I_{\text{con}}^{d} = I_{\text{g}}^{d} \\ I_{\text{con}}^{q} = I_{\text{g}}^{q} + \omega_0 C_{\text{f}} V_{\text{PCC}}^{d} \\ (V_{\text{g}}^{d})^2 + (V_{\text{g}}^{q})^2 = 1 \\ V_{\text{PCC}}^{d} = R_{\text{g}} I_{\text{g}}^{d} - \omega_0 L_{\text{g}} I_{\text{g}}^{q} + V_{\text{g}}^{d} \\ V_{\text{PCC}}^{q} = R_{\text{g}} I_{\text{g}}^{q} + \omega_0 L_{\text{g}} I_{\text{g}}^{d} + V_{\text{g}}^{q} \\ P = V_{\text{PCC}}^{d} I_{\text{g}}^{d} \end{cases} \tag{A.1}$$

$$\begin{cases} V_{\text{g}}^{d} = V_{\text{PCC}}^{d} - R_{\text{g}} I_{\text{g}}^{d} + \omega_0 L_{\text{g}} I_{\text{g}}^{q} \\ V_{\text{g}}^{q} = V_{\text{PCC}}^{q} - R_{\text{g}} I_{\text{g}}^{q} - \omega_0 L_{\text{g}} I_{\text{g}}^{d} \\ I_{\text{g}}^{d} = \dfrac{P}{V_{\text{PCC}}^{d}} \end{cases} \tag{A.2}$$

Stability Analysis of Converter-Rich Power Grids, First Edition. Jun Liang and Chuanyue Li.
© 2026 The Institute of Electrical and Electronics Engineers, Inc. Published 2026 by
John Wiley & Sons, Inc. All rights reserved.

A.1 State-space Matrix

A.1.1 Linearization of Active Power

For the *dq*-impedance derivation, the three-phase *abc*-system is presented in the *dq*-form. The power is therefore calculated as below:

$$P = \frac{3}{2}(V_{\mathrm{PCC}}^{d} I_{\mathrm{g}}^{d} + V_{\mathrm{PCC}}^{q} I_{\mathrm{g}}^{q}) \tag{A.3}$$

If small disturbances are applied based on small-signal method, Equation A.3 is rewritten as:

$$P + \tilde{P} = \frac{3}{2}[(V_{\mathrm{PCC}}^{d} + \tilde{v}_{\mathrm{PCC}}^{d})(I_{\mathrm{g}}^{d} + \tilde{i}_{\mathrm{g}}^{d}) + (V_{\mathrm{PCC}}^{q} + \tilde{v}_{\mathrm{PCC}}^{q})(I_{\mathrm{g}}^{q} + \tilde{i}_{\mathrm{g}}^{q})] \tag{A.4}$$

The linearized power is yielded by rearranging Equation A.4:

$$\tilde{P} \approx 1.5(I_{\mathrm{g}}^{d}\tilde{v}_{\mathrm{PCC}}^{d} + V_{\mathrm{PCC}}^{d}\tilde{i}_{\mathrm{g}}^{d} + I_{\mathrm{g}}^{q}\tilde{v}_{\mathrm{PCC}}^{q} + V_{\mathrm{PCC}}^{q}\tilde{i}_{\mathrm{g}}^{q}) \tag{A.5}$$

A.1.2 Linearization of *abc–dq* Transformation and the Phase-locked Loop

For the *dq*-impedance derivation, the three-phase system is presented in the *dq*-form, such as $\boldsymbol{v}_{\mathrm{con}}$, $\boldsymbol{v}_{\mathrm{PCC}}$, and $\boldsymbol{i}_{\mathrm{con}}$. The bold parameter stands for its *dq*-matrix parameters such as $\tilde{\boldsymbol{v}}_{\mathrm{PCC}} = \begin{bmatrix} \tilde{v}_{\mathrm{PCC}}^{d} \\ \tilde{v}_{\mathrm{PCC}}^{q} \end{bmatrix}$. The three parameters are explained below:

- $\boldsymbol{v}_{\mathrm{con}}$ is the output voltage of the converter;
- $\boldsymbol{v}_{\mathrm{PCC}}$ is the voltage after the *LC* filter, or the voltage of the connection point to the grid;
- $\boldsymbol{i}_{\mathrm{con}}$ is the output current of the converter.

The parameters in the *dq*-control frame are denoted as $\boldsymbol{v}_{\mathrm{con}}^{\mathrm{c}}$, $\boldsymbol{v}_{\mathrm{PCC}}^{\mathrm{c}}$, and $\boldsymbol{i}_{\mathrm{con}}^{\mathrm{c}}$. The *dq*-formed *abc*-parameters are equal to their *dq*-synchronized parameters after the *abc–dq* transformation at steady state but are different when a synchronized phase error $\tilde{\theta}_{\mathrm{PLL}}$ appears at the transformation. Their relation is derived below based on the linearization of the *abc–dq* transformation. The *abc–dq* transformation is shown below:

$$\mathbf{T} = \begin{bmatrix} \cos(0 + \tilde{\theta}_{\mathrm{PLL}}) & \sin(0 + \tilde{\theta}_{\mathrm{PLL}}) \\ -\sin(0 + \tilde{\theta}_{\mathrm{PLL}}) & \cos(0 + \tilde{\theta}_{\mathrm{PLL}}) \end{bmatrix} \approx \begin{bmatrix} 1 & \tilde{\theta}_{\mathrm{PLL}} \\ -\tilde{\theta}_{\mathrm{PLL}} & 1 \end{bmatrix} \tag{A.6}$$

Transformation from *dq* to *abc* yields:

$$\mathbf{T}^{-1} \approx \begin{bmatrix} 1 & -\tilde{\theta}_{\mathrm{PLL}} \\ \tilde{\theta}_{\mathrm{PLL}} & 1 \end{bmatrix} \tag{A.7}$$

The *abc–dq* transformation via $\mathbf{T}$ between $\boldsymbol{v}_{\mathrm{PCC}}$ and $\boldsymbol{v}_{\mathrm{PCC}}^{\mathrm{c}}$ based on the small-signal method is summarized below:

$$\begin{bmatrix} V_{\mathrm{PCC}}^{\mathrm{c},d} + \tilde{v}_{\mathrm{PCC}}^{\mathrm{c},d} \\ V_{\mathrm{PCC}}^{\mathrm{c},q} + \tilde{v}_{\mathrm{PCC}}^{\mathrm{c},q} \end{bmatrix} = \begin{bmatrix} 1 & \tilde{\theta}_{\mathrm{PLL}} \\ -\tilde{\theta}_{\mathrm{PLL}} & 1 \end{bmatrix} \begin{bmatrix} V_{\mathrm{PCC}}^{d} + \tilde{v}_{\mathrm{PCC}}^{d} \\ V_{\mathrm{PCC}}^{q} + \tilde{v}_{\mathrm{PCC}}^{q} \end{bmatrix} = \begin{bmatrix} V_{\mathrm{PCC}}^{d} + \tilde{v}_{\mathrm{PCC}}^{d} + V_{\mathrm{PCC}}^{q}\tilde{\theta}_{\mathrm{PLL}} + \tilde{\theta}_{\mathrm{PLL}}\tilde{v}_{\mathrm{PCC}}^{q} \\ V_{\mathrm{PCC}}^{q} + \tilde{v}_{\mathrm{PCC}}^{q} - V_{\mathrm{PCC}}^{d}\tilde{\theta}_{\mathrm{PLL}} - \tilde{\theta}_{\mathrm{PLL}}\tilde{v}_{\mathrm{PCC}}^{d} \end{bmatrix} \tag{A.8}$$

Equation A.8 is simplified below based on $V_{\mathrm{PCC}}^{\mathrm{c}} = V_{\mathrm{PCC}}$ and $\tilde{\theta}_{\mathrm{PLL}}\tilde{\boldsymbol{v}}_{\mathrm{PCC}} \approx 0$:

$$\begin{bmatrix} \tilde{v}_{\mathrm{PCC}}^{\mathrm{c},d} \\ \tilde{v}_{\mathrm{PCC}}^{\mathrm{c},q} \end{bmatrix} = \begin{bmatrix} \tilde{v}_{\mathrm{PCC}}^{d} + V_{\mathrm{PCC}}^{q}\tilde{\theta}_{\mathrm{PLL}} \\ \tilde{v}_{\mathrm{PCC}}^{q} - V_{\mathrm{PCC}}^{d}\tilde{\theta}_{\mathrm{PLL}} \end{bmatrix} \tag{A.9}$$

$$\begin{bmatrix} \tilde{i}_{\mathrm{con}}^{\mathrm{c},d} \\ \tilde{i}_{\mathrm{con}}^{\mathrm{c},q} \end{bmatrix} = \begin{bmatrix} \tilde{i}_{\mathrm{con}}^{d} + I_{\mathrm{con}}^{q}\tilde{\theta}_{\mathrm{PLL}} \\ \tilde{i}_{\mathrm{con}}^{q} - I_{\mathrm{con}}^{d}\tilde{\theta}_{\mathrm{PLL}} \end{bmatrix} \tag{A.10}$$

$$\begin{bmatrix} \tilde{v}_{\mathrm{con}}^{d} \\ \tilde{v}_{\mathrm{con}}^{q} \end{bmatrix} = \begin{bmatrix} \tilde{v}_{\mathrm{con}}^{\mathrm{c},d} - V_{\mathrm{con}}^{q}\tilde{\theta}_{\mathrm{PLL}} \\ \tilde{v}_{\mathrm{con}}^{\mathrm{c},q} + V_{\mathrm{con}}^{d}\tilde{\theta}_{\mathrm{PLL}} \end{bmatrix} \tag{A.11}$$

The relation between θ and $\tilde{v}_{\mathrm{PCC}}^{q}$ is shown below based on the phase-locked loop (PLL) and Equation A.9:

$$\underbrace{\tilde{v}_{\mathrm{PCC}}^{\mathrm{c},q}}_{\tilde{v}_{\mathrm{PCC}}^{q} - V_{\mathrm{PCC}}^{d}\tilde{\theta}_{\mathrm{PLL}}} \times \mathrm{PI}_{\mathrm{PLL}} \times \frac{1}{s} = \tilde{\theta}_{\mathrm{PLL}} \tag{A.12}$$

where $\mathrm{PI}_{\mathrm{PLL}} = k_{\mathrm{p}}^{\mathrm{PLL}} + \dfrac{k_{\mathrm{i}}^{\mathrm{PLL}}}{s}$.

Rearranging Equation A.12 yields:

$$\tilde{\theta}_{\mathrm{PLL}} = G_{\mathrm{PLL}}\tilde{v}_{\mathrm{PCC}}^{q} \tag{A.13}$$

where $G_{\mathrm{PLL}} = \dfrac{\mathrm{PI}_{\mathrm{PLL}}}{s + V_{\mathrm{PCC}}^{q}\mathrm{PI}_{\mathrm{PLL}}}$.

Substituting θ in Equation A.9 with Equation A.13 yields:

$$\tilde{\boldsymbol{v}}_{\mathrm{PCC}}^{\mathrm{c}} = \mathbf{G}_{\mathrm{ff}}\tilde{\boldsymbol{v}}_{\mathrm{PCC}} \tag{A.14}$$

where $\mathbf{G}_{\mathrm{ff}} = \begin{bmatrix} 1 & V_{\mathrm{PCC}}^{q} G_{\mathrm{PLL}} \\ 0 & 1 - V_{\mathrm{PCC}}^{d} G_{\mathrm{PLL}} \end{bmatrix}$.

$\tilde{\boldsymbol{i}}_{\mathrm{con}}^{\mathrm{c}}$ and $\tilde{\boldsymbol{v}}_{\mathrm{c}}^{\mathrm{c}}$ are found based on a similar derivation:

$$\tilde{\boldsymbol{i}}_{\mathrm{con}}^{\mathrm{c}} = \tilde{\boldsymbol{i}}_{\mathrm{con}} + \mathbf{H}_{\mathrm{cf}} \tilde{\boldsymbol{v}}_{\mathrm{PCC}} \tag{A.15}$$

where $\mathbf{H}_{\mathrm{cf}} = \begin{bmatrix} 0 & I_{\mathrm{con}}^{q} G_{\mathrm{PLL}} \\ 0 & -I_{\mathrm{con}}^{d} G_{\mathrm{PLL}} \end{bmatrix}$.

$$\tilde{\boldsymbol{v}}_{\mathrm{c}} = \tilde{\boldsymbol{v}}_{\mathrm{c}}^{\mathrm{c}} + \mathbf{G}_{\mathrm{cf}} \tilde{\boldsymbol{v}}_{\mathrm{PCC}} \tag{A.16}$$

where $\mathbf{G}_{\mathrm{cf}} = \begin{bmatrix} 0 & -V_{\mathrm{con}}^{q} G_{\mathrm{PLL}} \\ 0 & V_{\mathrm{con}}^{d} G_{\mathrm{PLL}} \end{bmatrix}$.

The linearization of the *abc–dq* transformation and the PLL is clarified in Equations A.14–A.16.

A.1.3 State-space Matrix

$$s\tilde{x}_{\mathrm{PLL}} = \tilde{v}_{\mathrm{PCC}}^{\mathrm{c},q} \tag{A.17}$$

Substituting $\tilde{v}_{\mathrm{PCC}}^{\mathrm{c},q}$ in Equation A.17 with Equation A.9 yields:

$$s\tilde{x}_{\mathrm{PLL}} = \tilde{v}_{\mathrm{PCC}}^{q} - V_{\mathrm{PCC}}^{d} \tilde{\theta}_{\mathrm{PLL}} \tag{A.18}$$

$$s\tilde{\theta}_{\mathrm{PLL}} = k_{\mathrm{p}}^{\mathrm{PLL}} \tilde{v}_{\mathrm{PCC}}^{\mathrm{c},q} + \frac{k_{\mathrm{i}}^{\mathrm{PLL}}}{s} \underbrace{\tilde{v}_{\mathrm{PCC}}^{\mathrm{c},q}}_{s\tilde{x}_{\mathrm{PLL}}} \tag{A.19}$$

Substituting $\tilde{v}_{\mathrm{PCC}}^{\mathrm{c},q}$ in Equation A.19 with Equation A.9 yields:

$$s\tilde{\theta}_{\mathrm{PLL}} = k_{\mathrm{p}}^{\mathrm{PLL}} \tilde{v}_{\mathrm{PCC}}^{q} - k_{\mathrm{p}}^{\mathrm{PLL}} V_{\mathrm{PCC}}^{d} \tilde{\theta}_{\mathrm{PLL}} + k_{\mathrm{i}}^{\mathrm{PLL}} \tilde{x}_{\mathrm{PLL}} \tag{A.20}$$

$$\begin{cases} s\tilde{x}_{\mathrm{p}} = \tilde{P}^{*} - \tilde{P} \\ s\tilde{x}_{\mathrm{v}} = \tilde{v}_{\mathrm{PCC}}^{\mathrm{c},d} - \tilde{v}_{\mathrm{PCC}}^{\mathrm{c},d,*} \end{cases} \tag{A.21}$$

Substituting $\tilde{P}$ and $\tilde{v}_{\mathrm{PCC}}^{\mathrm{c},d}$ in Equation A.21 with Equations A.5 and A.9 yields:

$$\begin{cases} s\tilde{x}_{\mathrm{p}} = \tilde{P}^{*} - 1.5I_{\mathrm{g}}^{d}\tilde{v}_{\mathrm{PCC}}^{d} - 1.5V_{\mathrm{PCC}}^{d}\tilde{i}_{\mathrm{g}}^{d} - 1.5I_{\mathrm{g}}^{q}\tilde{v}_{\mathrm{PCC}}^{q} - 1.5V_{\mathrm{PCC}}^{q}\tilde{i}_{\mathrm{g}}^{q} \\ s\tilde{x}_{\mathrm{v}} = \tilde{v}_{\mathrm{PCC}}^{d} + V_{\mathrm{PCC}}^{q}\tilde{\theta}_{\mathrm{PLL}} - \tilde{v}_{\mathrm{PCC}}^{\mathrm{c},d,*} \end{cases} \tag{A.22}$$

$$\begin{cases} s\tilde{x}_{\mathrm{id}} = \tilde{i}_{\mathrm{c}}^{\mathrm{c},d,*} - \tilde{i}_{\mathrm{con}}^{\mathrm{c},d} \\ s\tilde{x}_{\mathrm{iq}} = \tilde{i}_{\mathrm{c}}^{\mathrm{c},q,*} - \tilde{i}_{\mathrm{con}}^{\mathrm{c},q} \end{cases} \tag{A.23}$$

$$\begin{cases} \tilde{i}_{\mathrm{c}}^{\mathrm{c},d,*} = (\tilde{P}^{*} - \tilde{P})k_{\mathrm{p}}^{p} + \underbrace{(\tilde{P}^{*} - \tilde{P})}_{s\tilde{x}_{\mathrm{p}}}\dfrac{k_{\mathrm{i}}^{p}}{s} \\ \tilde{i}_{\mathrm{c}}^{\mathrm{c},q,*} = (\tilde{v}_{\mathrm{PCC}}^{\mathrm{c},d} - \tilde{v}_{\mathrm{PCC}}^{\mathrm{c},d,*})k_{\mathrm{p}}^{v} + \underbrace{(\tilde{v}_{\mathrm{PCC}}^{\mathrm{c},d} - \tilde{v}_{\mathrm{PCC}}^{\mathrm{c},d,*})}_{s\tilde{x}_{\mathrm{v}}}\dfrac{k_{\mathrm{i}}^{v}}{s} \end{cases} \tag{A.24}$$

$$\begin{cases} s\tilde{x}_{\mathrm{id}} = k_{\mathrm{p}}^{\mathrm{p}}\tilde{P}^{*} - k_{\mathrm{p}}^{\mathrm{p}}\tilde{P} + k_{\mathrm{i}}^{\mathrm{p}}\tilde{x}_{\mathrm{p}} - \tilde{i}_{\mathrm{con}}^{\mathrm{c},d} \\ s\tilde{x}_{\mathrm{iq}} = -k_{\mathrm{p}}^{v}\tilde{v}_{\mathrm{PCC}}^{\mathrm{c},d,*} + k_{\mathrm{p}}^{v}\tilde{v}_{\mathrm{PCC}}^{\mathrm{c},d} + k_{\mathrm{i}}^{v}\tilde{x}_{\mathrm{v}} - \tilde{i}_{\mathrm{con}}^{\mathrm{c},q} \end{cases} \tag{A.25}$$

Substituting $\tilde{P}$, $\tilde{v}_{\mathrm{PCC}}^{\mathrm{c},d}$, $\tilde{i}_{\mathrm{con}}^{\mathrm{c},d}$, and $\tilde{i}_{\mathrm{con}}^{\mathrm{c},q}$ in Equation A.25 with Equations A.5, A.9, and A.10 yields:

$$\begin{cases} s\tilde{x}_{\mathrm{id}} = k_{\mathrm{p}}^{\mathrm{p}}\tilde{P}^{*} - 1.5k_{\mathrm{p}}^{\mathrm{p}}I_{\mathrm{g}}^{d}\tilde{v}_{\mathrm{PCC}}^{d} - 1.5k_{\mathrm{p}}^{\mathrm{p}}V_{\mathrm{PCC}}^{d}\tilde{i}_{\mathrm{g}}^{d} - \\ 1.5k_{\mathrm{p}}^{\mathrm{p}}I_{\mathrm{g}}^{q}\tilde{v}_{\mathrm{PCC}}^{q} - 1.5k_{\mathrm{p}}^{\mathrm{p}}V_{\mathrm{PCC}}^{q}\tilde{i}_{\mathrm{g}}^{q} + k_{\mathrm{i}}^{\mathrm{p}}\tilde{x}_{\mathrm{p}} - \tilde{i}_{\mathrm{con}}^{d} - I_{\mathrm{con}}^{q}\tilde{\theta}_{\mathrm{PLL}} \\ s\tilde{x}_{\mathrm{iq}} = -k_{\mathrm{p}}^{\mathrm{v}}\tilde{v}_{\mathrm{PCC}}^{\mathrm{c},d,*} + k_{\mathrm{p}}^{\mathrm{v}}\tilde{v}_{\mathrm{PCC}}^{d} + k_{\mathrm{i}}^{\mathrm{v}}\tilde{x}_{\mathrm{v}} - \tilde{i}_{\mathrm{con}}^{q} + (k_{\mathrm{p}}^{\mathrm{v}}V_{\mathrm{PCC}}^{q} + I_{\mathrm{con}}^{d})\tilde{\theta}_{\mathrm{PLL}} \end{cases} \tag{A.26}$$

$$\begin{cases} \tilde{v}_{\mathrm{con}}^{d} = \tilde{v}_{\mathrm{PCC}}^{\mathrm{c},d} - \omega_0 L_{\mathrm{f}}\tilde{i}_{\mathrm{con}}^{\mathrm{c},q} + k_{\mathrm{p}}^{\mathrm{c}}(\tilde{i}_{\mathrm{c}}^{\mathrm{c},d,*} - \tilde{i}_{\mathrm{con}}^{\mathrm{c},d}) + \underbrace{(\tilde{i}_{\mathrm{c}}^{\mathrm{c},d,*} - \tilde{i}_{\mathrm{con}}^{\mathrm{c},d})}_{s\tilde{x}_{\mathrm{id}}}\dfrac{k_{\mathrm{i}}^{\mathrm{c}}}{s} \\ \tilde{v}_{\mathrm{con}}^{q} = \tilde{v}_{\mathrm{PCC}}^{\mathrm{c},q} + \omega_0 L_{\mathrm{f}}\tilde{i}_{\mathrm{con}}^{\mathrm{c},d} + k_{\mathrm{p}}^{\mathrm{c}}(\tilde{i}_{\mathrm{c}}^{\mathrm{c},q,*} - \tilde{i}_{\mathrm{con}}^{\mathrm{c},q}) + \underbrace{(\tilde{i}_{\mathrm{c}}^{\mathrm{c},q,*} - \tilde{i}_{\mathrm{con}}^{\mathrm{c},q})}_{s\tilde{x}_{\mathrm{iq}}}\dfrac{k_{\mathrm{i}}^{\mathrm{c}}}{s} \end{cases} \tag{A.27}$$

Substituting $\tilde{v}_{\text{con}}^{d}$ and $\tilde{v}_{\text{con}}^{q}$ in Equation A.11 with Equation A.27 yields:

$$\begin{cases} \tilde{v}_{\text{con}}^{\text{c},d} = \tilde{v}_{\text{PCC}}^{\text{c},d} - \omega_0 L_{\text{f}} \tilde{i}_{\text{con}}^{\text{c},q} + k_{\text{p}}^{\text{c}}(\tilde{i}_{\text{c}}^{\text{c},d,*} - \tilde{i}_{\text{con}}^{\text{c},d}) + k_{\text{i}}^{\text{c}}\tilde{x}_{\text{id}} - V_{\text{con}}^{q}\tilde{\theta}_{\text{PLL}} \\ \tilde{v}_{\text{con}}^{\text{c},q} = \tilde{v}_{\text{PCC}}^{\text{c},q} + \omega_0 L_{\text{f}} \tilde{i}_{\text{con}}^{\text{c},d} + k_{\text{p}}^{\text{c}}(\tilde{i}_{\text{c}}^{\text{c},q,*} - \tilde{i}_{\text{con}}^{\text{c},q}) + k_{\text{i}}^{\text{c}}\tilde{x}_{\text{iq}} + V_{\text{con}}^{d}\tilde{\theta}_{\text{PLL}} \end{cases} \tag{A.28}$$

Deriving $\tilde{v}_{\text{con}}^{\text{c},d}$:

$$\begin{cases} \tilde{v}_{\text{con}}^{\text{c},d} = \tilde{v}_{\text{PCC}}^{d} + V_{\text{PCC}}^{q}\tilde{\theta}_{\text{PLL}} - \omega_0 L_{\text{f}}(\tilde{i}_{\text{con}}^{q} - I_{\text{con}}^{d}\tilde{\theta}_{\text{PLL}}) \\ \quad + k_{\text{p}}^{\text{c}}(\tilde{i}_{\text{c}}^{\text{c},d,*} - \tilde{i}_{\text{con}}^{\text{c},d}) + k_{\text{i}}^{\text{c}}\tilde{x}_{\text{id}} - V_{\text{con}}^{q}\tilde{\theta}_{\text{PLL}} \\ k_{\text{p}}^{\text{i}}(\tilde{i}_{\text{c}}^{\text{c},d,*} - \tilde{i}_{\text{con}}^{\text{c},d}) = \\ \quad k_{\text{p}}^{\text{c}}k_{\text{p}}^{\text{p}}\tilde{P}^{*} - 1.5k_{\text{p}}^{\text{c}}k_{\text{p}}^{\text{p}}I_{\text{g}}^{d}\tilde{v}_{\text{PCC}}^{d} - 1.5k_{\text{p}}^{\text{c}}k_{\text{p}}^{\text{p}}V_{\text{PCC}}^{d}\tilde{i}_{\text{g}}^{d} - \\ \quad 1.5k_{\text{p}}^{\text{c}}k_{\text{p}}^{\text{p}}I_{\text{g}}^{q}\tilde{v}_{\text{PCC}}^{q} - 1.5k_{\text{p}}^{\text{c}}k_{\text{p}}^{\text{p}}V_{\text{PCC}}^{q}\tilde{i}_{\text{g}}^{q} + k_{\text{p}}^{\text{c}}k_{\text{i}}^{\text{p}}\tilde{x}_{\text{p}} - k_{\text{p}}^{\text{c}}\tilde{i}_{\text{con}}^{d} - k_{\text{p}}^{\text{c}}I_{\text{con}}^{q}\tilde{\theta}_{\text{PLL}} \end{cases} \tag{A.29}$$

$$\begin{cases} \tilde{v}_{\text{con}}^{\text{c},d} = \\ \quad k_{\text{p}}^{\text{c}}k_{\text{p}}^{\text{p}}\tilde{P}^{*} + (1 - 1.5k_{\text{p}}^{\text{c}}k_{\text{p}}^{\text{p}}I_{\text{g}}^{d})\tilde{v}_{\text{PCC}}^{d} - 1.5k_{\text{p}}^{\text{c}}k_{\text{p}}^{\text{p}}V_{\text{PCC}}^{d}\tilde{i}_{\text{g}}^{d} \\ \quad -1.5k_{\text{p}}^{\text{c}}k_{\text{p}}^{\text{p}}I_{\text{g}}^{q}\tilde{v}_{\text{PCC}}^{q} - 1.5k_{\text{p}}^{\text{c}}k_{\text{p}}^{\text{p}}V_{\text{PCC}}^{q}\tilde{i}_{\text{g}}^{q} + k_{\text{p}}^{\text{c}}k_{\text{i}}^{\text{p}}\tilde{x}_{\text{p}} - k_{\text{p}}^{\text{c}}\tilde{i}_{\text{con}}^{d} \\ \quad -\omega_0 L_{\text{f}}\tilde{i}_{\text{con}}^{q} + k_{\text{i}}^{\text{i}}\tilde{x}_{\text{id}} + (V_{\text{PCC}}^{q} + \omega_0 L_{\text{f}} I_{\text{con}}^{d} - k_{\text{p}}^{\text{c}}I_{\text{con}}^{q} - V_{\text{con}}^{q})\tilde{\theta}_{\text{PLL}} \end{cases} \tag{A.30}$$

Deriving $\tilde{v}_{\text{con}}^{\text{c},q}$:

$$\begin{cases} \tilde{v}_{\text{con}}^{\text{c},q} = \tilde{v}_{\text{PCC}}^{q} - V_{\text{PCC}}^{d}\tilde{\theta}_{\text{PLL}} + \omega_0 L_{\text{f}}(\tilde{i}_{\text{con}}^{d} + I_{\text{con}}^{q}\tilde{\theta}_{\text{PLL}}) + k_{\text{p}}^{\text{c}}(\tilde{i}_{\text{c}}^{\text{c},q,*} \\ \quad - \tilde{i}_{\text{con}}^{\text{c},q}) + k_{\text{i}}^{\text{c}}\tilde{x}_{\text{iq}} + V_{\text{con}}^{d}\tilde{\theta}_{\text{PLL}} \\ k_{\text{p}}^{\text{c}}(\tilde{i}_{\text{c}}^{\text{c},q,*} - \tilde{i}_{\text{con}}^{\text{c},q}) = \\ \quad - k_{\text{p}}^{\text{c}}k_{\text{p}}^{\text{v}}\tilde{v}_{\text{PCC}}^{\text{c},d,*} + k_{\text{p}}^{\text{c}}k_{\text{p}}^{\text{v}}\tilde{v}_{\text{PCC}}^{d} + k_{\text{p}}^{\text{c}}k_{\text{i}}^{\text{v}}\tilde{x}_{\text{v}} \\ \quad - k_{\text{p}}^{\text{c}}\tilde{i}_{\text{con}}^{q} + (k_{\text{p}}^{\text{c}}k_{\text{p}}^{\text{v}}V_{\text{PCC}}^{q} + k_{\text{p}}^{\text{c}}I_{\text{con}}^{d})\tilde{\theta}_{\text{PLL}} \end{cases} \tag{A.31}$$

$$\begin{cases} \tilde{v}_{\text{con}}^{\text{c},q} = \\ \quad - k_{\text{p}}^{\text{c}}k_{\text{p}}^{\text{v}}\tilde{v}_{\text{PCC}}^{\text{c},d,*} + \tilde{v}_{\text{PCC}}^{q} + k_{\text{p}}^{\text{c}}k_{\text{p}}^{\text{v}}\tilde{v}_{\text{PCC}}^{d} + k_{\text{p}}^{\text{c}}k_{\text{i}}^{\text{v}}\tilde{x}_{\text{v}} - k_{\text{p}}^{\text{c}}\tilde{i}_{\text{con}}^{q} + (k_{\text{p}}^{\text{c}}k_{\text{p}}^{\text{v}}V_{\text{PCC}}^{q} \\ \quad + k_{\text{p}}^{\text{c}}I_{\text{con}}^{d} - V_{\text{PCC}}^{d} + \omega_0 L_{\text{f}} I_{\text{con}}^{q} + V_{\text{con}}^{d})\tilde{\theta}_{\text{PLL}} + \omega_0 L_{\text{f}}\tilde{i}_{\text{con}}^{d} + k_{\text{i}}^{\text{c}}\tilde{x}_{\text{iq}} \end{cases} \tag{A.32}$$

$$\begin{cases}(1+0.5T_{\text{del}}s)\tilde{x}_d^{\text{PWM}} = \tilde{v}_{\text{con}}^{\text{c},d} \\ (1+0.5T_{\text{del}}s)\tilde{x}_q^{\text{PWM}} = \tilde{v}_{\text{con}}^{\text{c},q}\end{cases} \tag{A.33}$$

$$\begin{cases}s\tilde{x}_d^{\text{PWM}} = \dfrac{\tilde{v}_{\text{con}}^{\text{c},d}}{0.5T_{\text{del}}} - \dfrac{\tilde{x}_d^{\text{PWM}}}{0.5T_{\text{del}}} \\ s\tilde{x}_q^{\text{PWM}} = \dfrac{\tilde{v}_{\text{con}}^{\text{c},q}}{0.5T_{\text{del}}} - \dfrac{\tilde{x}_q^{\text{PWM}}}{0.5T_{\text{del}}}\end{cases} \tag{A.34}$$

$$\begin{cases}\tilde{v}_{\text{cd}} = \dfrac{1-0.5T_{\text{del}}s}{1+0.5T_{\text{del}}s}\tilde{v}_{\text{con}}^{\text{c},d} = \left(\dfrac{2}{1+0.5T_{\text{del}}s} - 1\right)\tilde{v}_{\text{con}}^{\text{c},d} \\ \tilde{v}_{\text{cq}} = \dfrac{1-0.5T_{\text{del}}s}{1+0.5T_{\text{del}}s}\tilde{v}_{\text{con}}^{\text{c},q} = \left(\dfrac{2}{1+0.5T_{\text{del}}s} - 1\right)\tilde{v}_{\text{con}}^{\text{c},q}\end{cases} \tag{A.35}$$

$$\begin{cases}\tilde{v}_{\text{cd}} = \dfrac{2}{1+0.5T_{\text{del}}s}\underbrace{\tilde{v}_{\text{con}}^{\text{c},d}}_{(1+0.5T_{\text{del}}s)\tilde{x}_d^{\text{PWM}}} - \tilde{v}_{\text{con}}^{\text{c},d} \\ \tilde{v}_{\text{cq}} = \dfrac{2}{1+0.5T_{\text{del}}s}\underbrace{\tilde{v}_{\text{con}}^{\text{c},q}}_{(1+0.5T_{\text{del}}s)\tilde{x}_q^{\text{PWM}}} - \tilde{v}_{\text{con}}^{\text{c},q}\end{cases} \tag{A.36}$$

$$\begin{cases}\tilde{v}_{\text{cd}} = 2\tilde{x}_d^{\text{PWM}} - \tilde{v}_{\text{con}}^{\text{c},d} \\ \tilde{v}_{\text{cq}} = 2\tilde{x}_q^{\text{PWM}} - \tilde{v}_{\text{con}}^{\text{c},q}\end{cases} \tag{A.37}$$

$$\begin{cases}L_{\text{f}}s\tilde{i}_{\text{con}}^d + R_{\text{f}}\tilde{i}_{\text{con}}^d - \omega_0 L_{\text{f}}\tilde{i}_{\text{con}}^q = \tilde{v}_{\text{cd}} - \tilde{v}_{\text{PCC}}^d \\ L_{\text{f}}s\tilde{i}_{\text{con}}^q + R_{\text{f}}\tilde{i}_{\text{con}}^q + \omega_0 L_{\text{f}}\tilde{i}_{\text{con}}^d = \tilde{v}_{\text{cq}} - \tilde{v}_{\text{PCC}}^q\end{cases} \tag{A.38}$$

$$\begin{cases}s\tilde{i}_{\text{con}}^d = \dfrac{\tilde{v}_{\text{cd}}}{L_{\text{f}}} - \dfrac{\tilde{v}_{\text{PCC}}^d}{L_{\text{f}}} - \dfrac{R_{\text{f}}\tilde{i}_{\text{con}}^d}{L_{\text{f}}} + \omega_0\tilde{i}_{\text{con}}^q \\ s\tilde{i}_{\text{con}}^q = \dfrac{\tilde{v}_{\text{cq}}}{L_{\text{f}}} - \dfrac{\tilde{v}_{\text{PCC}}^q}{L_{\text{f}}} - \dfrac{R_{\text{f}}\tilde{i}_{\text{con}}^q}{L_{\text{f}}} - \omega_0\tilde{i}_{\text{con}}^d\end{cases} \tag{A.39}$$

Substituting $\tilde{v}_{\text{cd}}$ and $\tilde{v}_{\text{cq}}$ in Equation A.39 with Equation A.37 yields:

$$\begin{cases} s\tilde{i}_{\text{con}}^{d} = \dfrac{2\tilde{x}_{d}^{\text{PWM}}}{L_{\text{f}}} - \dfrac{\tilde{v}_{\text{con}}^{\text{c},d}}{L_{\text{f}}} - \dfrac{\tilde{v}_{\text{PCC}}^{d}}{L_{\text{f}}} - \dfrac{R_{\text{f}}\tilde{i}_{\text{con}}^{d}}{L_{\text{f}}} + \omega_0\tilde{i}_{\text{con}}^{q} \\ s\tilde{i}_{\text{con}}^{q} = \dfrac{2\tilde{x}_{q}^{\text{PWM}}}{L_{\text{f}}} - \dfrac{\tilde{v}_{\text{con}}^{\text{c},q}}{L_{\text{f}}} - \dfrac{\tilde{v}_{\text{PCC}}^{q}}{L_{\text{f}}} - \dfrac{R_{\text{f}}\tilde{i}_{\text{con}}^{q}}{L_{\text{f}}} - \omega_0\tilde{i}_{\text{con}}^{d} \end{cases} \tag{A.40}$$

$$\begin{cases} L_{\text{g}}s\tilde{i}_{\text{g}}^{d} + R_{\text{g}}\tilde{i}_{\text{g}}^{d} - \omega_0 L_{\text{g}}\tilde{i}_{\text{g}}^{q} = \tilde{v}_{\text{PCC}}^{d} - \tilde{v}_{\text{gd}} \\ L_{\text{g}}s\tilde{i}_{\text{g}}^{q} + R_{\text{g}}\tilde{i}_{\text{g}}^{q} + \omega_0 L_{\text{g}}\tilde{i}_{\text{g}}^{d} = \tilde{v}_{\text{PCC}}^{q} - \tilde{v}_{\text{gq}} \end{cases} \tag{A.41}$$

$$\begin{cases} s\tilde{i}_{\text{g}}^{d} = \dfrac{\tilde{v}_{\text{PCC}}^{d}}{L_{\text{g}}} - \dfrac{\tilde{v}_{\text{gd}}}{L_{\text{g}}} - \dfrac{R_{\text{g}}\tilde{i}_{\text{g}}^{d}}{L_{\text{g}}} + \omega_0\tilde{i}_{\text{g}}^{q} \\ s\tilde{i}_{\text{g}}^{q} = \dfrac{\tilde{v}_{\text{PCC}}^{q}}{L_{\text{g}}} - \dfrac{\tilde{v}_{\text{gq}}}{L_{\text{g}}} - \dfrac{R_{\text{g}}\tilde{i}_{\text{g}}^{q}}{L_{\text{g}}} - \omega_0\tilde{i}_{\text{g}}^{d} \end{cases} \tag{A.42}$$

$$\begin{cases} C_{\text{f}}s\tilde{v}_{\text{PCC}}^{d} - \omega_0 C_{\text{f}}\tilde{v}_{\text{PCC}}^{q} = \tilde{i}_{\text{con}}^{d} - \tilde{i}_{\text{g}}^{d} \\ C_{\text{f}}s\tilde{v}_{\text{PCC}}^{q} + \omega_0 C_{\text{f}}\tilde{v}_{\text{PCC}}^{d} = \tilde{i}_{\text{con}}^{q} - \tilde{i}_{\text{g}}^{q} \end{cases} \tag{A.43}$$

$$\begin{cases} s\tilde{v}_{\text{PCC}}^{d} = \dfrac{\tilde{i}_{\text{con}}^{d}}{C_{\text{f}}} - \dfrac{\tilde{i}_{\text{g}}^{d}}{C_{\text{f}}} + \omega_0\tilde{v}_{\text{PCC}}^{q} \\ s\tilde{v}_{\text{PCC}}^{q} = \dfrac{\tilde{i}_{\text{con}}^{q}}{C_{\text{f}}} - \dfrac{\tilde{i}_{\text{g}}^{q}}{C_{\text{f}}} - \omega_0\tilde{v}_{\text{PCC}}^{d} \end{cases} \tag{A.44}$$

In summary, the state-space matrix **A** is defined by Equations A.18, A.20, A.22, A.26, A.34, A.40, A.42, and A.44.

A.2 Impedance Representation

For example, the *abc–dq* transformation of v_{PCC} is given below:

$$\begin{bmatrix} \tilde{v}_{\text{PCC}}^{d} + V_{\text{PCC}}^{q}\tilde{\theta}_{\text{PLL}} \\ \tilde{v}_{\text{PCC}}^{q} - V_{\text{PCC}}^{d}\tilde{\theta}_{\text{PLL}} \end{bmatrix} = \begin{bmatrix} \tilde{v}_{\text{PCC}}^{\text{c},d} \\ \tilde{v}_{\text{PCC}}^{\text{c},q} \end{bmatrix} \tag{A.46}$$

The phase disturbance $\tilde{\theta}_{\text{PLL}}$ comes from the PLL and grid resistance compensation, which is summarized below:

$$\underbrace{\tilde{v}_{\text{PCC}}^{\text{c},q}}_{\tilde{v}_{\text{PCC}}^{q}-V_{\text{PCC}}^{d}\tilde{\theta}_{\text{PLL}}} \times \underbrace{\text{PI}_{\text{PLL}}}_{k_{\text{p}}^{\text{PLL}}+k_{\text{i}}^{\text{PLL}}/s} \times \frac{1}{s} = \tilde{\theta}_{\text{PLL}} \tag{A.47}$$

Rearranging Equation A.47 yields:

$$G_{\text{PLL}}\tilde{v}_{\text{PCC}}^{q} + G_{\text{PLL}}\tilde{i}_{\text{g}}^{q} = \tilde{\theta}_{\text{PLL}} \tag{A.48}$$

where $G_{\text{PLL}} = \dfrac{k_{\text{p}}^{\text{PLL}} + k_{\text{i}}^{\text{PLL}}/s}{s + (k_{\text{p}}^{\text{PLL}} + k_{\text{i}}^{\text{PLL}}/s)(V_{\text{PCC}}^{d} + R_{\text{v}}s/(s+\omega_{\text{c}})I_{\text{g}}^{d})}$

Substituting $\tilde{\theta}$ in Equation A.46 with Equation A.48 gives:

$$\mathbf{G}_{\text{fg}}\tilde{\boldsymbol{v}}_{\text{PCC}} + \mathbf{H}_{\text{fg}}\tilde{\boldsymbol{i}}_{\text{g}} = \tilde{\boldsymbol{v}}_{\text{PCC}}^{\text{c}} \tag{A.49}$$

where $\mathbf{G}_{\text{fg}} = \begin{bmatrix} 1 & V_{\text{PCC}}^{q}G_{\text{PLL}} \\ 0 & 1 - V_{\text{PCC}}^{d}G_{\text{PLL}} \end{bmatrix}$, $\mathbf{H}_{\text{fg}} = \begin{bmatrix} 0 & V_{\text{PCC}}^{q}G_{\text{PLL}} \\ 0 & -V_{\text{PCC}}^{d}G_{\text{PLL}} \end{bmatrix}$.

$\tilde{\boldsymbol{i}}_{\text{con}}^{\text{c}}$ and $\tilde{\boldsymbol{v}}_{\text{c}}^{\text{c}}$ are found based on a similar derivation:

$$\tilde{\boldsymbol{i}}_{\text{con}} + \mathbf{G}_{\text{cf}}\tilde{\boldsymbol{v}}_{\text{PCC}} + \mathbf{H}_{\text{cf}}\tilde{\boldsymbol{i}}_{\text{g}} = \tilde{\boldsymbol{i}}_{\text{con}}^{\text{c}} \tag{A.50}$$

where $\mathbf{G}_{\text{cf}} = \begin{bmatrix} 0 & I_{\text{con}}^{q}G_{\text{PLL}} \\ 0 & -I_{\text{con}}^{d}G_{\text{PLL}} \end{bmatrix}$, $\mathbf{H}_{\text{cf}} = \begin{bmatrix} 0 & I_{\text{con}}^{q}G_{\text{PLL}} \\ 0 & -I_{\text{con}}^{d}G_{\text{PLL}} \end{bmatrix}$.

$$\tilde{\boldsymbol{v}}_{\text{c}}^{\text{c}} + \mathbf{G}_{\text{fc}}\tilde{\boldsymbol{v}}_{\text{PCC}} + \mathbf{H}_{\text{fc}}\tilde{\boldsymbol{i}}_{\text{g}} = \tilde{\mathbf{v}}_{\text{c}} \tag{A.51}$$

where $\mathbf{G}_{\text{fc}} = \begin{bmatrix} 0 & -V_{\text{con}}^{q}G_{\text{PLL}} \\ 0 & V_{\text{con}}^{d}G_{\text{PLL}} \end{bmatrix}$, $\mathbf{H}_{\text{fc}} = \begin{bmatrix} 0 & -V_{\text{con}}^{q}G_{\text{PLL}} \\ 0 & V_{\text{con}}^{d}G_{\text{PLL}} \end{bmatrix}$.

The linearization of the *abc–dq* transformation is clarified in Equations A.49–A.51.

The impedance of a voltage source converter (VSC) includes the control system and its *LC* filter. The control system is represented below:

$$\mathbf{PI}_{\text{c}}(\mathbf{PI}_{\text{pv}}(\tilde{\mathbf{r}}^{*} - [\tilde{P}\,; -\tilde{v}_{\text{PCC}}^{\text{c},d}] - \tilde{\boldsymbol{i}}_{\text{con}}^{\text{c}})) + \tilde{\boldsymbol{v}}_{\text{PCC}}^{\text{c}} + \mathbf{G}_{\text{dec}}\tilde{\boldsymbol{i}}_{\text{con}}^{\text{c}} = \tilde{\boldsymbol{v}}_{\text{con}}^{\text{c}} \tag{A.52}$$

where $\tilde{\mathbf{r}}^* = \begin{bmatrix} P^* \\ -v^*_{\text{mag}} \end{bmatrix}$, $\mathbf{G}_{\text{dec}} = \begin{bmatrix} 0 & -L_f\omega_0 \\ L_f\omega_0 & 0 \end{bmatrix}$, $\mathbf{PI}_c = \begin{bmatrix} k_p^c + k_i^c/s & 0 \\ 0 & k_p^c + k_i^c/s \end{bmatrix}$, and $\mathbf{PI}_{pv} = \begin{bmatrix} k_p^p + k_i^p/s & 0 \\ 0 & k_p^v + k_i^v/s \end{bmatrix}$.

The linearized active power in dq-frame is shown below:

$$\tilde{P} \approx 1.5(I_g^d\tilde{v}_{\text{PCC}}^d + V_{\text{PCC}}^d\tilde{i}_g^d + I_g^q\tilde{v}_{\text{PCC}}^q + V_{\text{PCC}}^q\tilde{i}_g^q) \tag{A.53}$$

Substituting $\tilde{P}$ and $\tilde{v}_{\text{PCC}}^{c,d}$ in Equation A.52 with Equations A.49 and A.53 yields:

$$\mathbf{PI}_c\mathbf{PI}_{pv}\tilde{\mathbf{r}}^* + \mathbf{G}_{rv}\tilde{\boldsymbol{v}}_{\text{PCC}} + \mathbf{G}_{ri}\tilde{\boldsymbol{i}}_g - \mathbf{PI}_c\tilde{\boldsymbol{i}}_{\text{con}}^c + \tilde{\boldsymbol{v}}_{\text{PCC}}^c + \mathbf{G}_{\text{dec}}\tilde{\boldsymbol{i}}_{\text{con}}^c = \tilde{\boldsymbol{v}}_{\text{con}}^c \tag{A.54}$$

where $\text{PI}_i = k_p^c + k_i^c/s$, $\text{PI}_p = k_p^p + k_i^p/s$, $\text{PI}_v = k_p^v + k_i^v/s$,
$\mathbf{G}_{rv} = \begin{bmatrix} -1.5\text{PI}_p I_g^d & -1.5\text{PI}_i\text{PI}_p I_g^q \\ \text{PI}_i\text{PI}_v & \text{PI}_i\text{PI}_v V_{\text{PCC}}^q G_{\text{PLL}} \end{bmatrix}$, and $\mathbf{G}_{ri} = \begin{bmatrix} -1.5\text{PI}_i\text{PI}_p V_{\text{PCC}}^d & -1.5\text{PI}_i\text{PI}_p V_{\text{PCC}}^q \\ 0 & \text{PI}_i\text{PI}_v V_{\text{PCC}}^q G_{\text{PLL}} \end{bmatrix}$.

Substituting $\tilde{\boldsymbol{i}}_{\text{con}}^c$, $\tilde{\boldsymbol{v}}_{\text{PCC}}^c$, and $\tilde{\boldsymbol{v}}_{\text{con}}^c$ in Equation A.54 with Equations A.49–A.51 and taking the time T_{del} including control delay and the dead time of pulse width modulation (PWM) into account yields:

$$\begin{aligned} &\mathbf{G}_{\text{del}}\{\mathbf{PI}_c\mathbf{PI}_{pv}\tilde{\mathbf{r}}^* + [\mathbf{G}_{rv} + (\mathbf{G}_{\text{dec}} - \mathbf{PI}_c)\mathbf{G}_{cf} + \mathbf{G}_{fg} + \mathbf{G}_{fc}]\tilde{\boldsymbol{v}}_{\text{PCC}} \\ &+[\mathbf{G}_{ri} + (\mathbf{G}_{\text{dec}} - \mathbf{PI}_c)\mathbf{H}_{cf} + \mathbf{H}_{fg} + \mathbf{H}_{fc}]\tilde{\boldsymbol{i}}_g + (\mathbf{G}_{\text{dec}} - \mathbf{PI}_c)\tilde{\boldsymbol{i}}_{\text{con}}\} = \tilde{\boldsymbol{v}}_{\text{con}} \end{aligned} \tag{A.55}$$

where $\mathbf{G}_{\text{del}} = \begin{bmatrix} (1-0.5T_{\text{del}}s)/(1+0.5T_{\text{del}}s) & 0 \\ 0 & (1-0.5T_{\text{del}}s)/(1+0.5T_{\text{del}}s) \end{bmatrix}$.

The voltage and current relations on the LC filter can be found below:

$$\tilde{\boldsymbol{v}}_{\text{con}} = \tilde{\boldsymbol{v}}_{\text{PCC}} + \mathbf{Z}_f\tilde{\boldsymbol{i}}_{\text{con}} \tag{A.56}$$

$$\tilde{\boldsymbol{i}}_{\text{con}} = \mathbf{Y}_c\tilde{\boldsymbol{v}}_{\text{PCC}} + \tilde{\boldsymbol{i}}_g \tag{A.57}$$

where $\mathbf{Y}_c = \begin{bmatrix} C_f s & -\omega_0 C_f \\ \omega_0 C_f & C_f s \end{bmatrix}$, $\mathbf{Z}_f = \begin{bmatrix} L_f s + R_f & -\omega_0 L_f \\ \omega_0 L_f & L_f s + R_f \end{bmatrix}$.

Substituting $\tilde{\boldsymbol{i}}_{\text{con}}$ in Equation A.56 with Equation A.57 yields:

$$\tilde{\boldsymbol{v}}_{\text{con}} = \tilde{\boldsymbol{v}}_{\text{PCC}} + \mathbf{Z}_f\mathbf{Y}_c\tilde{\boldsymbol{v}}_{\text{PCC}} + \mathbf{Z}_f\tilde{\boldsymbol{i}}_g \tag{A.58}$$

Substituting $\tilde{\boldsymbol{i}}_{\text{con}}$ and $\tilde{\boldsymbol{v}}_c$ in Equation A.55 with Equations A.57 and A.58 yields:

$$\tilde{\boldsymbol{v}}_{\text{PCC}} + \mathbf{Z}_f\mathbf{Y}_c\tilde{\boldsymbol{v}}_{\text{PCC}} + \mathbf{Z}_f\tilde{\boldsymbol{i}}_g = \mathbf{G}_{\text{del}}\mathbf{PI}_c\mathbf{PI}_{pv}\tilde{\mathbf{r}}^* + \mathbf{G}_v\tilde{\boldsymbol{v}}_{\text{PCC}} + \mathbf{G}_i\tilde{\boldsymbol{i}}_g \tag{A.59}$$

$\mathbf{G}_{\text{ref}}$ is the gain factor function of the VSC reference. $\mathbf{G}_{\text{ref}}$ is the desired admittance of the VSC. $\mathbf{Z}_{\text{o}}$ and $\mathbf{G}_{\text{ref}}$ are defined via rearranging Equation A.59:

$$\underbrace{(\mathbf{G}_{\text{v}} - \mathbf{Z}_{\text{f}}\mathbf{Y}_{\text{c}} - \mathbf{I})^{-1}(\mathbf{G}_{\text{i}} - \mathbf{Z}_{\text{f}})}_{\mathbf{Z}_{\text{O}}}\tilde{\boldsymbol{i}}_{\text{g}} = \underbrace{(\mathbf{G}_{\text{v}} - \mathbf{Z}_{\text{f}}\mathbf{Y}_{\text{c}} - \mathbf{I})^{-1}(\mathbf{G}_{\text{del}}\mathbf{PI}_{\text{c}}\mathbf{PI}_{\text{pv}})}_{\mathbf{G}_{\text{ref}}}\tilde{\mathbf{r}}^{*} - \tilde{\boldsymbol{v}}_{\text{PCC}} \tag{A.60}$$

where $\mathbf{G}_{\text{v}} = \mathbf{G}_{\text{del}}[\mathbf{G}_{\text{rv}} + (\mathbf{G}_{\text{dec}} - \mathbf{PI}_{\text{c}})\mathbf{G}_{\text{cf}} + \mathbf{G}_{\text{fg}} + \mathbf{G}_{\text{fc}} + (\mathbf{G}_{\text{dec}} - \mathbf{PI}_{\text{c}})\mathbf{Y}_{\text{c}}]$, $\mathbf{G}_{\text{i}} = \mathbf{G}_{\text{del}}[\mathbf{G}_{\text{ri}} + (\mathbf{G}_{\text{dec}} - \mathbf{PI}_{\text{c}})\mathbf{H}_{\text{cf}} + \mathbf{H}_{\text{fg}} + \mathbf{H}_{\text{fc}} + \mathbf{G}_{\text{dec}} - \mathbf{PI}_{\text{c}}]$.

Therefore, the whole VSC admittance $\mathbf{Y}_{\text{o}}$ and impedance $\mathbf{Z}_{\text{o}}(= \mathbf{Y}_{\text{o}}^{-1})$ are fully derived for the VSC system with the proposed grid impedance compensation within the PLL.

	1	2	3	4	5	6
$\tilde{i}_{\text{g}}^{d}$	$-\frac{R_{\text{g}}}{L_{\text{g}}}$	ω_0	$\frac{1}{L_{\text{g}}}$			
$\tilde{i}_{\text{g}}^{q}$	$-\omega_0$	$-\frac{R_{\text{g}}}{L_{\text{g}}}$		$\frac{1}{L_{\text{g}}}$		
$\tilde{v}_{\text{PCC}}^{d}$	$-\frac{1}{C_{\text{f}}}$			ω_0	$\frac{1}{C_{\text{f}}}$	
$\tilde{v}_{\text{PCC}}^{q}$		$-\frac{1}{C_{\text{f}}}$	$-\omega_0$			$\frac{1}{C_{\text{f}}}$
$\tilde{i}_{\text{con}}^{d}$	$\frac{-1.5k_p^i k_p^p V_{\text{PCC}}^d}{-L_{\text{f}}}$	$\frac{-1.5k_p^i k_p^p V_{\text{PCC}}^q}{-L_{\text{f}}}$	$\frac{1-1.5k_p^i k_p^p I_{\text{g}}^d}{-L_{\text{f}}} - \frac{1}{L_{\text{f}}}$	$\frac{-1.5k_p^i k_p^p I_{\text{g}}^q}{-L_{\text{f}}}$	$\frac{-k_p^i}{-L_{\text{f}}} - \frac{R_{\text{f}}}{L_{\text{f}}}$	$\frac{-\omega_0 L_{\text{f}}}{-L_{\text{f}}} + \omega_0$
$\tilde{i}_{\text{con}}^{q}$			$\frac{k_p^i k_p^v}{-L_{\text{f}}}$	$\frac{1}{-L_{\text{f}}} - \frac{1}{L_{\text{f}}}$	$\frac{\omega_0 L_{\text{f}}}{-L_{\text{f}}} - \omega_0$	$\frac{-k_p^i}{-L_{\text{f}}} - \frac{R_{\text{f}}}{L_{\text{f}}}$
$\tilde{\theta}_{\text{PLL}}$				k_p^{PLL}		
$\tilde{x}_{\text{PLL}}$				$\mathbf{1}$		
$\tilde{x}_{\text{p}}$	$-1.5V_{\text{PCC}}^d$	$-1.5V_{\text{PCC}}^q$	$-1.5I_{\text{g}}^d$	$-1.5I_{\text{g}}^q$		
$\tilde{x}_{\text{v}}$			$\mathbf{1}$			
$\tilde{x}_{\text{id}}$	$-1.5k_p^p V_{\text{PCC}}^d$	$-1.5k_p^p V_{\text{PCC}}^q$	$-1.5k_p^p I_{\text{g}}^d$	$-1.5k_p^p I_{\text{g}}^q$	$\mathbf{-1}$	
$\tilde{x}_{\text{iq}}$			k_p^v			$\mathbf{-1}$
$\tilde{x}_d^{\text{PWM}}$	$\frac{-1.5k_p^i k_p^p V_{\text{PCC}}^d}{0.5T_{\text{del}}}$	$\frac{-1.5k_p^i k_p^p V_{\text{PCC}}^q}{0.5T_{\text{del}}}$	$\frac{1-1.5k_p^i k_p^p I_{\text{g}}^d}{0.5T_{\text{del}}}$	$\frac{-1.5k_p^i k_p^p I_{\text{g}}^q}{0.5T_{\text{del}}}$	$\frac{-k_p^i}{0.5T_{\text{del}}}$	$\frac{-\omega_0 L_{\text{f}}}{0.5T_{\text{del}}}$
$\tilde{x}_q^{\text{PWM}}$			$\frac{k_p^i k_p^v}{0.5T_{\text{del}}}$	$\frac{1}{0.5T_{\text{del}}}$	$\frac{\omega_0 L_{\text{f}}}{0.5T_{\text{del}}}$	$\frac{-k_p^i}{0.5T_{\text{del}}}$

$$
\left[\begin{array}{lcccccccc}
 & 7 & 8 & 9 & 10 & 11 & 12 & 13 & 14 \\
\tilde{i}_{\mathrm{g}}^{d} & & & & & & & & \\
\tilde{i}_{\mathrm{g}}^{q} & & & & & & & & \\
\tilde{v}_{\mathrm{PCC}}^{d} & & & & & & & & \\
\tilde{v}_{\mathrm{PCC}}^{q} & & & & & & & & \\
\tilde{i}_{\mathrm{con}}^{d} & \dfrac{V_{\mathrm{PCC}}^{q}+\omega_0 L_{\mathrm{f}} I_{\mathrm{con}}^{d}-k_p^{i} I_{\mathrm{con}}^{q}-V_{\mathrm{con}}^{q}}{-L_{\mathrm{f}}} & & \dfrac{k_p^{i} k_p^{p}}{-L_{\mathrm{f}}} & & \dfrac{k_i^{i}}{-L_{\mathrm{f}}} & & \dfrac{2}{L_{\mathrm{f}}} & \\
\tilde{i}_{\mathrm{con}}^{q} & \dfrac{k_p^{i} k_p^{v} V_{\mathrm{PCC}}^{q}+k_p^{i} I_{\mathrm{con}}^{d}-V_{\mathrm{PCC}}^{d}+\omega_0 L_{\mathrm{f}} I_{\mathrm{con}}^{q}+V_{\mathrm{con}}^{d}}{-L_{\mathrm{f}}} & & & \dfrac{k_p^{i} k_i^{v}}{-L_{\mathrm{f}}} & & \dfrac{k_i^{i}}{-L_{\mathrm{f}}} & & \dfrac{2}{L_{\mathrm{f}}} \\
\tilde{\theta}_{\mathrm{PLL}} & -k_p^{\mathrm{PLL}} V_{\mathrm{PCC}}^{d} & k_i^{\mathrm{PLL}} & & & & & & \\
\tilde{x}_{\mathrm{PLL}} & -V_{\mathrm{PCC}}^{d} & & & & & & & \\
\tilde{x}_{\mathrm{p}} & & & & & & & & \\
\tilde{x}_{\mathrm{v}} & V_{\mathrm{PCC}}^{q} & & & & & & & \\
\tilde{x}_{\mathrm{id}} & -I_{\mathrm{con}}^{q} & & k_i^{p} & & & & & \\
\tilde{x}_{\mathrm{iq}} & (k_p^{v} V_{\mathrm{PCC}}^{q}+I_{\mathrm{con}}^{d}) & & & k_i^{v} & & & & \\
\tilde{x}_{d}^{\mathrm{PWM}} & \dfrac{V_{\mathrm{PCC}}^{q}+\omega_0 L_{\mathrm{f}} I_{\mathrm{con}}^{d}-k_p^{i} I_{\mathrm{con}}^{q}-V_{\mathrm{con}}^{q}}{0.5T_{\mathrm{del}}} & & \dfrac{k_p^{i} k_p^{p}}{0.5T_{\mathrm{del}}} & & \dfrac{k_i^{i}}{0.5T_{\mathrm{del}}} & & -\dfrac{1}{0.5T_{\mathrm{del}}} & \\
\tilde{x}_{q}^{\mathrm{PWM}} & \dfrac{k_p^{i} k_p^{v} V_{\mathrm{PCC}}^{q}+k_p^{i} I_{\mathrm{con}}^{d}-V_{\mathrm{PCC}}^{d}+\omega_0 L_{\mathrm{f}} I_{\mathrm{con}}^{q}+V_{\mathrm{con}}^{d}}{0.5T_{\mathrm{del}}} & & & \dfrac{k_p^{i} k_i^{v}}{0.5T_{\mathrm{del}}} & & \dfrac{k_i^{i}}{0.5T_{\mathrm{del}}} & & -\dfrac{1}{0.5T_{\mathrm{del}}}
\end{array}\right]
\tag{A.45}
$$

Index

Note: Page numbers in *italics* and **bold** refers to figures and tables respectively.

a

abc–dq transformation 50–53, 56, 104
$\alpha\beta$ impedance 84–85, *85*
  measurement of 135–137, *136–137*
alternating current (AC) 162
  grid 145
  impedance measurements 127–130, *128*, *131*
  impedance measurements, noise on 133–135, **135**, *135*
argument principle 44, *44*
artificial neural network (ANN)-based estimation method 205, 209

b

balanced control system
  defined 119
  in mirror-frequency effect 119–122, *121–122*
  unbalanced control system and 122–123
band-pass (BP) filter 93–94
battery technology 10
Bode plot analysis 41–43, *43*

c

capacitor-clamped multilevel converter 3
CCA. *See* critical clearing angle (CCA)
CCT. *See* critical clearing time (CCT)
chemical energy storage 10
CIG. *See* converter interfaced generation (CIG)
CIGRE Committee 24, 26
closed-loop cutoff frequency 177
control conflict analysis 171–172
control interaction
  analysis of *17*
  of converter interfaced generation 27–31
  with mechanical mass system of turbine-generators 28–29, *28–29*
  against network 24
  against other devices 24
  with other power electronic devices 30–31
  from power electronic devices 27
  with series capacitors 29–30
  of voltage source converter 28–31
  with weak power system 31–33, *33*
control system matrix 120
conventional solution 148
conventional stability classification 13. *See also* power system
conventional synchronous generators based power systems 185–189, *186*
converter(s). *See also specific converters*
  energy storage system 9–10, *9*
  high-voltage direct current 1–5

Stability Analysis of Converter-Rich Power Grids, First Edition. Jun Liang and Chuanyue Li.
© 2026 The Institute of Electrical and Electronics Engineers, Inc. Published 2026 by John Wiley & Sons, Inc. All rights reserved.

converter(s). *See also specific converters* (*Cont.*)
mirror-frequency effect of 107–115
output current 139
renewable energy 5–9
stability 19, 26
system 20, *21*
voltage signal 107
converter-integrated power systems
case study 100–103, **100**, *100–102*
in high-order system 97–104
impedance interaction analysis for 89–104
impedance interaction criterion 97–99, *98–99*
negative capacitors 91–97, *92–96*, **96**
negative inductors 91–97, *92–96*, **96**
negative resistor 93–97, *93–96*, **96**
negative resistor criterion 89–91, *90*
2×2 matrix 103–104
converter interfaced generation (CIG) 26
control interaction of 27–31
past experiences from power electronic devices 27
VSCs control interaction 28–31
critical clearing angle (CCA) 204
critical clearing time (CCT) 204
current control 153
causing PCC voltage fluctuation 142–144, *142*, *144*
coordinated tuning of 156, *157*
phase-locked loop and 139–142, *140*, *141*, 149–152, *151*, *152*
current-injection method 128, 130
current reference 139
current-source-based GFM control 173–174
current source converter 2
cutoff frequencies 177–178, *178*, *179*

d

DDSSO. *See* device dependent subsynchronous oscillations (DDSSO)
device dependent subsynchronous oscillations (DDSSO) 22
electrical resonance 23
torsional resonance 23
DFIGs. *See* doubly fed induction generators (DFIGs)
diode-clamped converters 3
direct current (DC)
impedance measurement 126–127, *127*, *128*
impedance measurement, impact of noise on 131–133, *131*, **132**, *133*
dominant instability, solution for 146–149, *147–149*
doubly fed induction generators (DFIGs) 19, 30
dq–abc transformation 123
dq–$\alpha\beta$ transformation 109, 111–112, 114
dq-frame control 166
dq measurement 137
dq-sequence impedance 82–83, *82*
impedance measurement of 135–137, *136–137*
stability criteria for 83–84

e

EAC. *See* equal area criterion (EAC)
EEAC. *See* extended equal area criterion (EEAC)
electrical quantities 117–119
electrical resonance 23, 26
electrochemical double layer capacitors 10
electrochemical energy storage system 10
elimination methods
AC impedance and 133–135, **135**, *135*
DC impedance and 131–133, *131*, **132**, *133*
energy storage system 9–10, *9*
chemical 10
electrochemical 10
mechanical 9
thermal 10
equal area criterion (EAC) 18, 185–187
Euler's theorem 51
excitation control 189
extended equal area criterion (EEAC) 187

f

FACTS. *See* flexible alternating current transmission systems (FACTS)
fast Fourier transform (FFT) 132
feedback systems
Bode plot analysis for 41–43, *43*
mathematical mechanism of system stability 37–39, *38*

Nyquist plot analysis for 43–46, *44*, *45*
pole map for 39–41, *41*
stability criteria for 37–46
FFT. *See* fast Fourier transform (FFT)
flexible alternating current transmission systems (FACTS) 31
frame transformation
linearization of 54–55
modeling 55–60, *55*
phase-locked loops 56–57, *56*
power-frequency droop control 57–60, *58*
for single-phase systems *49*, 50–52
for three-phase systems *49*, 52–54
frequency
alignment methods 47–48
decoupling 51–53
frequency-domain inertia design approach 199
stability 26

g

Generalized Nyquist Criterion 78–79, *78*, *79*
grid-connected converters
CIG, control interaction of 27–31
modeling of frame transformation 55–60, *55*
oscillation frequency of electrical quantities 20–21, *21*
power system stability, classification of 13–18, 24–27
subsynchronous oscillation 22–24
VSC, control configurations of 47–48, *48*
VSC control systems, linearization of 48–55
VSC-induced oscillation events 18–20, **19**
VSC system, modeling and stability analysis of 60–67
weak-grid caused instabilities 31–33
grid-connected PV systems 7, *8*
grid following (GFL) 170
grid-following (GFL) converters
control structure 191–192, *192*
current control and PCC 142–144, *142*, *144*
current control and PLL 139–142, *140*, *141*, 149–152
with cutoff frequencies 177–178, *178*, *179*
dominant instability mechanism 146–149, *147–149*
GFM converters and 175–180, **175**
under grid strength conditions 176–177, *176*, *177*
outer loop, instability cause of 144–146
PLL-based control, instability causes of 139–146
PLL-based current control, tuning method for 146–154
PLL-based PV control systems 154–158
simulation verifications 196–198, *196–198*
transient stability analysis 189–196, *191*, *193–196*
tuning strategies 153, *154*
using virtual impedance 178–180, *180*, *181*
grid-forming (GFM) 159
grid-forming (GFM) converters 191
comparative analysis of stability characteristics 167
control structure 200–201, *201*
current-source-based GFM control 173–174, *174*
cutoff frequencies of inner current control loop 177–178, *178*, *179*
definition and variants of 162–166, *163–164*
development of 159–162, *162*
GFL and 175–180, **175**
GFM capabilities 162
GFM control, definition and variants of 162–166, *163–164*
GFM control, development of 161–162, *162*
grid strength conditions 176–177, *176*, *177*
instability in grids 167–168, *169*
instability in strong grids 167–168
PLL-based grid-supporting converter 170–171, **172**, *173*, *174*
P–ω droop and inner current control in 171–172, **172**
and SGs in strong and weak grids 167
simulation verifications 206–211, *207–210*, **211**
synchronous generators and 159–161, *160–161*, 169–170, *170*
transient stability of 198–211, *199*, 201–206, *203–207*
virtual impedance, stability enhancement using 178–180, *180*, *181*
grid impedance 176

h

Hellsjön project 3
high-pass (HP) filter 93–94
high-voltage direct current (HVDC) 166
  LCC-HVDC 1–2, *2*
  VSC-HVDC 3–5, *5*
HVDC. *See* high voltage direct current (HVDC)

i

IEEE Committee 24, 26
IGBTs. *See* insulated gate bipolar transistors (IGBTs)
IGE. *See* induction generator effect (IGE)
impedance analysis 176
impedance calculation method 132
impedance interaction analysis
  case study 100–103, **100**, *100–102*
  for converter-integrated power systems 89–104
  in high-order system 97–104
  impedance interaction criterion 97–99, *98–99*
  negative inductors and capacitors 91–93, *92*
  negative resistor, inductor, and capacitor 93–97, *93–96*, **96**
  negative resistor criterion 89–91, *90*
  2×2 matrix 103–104
impedance interaction criterion 97–98
  modified 98–99, *98–100*
impedance measurement techniques 125–126
  AC 127–130, *128*, *131*
  DC 126–127, *127*, *128*
  of *dq*-sequence impedance or $\alpha\beta$ impedance 135–137, *136–137*
  mechanism of 125–126, *126*
  noise impact on accuracy and elimination methods 130–135
impedance stability analysis method 75
  $\alpha\beta$ impedance 84–85, *85*
  Bode plot, criterion in 80–81, *81*
  converter impedance 76–77
  dq-sequence impedance 82–84, *82*
  Generalized Nyquist Criterion, criterion in 78–79, *78*, *79*
  Generalized Nyquist Criterion and Bode criterion 77–81
  Norton and Thevenin equivalent circuits 75–76, *75*
induction generator effect (IGE) 23, 30
inner-loop current control 168
insulated gate bipolar transistors (IGBTs) 3
integral–proportional (IP) controller 147

k

Kunbei-Liuzhou-Longmen (KBL) Hybrid Multi-terminal Ultra HVDC project 4

l

LCC-HVDC 1–2, *2*
LCCs. *See* line-commutated converters (LCCs)
left half-plane (LHP) 39, 140, 180
LHP. *See* left half-plane (LHP)
linearize sinusoidal functions 51–53
linear time-invariant system 103–104
line-commutated converters (LCCs) 1, 3
loop transfer function 37, 42, 45
low-pass (LP) filter 93–94
Lyapunov-based techniques 185
Lyapunov direct methods 187

m

mathematical mechanism, of system stability 37–39, *38*
MATLAB Neural Net Fitting toolbox 209
maximum power point tracking (MPPT) 7
mechanical energy storage system 9
MFE. *See* mirror-frequency effect (MFE)
minimum-phase transfer function 45
mirror-frequency effect (MFE) 107
  balanced control systems in 119–122, *121–122*
  of converter systems 107–115
  mirror frequency in three-phase system 109–115, *110*
  in single-phase control 108–109, *108–109*
  unbalanced control systems in 119–122, *121–122*
  understanding of 111–115, *115*
mirror-frequency oscillation
  converter control 122, *123*
  across different electrical quantities 117–119
  due to converters 107–124

mirror-frequency effect of converter systems 107–115
positive/negative sequence oscillations and 115–117, *117*
transformation mechanism of 109–114, *115*
vector-form 113
MMC. *See* modular multilevel converter (MMC)
mode-switching control strategy 199
modular multilevel converter (MMC) 3–5, *4*, 20
MPPT. *See* maximum power point tracking (MPPT)
Murray link HVDC Light Interconnection 4

n

negative capacitors 91–97, *92–96*, **96**
negative inductors 91–97, *92–96*, **96**
negative resistor 93–97, *93–96*, **96**
negative resistor criterion 89–91, *90*
negative sequence oscillations 115–117, *117*
neutral point clamped (NPC) converter 3, *4*
noise
on AC impedance 133–135, **135**, *135*
on DC impedance 131–133, *131*, **132**, *133*
on impedance and elimination methods 130–135
Norton and Thevenin equivalent circuits 75–76, *75*
Nyquist path 44–45
Nyquist plot analysis, stability criterion via 43–46, *44*, *45*

o

OMIB. *See* one-machine infinite-bus (OMIB)
one-machine infinite-bus (OMIB) 187
oscillation frequency 20–21, *21*
outer loop
instability cause of 144–146
tuning strategy 154–156, *155–156*
overdamped system 41

p

PEBS. *See* potential energy boundary surface (PEBS)
PEDIs. *See* power electronic device interactions (PEDIs)
phase-locked loop (PLL) 47, 63–64, **64**, *64*, 122–124
-based control 139–146
control conflict analysis 171–172
converter control 163
converter system 130
coordinated tuning of 156, *157*
current control and 139–142, *140*, *141*, 149–152, *151*, *152*
current-source-based GFM control, challenges of 173–174
for grid-connected VSC *56*
grid-supporting and GFM control strategies 172–173, **172**, *173*, *174*
-based grid-supporting converter *167*, *168*, 170–171
modeling of 56–57, *56*
PLL-based current control, tuning method for 146, 153–154, *153–154*
PLL-based PV control systems 154–158
P–ω droop-based GFM controls and 170–174
phase portrait analysis 185, 187–188
phase tracking process of PLL 140
photovoltaic (PV) systems 7
central configuration 8
extended configuration 8
grid-connected PV systems 7–9, *8*
string configuration 8
VSC technology 9
PLL. *See* phase-locked loop (PLL)
PNF. *See* positive and negative-frequency (PNF)
point of common coupling (PCC) voltage 48, 122, 139, 150, 164
dominant factor of current control 142–144, *142*, *144*
fluctuation 151
pole map, stability criterion via 39–41, *41*
positive and negative-frequency (PNF) 82, 110, 120
positive/negative sequence oscillations 115–117, *117*
potential energy boundary surface (PEBS) 187
power and voltage (PV) control 144
power electronic device interactions (PEDIs) 22
power electronic (PE) devices 24
power-frequency (P–f) droop 47
power-single-frequency oscillation 118
power systems 159
classification of 19, 24–27, *25*
converter-driven stability 26
converter-integrated. *See* converter-integrated power system
frequency stability 26

power systems (*Cont.*)
instabilities 18
resonance stability 26
rotor angle stability 25
stability 13
steady-state stability condition (without damping) 14–16
steady-state stability condition (damping) 16
subsynchronous resonance 16–17
transient stability 17–18, *18*
voltage stability 25
PR. *See* proportional-resonant (PR)
proportional–integral (PI) controller 49, 139
proportional-resonant (PR) controller 49
proposed solution 148
pulse width modulation (PWM) 3, 48
PWM. *See* pulse width modulation (PWM)
P–*ω* droop-based GFM control 165–166
control conflict analysis 171–172
current-source-based GFM control 173–174
grid-supporting and GFM control strategies 172–173, **172**, *173*, *174*
PLL-based grid-supporting control and 170–174
PLL-based grid-supporting converter *167*, *168*, 170–171
P–ω droop control 57–60, 66–67, 123–124, 167
in GFM converters 171–172

r

rechargeable batteries 10
redox flow batteries 10
renewable energy
solar photovoltaic 7–9, *8*
wind generation 5–7, *6*
Réseau de Transport d'Électricité (RTE) 31
resonance stability 19, 26, 27
RHP. *See* right half-plane (RHP)
right half-plane (RHP) 39, 41, 140, 168
RLC circuit
negative capacitors 91–97, *92–96*, **96**
negative inductors 91–97, *92–96*, **96**
negative resistor 93–97, *93–96*, **96**
RMS. *See* root mean square (RMS)
root mean square (RMS) 20, 118
rotor angle stability 25
Routh array 39
Routh-Hurwitz criterion 39–40
RTE. *See* Réseau de Transport d'Électricité (RTE)

s

SCR. *See* short-circuit ratio (SCR)
sensitivity stability analysis 70–71
separation process 109, 114
series capacitor SSO 30
SGs. *See* synchronous generators (SGs)
shaft torque amplification 23
short-circuit ratio (SCR) 14–16, 31, 140, 168
simulation verifications 196–198, *196–198*
single-machine infinite-bus (SMIB) context 186
single-phase control, mirror-frequency effect in 108–109, *108–109*
single-phase PLL 64–66, *65*, **66**, *66*
single-phase systems *49*, 50–51
frequency decoupling 51
linearize sinusoidal functions 51–52
small-signal modeling 51
SMs. *See* sub-modules (SMs)
solar energy 7
solar photovoltaic system 7–9, *8*
central configuration 8
configurations of 8, *8*
extended configuration 8
grid-connected PV systems 7–8, *8*
string configuration 8
SSCI. *See* subsynchronous control interaction (SSCI)
SSO. *See* subsynchronous oscillation (SSO)
SSR. *See* subsynchronous resonance (SSR)
SSTI. *See* subsynchronous torsional interaction (SSTI)
STATCOMs. *See* static synchronous compensator (STATCOMs)
state-space stability analysis method 69
demonstration of 71–74, *74*
sensitivity stability analysis 70–71
system of 70
static synchronous compensator (STATCOMs) 20
static var compensator (SVC) 27
steady-state power limit 14–15
steady-state stability condition
without damping characteristics 14–16
with damping characteristics 16

sub-modules (SMs) 3
subsynchronous control interaction (SSCI) 23, 27, 28, 30
subsynchronous oscillation (SSO) 22–23
  classification of *22*
  power electronic device interactions 24
  subsynchronous resonance 23–24
subsynchronous resonance (SSR) 16–17, 22
  DDSSO 23–24
  electrical resonance 23
  induction generator effect 23
  shaft torque amplification 23
  torsional interaction 23
  torsional resonance 23
subsynchronous torsional interaction (SSTI) 23, 28–29
supercapacitors 10
SVC. *See* static var compensator (SVC)
swing equation 160, 165, 185, 200
synchronous generators (SGs) 16–17, 159
  connection with 159–161, *160–161*
  stability under weak grid conditions 169–170, *170*

t

TES. *See* thermal energy storage (TES)
thermal energy storage (TES) 10
Thevenin equivalent circuit 125–126
three-phase systems *49*, 52
  frequency decoupling 52–53
  linearize sinusoidal functions 53
  MFE, detailed understanding of 111–115, *115*
  mirror frequency in 109–111, *110*
  single-phase *vs.* three-phase systems 53–54
time-domain simulation 186
torsional interaction 23
torsional resonance 24, 26, 27
Trans Bay Cable project 4
transfer function 37
transformation matrix 54–55
transient stability 17–18, *18*
  analysis approaches 186–188
  of conventional synchronous generators based power systems 185–189, *186*
  enhancement methods 188–189, *188*, *189*, *190*
  enhancement strategies 199
  equal area criterion 186–187
  of grid-following converters 189–198
  of grid-forming converters 198–211, 201–206, *203–207*
  Lyapunov direct methods 187
  phase portrait method 187–188
tuning method
  current control and PLLs 149–152, *151*, *152*
  dominant instability 146–149, *147–149*
  outer-loop tuning strategy 154–156, *155–156*
  for PLL-based current control 146, 153–154, *153–154*, 156, *157*
  for PLL-based PV control systems 154, 156–158, *157–158*
two-level converter 3–4, *4*

u

unbalanced control system 124
  balanced control system and 122–123
  defined 119
  in mirror-frequency effect 119–122, *121–122*
underdamped system 41
unstable equilibrium point (UEP) method 187

v

vector-form mirror-frequency oscillations 113
Viking Link 4
virtual impedance, stability enhancement using 178–180, *180*, *181*
virtual oscillator-based GFM control 166
virtual synchronous generator/machine (VSG/M) 161
voltage
  stability 25
  voltage-injection method 129–130
voltage source converter (VSC) 3, 28, 47, 60–67, 107, 159
  AC-DC transformation 49–50
  control 126
  control interaction of 28–31
  DC-AC transformation 49–50
  frame transformation 54–55
  induced oscillation events 18–20, **19**
  linearization control systems 48–50
  mechanical mass system 28–29, *28*, *29*
  modeling of 60–62

voltage source converter (VSC) (*Cont.*)
oscillation frequency of electrical quantities 20–21, *21*
parameters 49–50
PI-based control system 49–50
PLL, stability analysis of *56*, 63–64, **64**, *64*
power electronic devices 30–31
proportional-integral 49
proportional-resonant 49
P–ω droop control 66–67, **67**, *67*
series capacitors 29–30
single-phase PLL 64–66, *65*, **66**, *66*
single-phase system and three-phase system *49*
transfer-function-based stability analysis method 62–63
weak-grid instabilities 32–33
weak power system 31
VSC. *See* voltage source converter (VSC)
VSC-HVDC
development of *5*
modular multilevel converter (MMC) 3–5, *4*
neutral point clamped (NPC) converter 3, *4*
two-level converter 3–4, *4*
VSG/M. *See* virtual synchronous generator/machine (VSG/M)

W

weak-grid caused instabilities 31–32
experiences with 32
VSC 32–33, *33*
wind generation 5–7, *6*
wind turbine
back-to-back converters 6–7, *6*
drawbacks 5
fixed speed and connects directly to grid 6, *6*
power electronic-controlled variable resistor 6, *6*
synchronous generator *6*, 7
types of 6–7, *6*